NEUROPEPTIDE RESEARCH TRENDS

NEUROPEPTIDE RESEARCH TRENDS

BERNICE A. LEVINE
EDITOR

Nova Biomedical Books
New York

For permission to use material from this book please contact us:
Telephone 631-231-7269; Fax 631-231-8175
Web Site: http://www.novapublishers.com

Library of Congress Cataloging-in-Publication Data

Neuropeptide research trends / Bernice A. Levine, editor.
p. ; cm.
Includes bibliographic references and index.
ISBN 13: 978-1-60021-640-4
ISBN 10: 1-60021-640-4
1.Neuropeptides- -Physiological effect. I. Levine, Bernice A.-
[DNLM: 1. Neuropeptides- -physiology. WL 104 N4942076 2007]
QP552.N39N4745 2007
572'.65- -dc22

2007006568

Published by Nova Science Publishers, Inc. ✣ New York

Contents

Preface

The explosion of research activity in the field of neuropeptides has led to the idenfication of numerous naturally occurring endogenous peptides which act as neurotransmitters, neuromodulators, or trophic factors, to mediate nervous system functions. Increasing numbers of non-peptide ligands of neuropeptide receptors have been developed, which act as agonists or antagonists in peptidergic systems.

The scope of this new book includes gene regulation of peptide expression, peptide receptor subtypes, transgenic and knockout mice with mutations in genes for neuropeptides and peptide receptors, neuroanatomy, physiology, behaviour, neurotrophic factors, preclinical drug evaluation, clinical studies, and clinical trials.

The tackykinin family are a group of bioactive neuropeptides. These originally only included substance P, neurokinin A and neurokinin B. Recent discoveries of new tachykinins, hemokinin-1, endokinin A and endokinin B have prompted us to consider their possible roles within the central nervous system. Here we discuss their differences and similarities to the original tachykinins, and whether there is any potential for their use to develop more targeted drugs to combat conditions such as neurodegenerative and psychiatric disorders where the tachykinins have been implicated.

Chapter I - Neurotensin is a tridecapeptide originally extracted from bovine hypothalamus by Carraway and Leeman (1973), and later found in a variety of peripheral tissues with the highest levels determined in the jejuno-ileal portion of the gastrointestinal tract. Neurotensin possess a broad spectrum of physiological activity, and could act as a neurotransmitter and neuromodulator in the central nervous system and as a local hormone in gastrointestinal tract. Importantly, there is increasing evidence to suggest that neurotensin plays a pivotal role in regulation of the cardiovascular system. Indeed, neurotensin-containing neural fibers are found in close contact with atrial and ventricular cardiomyocytes, sinoatrial and atrioventricular nodal cells, neurons of intracardiac ganglia, as well as coronary vessels in humans and various animal species. The density of neurotensin-immunoreactive myocardial innervation is reduced in cardiac disease. Neurotensin produces a variety of cardiovascular actions including potent effects on blood pressure, heart rate, myocardial contractility, coronary vascular tone, venous smooth muscle tone as well as regional blood flow in the gastrointestinal tract, skin and adipose tissue. Neurotensin also modifies the electrophysiological properties of the heart and exerts a protective effect in vagally-induced cardiac arrhythmias. Moreover, neurotensin could trigger cardiovascular reflexes by stimulating primary visceral afferents synaptically connected with preganglionic sympathetic neurons in the spinal cord. Neurotensin-induced cardiovascular effects are strongly influenced by species-dependent differences, and rapidly subjected to tachyphylaxis. Structural determinants of biological activity of neurotensin reside primarily in the C-terminal portion of its molecule. Indeed, C-terminal neurotensin hexapeptide fragment, neurotensin (8-13), is shown to be equipotent to intact neurotensin in eliciting cardiovascular effects, as well as in binding to the purified neurotensin receptor. In contrast, short N-terminal neurotensin fragments, neurotensin (1-6), (1-8) and (1-10), do not possess cardiovascular activity, and exhibit very low affinity for the neurotensin receptor. Neurotensin-induced cardiovascular responses are mediated via direct activation of neurotensin receptors, as well as through stimulation of release of various endogenous biologically active substances such as histamine, serotonin, catecholamines and prostaglandins. To date, three subtypes of neurotensin receptors (NTS_1, NTS_2 and NTS_3) have been identified and found to be expressed in the brain and various peripheral tissues including the myocardium. NTS_1, a G-protein-coupled receptor with seven putative transmembrane domains, represents a high-affinity ($K_d \sim 0.1$-0.3 nM) neurotensin binding site coupled to phospholipase C-inositoltrisphosphate transduction pathway. The NTS_1 receptor seems to play a pivotal role in the mechanism of neurotensin-induced cardiovascular responses, whereas a putative significance of NTS_2 and NTS_3 receptor subtypes remains to be determined.

Chapter II - INTRODUCTION: *Helicobacter pylori (Hp)* causes chronic active gastritis and diffuse inflammation of gastric mucosa. While the majority of infected individuals remain asymptomatic, a subset of patients develop severe gastroduodenal pathologies, such as peptic ulcer disease, gastric cancer and low-grade gastric B cell lymphoma. We aimed to determine if previously reported *Hp* induced gastrin-somatostatin link impairment is a universal *Hp* related phenomena or if it depends on the disease outcome.

AIMS and METHODS: Gastrin and somatostatin (SST) protein and mRNA expression were quantified using ELISA and RT-PCR in gastric mucosa biopsies obtained during endoscopy from 70 *Hp* infected individuals and 25 *Hp* negative controls. Based on endoscopic findings and histological examination *Hp* infected patients were further divided into duodenal ulcer (DU, n=17), gastric cancer (GC, n=22) and non-ulcer dyspepsia (NUD, n=31) group.

RESULTS: Significantly lower antral gastrin levels were detected in *Hp* infected gastric cancer patients compared to both *Hp*+DU and *Hp*+NUD patients and controls. No significant difference in antral gastrin mRNA levels was detected. Comparable and disease outcome unrelated antral SST levels were detected in all *Hp* infected individuals. However, these were significantly lower when compared to *Hp* negative controls ($p<0.05$). The same trend was observed in corpus mucosa, but with no statistical significance ($p>0.05$). Antral SST gene expression is lower in all *Hp* infected patients regardless of the disease outcome, while in corpus only NUD patients exhibited significantly lower SST mRNA levels when compared to all other groups and controls.

CONCLUSION: Based on the results of our study and available data in the literature we concluded that decreased gastric somatostatin synthesis and secretion is a universal

Helicobacter pylori related phenomenon related to the presence of the infection and not to the disease outcome. On the other hand gastrin levels in *Helicobacter pylori* positive individuals depend on the disease outcome and in gastric cancer patients are probably influenced by more than one mechanism out of which some are most likely *Hp* independent.

Chapter III - Neuropeptides, which show a widespread distribution in both the central and peripheral nervous systems, are neuroactive substances that acting as neurotransmitters, neuromodulators (paracrine and autocrine actions) and neurohormones are involved in numerous physiological actions. Thus, for example, neuropeptides have been implicated in inflammatory and immune processes, as well as in angiogenesis, mitogenesis, cancer and metastasis.

In this sense, it is known that after binding to the neurokinin (NK)-1 receptor, substance P, a neuropeptide that belongs to the tachykinin family of peptides and that shows a widespread distribution in the nervous system, promotes proliferation of cancer cells (neuroblastoma, glioma, retinoblastoma, melanoma...) favoring tumour growth. It is also known that in some tumours, tachykinins facilitate tumour metastasis. Moreover, cancer cells express more NK-1 receptors than normal cells and NK-1 receptor antagonists exert an antitumoural action on numerous cancer cell lines, being the cell death observed produced by apoptosis. These data suggest that the NK-1 receptor is a new and promising target in the treatment of cancer and that the development of antagonist molecules of the NK-1 receptor represents an important opportunity for exploiting these molecules as novel therapeutic agents.

According to this, in this review the following aspects will be discussed: 1) Substance P/NK-1 receptor system: a) presence of substance P in human tumours; b) presence of NK-1 receptors in human tumours; 2) Cancer progression and mitogenic action of substance P: a) mitogenic action of substance P; b) angiogenesis; c) metastasis; d) substance P, inflammation, pain and cancer; e) substance P, emotional behaviour and cancer; and 3) Future perspectives: the NK-1 receptor as a new target in the treatment of cancer: a) NK-1 receptor antagonists; b) antitumoural action of NK-1 receptor antagonists.

Chapter IV - Recently, several neuropeptides have been discovered as the ligand for orphan receptors in the brain, some of which are appetite-regulating peptides that induce satiety or hunger. Since feeding is closely related to digestive function, the effects of these peptides on gastric acid secretion, bile secretion and gastric motility are well investigated. Orexigenic (appetite-enhancing) peptides generally facilitate gastric acid secretion and also facilitate phasic contractions of the distal stomach, which enhances the ability to churn gastric contents to quicken gastric emptying. These gastric functions are particularly important for digestion. In addition, accommodation of food is also important for smooth digestion. The proximal region of the stomach, which serves as a reservoir, relaxes to accommodate swallowed food. Reflex relaxations, induced by the activation of esophageal and/or pharyngo-laryngeal receptors before food reaches the stomach, are well known as so-called "receptive relaxation"; however, the relationship between feeding-related peptides and gastric relaxation of the proximal stomach has never been clarified. Gastric relaxation is achieved by the intramural nervous system and extrinsic nervous system. The vagus nerves, whose cell bodies are in the dorsal motor nucleus of the vagus (DMV), have a major role in the extrinsic control of gastric motility. Neural circuits in the dorsal vagal complex (DVC), which consists of the area postrema, the nucleus of the solitary tract and the DMV, form output signals for regulating gastric motor activities; therefore, the DVC must be the main target of feeding-related peptides to regulate gastric motility. Recently, our studies revealed that some orexigenic neuropeptides induced gastric relaxation of the proximal stomach by way of the DVC. The administration of orexin-A into the fourth ventricle induced gastric relaxation of the proximal stomach, as well as enhanced gastric contractility of the distal stomach. Neuropeptide Y (NPY), which is a classical but substantial orexigenic neuropeptide, also induced relaxation in the proximal stomach via Y1 receptors (Y1R) situated in the DVC. Since several studies revealed that the Y1R subtype plays a prominent role in mediating feeding induced by NPY, gastric relaxation induced by NPY might be closely associated with food intake. Ghrelin, which was originally isolated from the stomach and has orexigenic effects, induced gastric relaxation of the proximal stomach. These observations strongly suggest that central orexigenic neuropeptides facilitate accommodation by the stomach of a meal without an increase in intragastric pressure. Fasting is usually followed by feeding, therefore, gastric relaxation achieved by the actions of orexigenic neuropeptides, which are released by hunger, to enhance the easy accommodation of food is rational. In this report it is hypothesized that orexigenic neuropeptides induce anticipatory gastric relaxation to compensate for hyperphagia. Centrally released neuropeptides might cooperate with the acute reflex relaxation just after swallowing to facilitate reservoir function of the stomach.

Chapter V - Calcitonin gene-related peptide (CGRP) is a potent vasodilatory and cardiotonic neuropeptide with widespread distribution in the central and peripheral nervous systems. CGRP-positive nerve terminals have been detected in the heart and vasculature, particularly in arteries. The perivascular sensory nerve endings have been suggested to be the major source of circulating CGRP. In the present study, we investigated the impact of type 2 diabetes on CGRP concentrations in the heart, selected arteries and plasma in relation to basal cardiovascular parameters, i.e. resting heart rate, systolic and diastolic blood pressures. Type 2 diabetes was induced in 2-day-old male and female rats by a single intraperitoneal streptozotocin injection (STZ; 100 mg/kg; n2-STZ model). Diabetic state closely resembling human type 2 diabetes was verified by determinations of blood glucose levels, glucose tolerance tests, plasma and pancreatic insulin concentrations, and immunohistochemical analysis of the pancreatic tissue. CGRP concentrations were measured in the heart compartments, thoracic and abdominal aortae, mesenteric and femoral arteries, and in the plasma using radioimmunoassay diagnostic kits. At the age of 3 months, diabetic animals displayed significantly higher blood glucose levels, lower plasma insulin concentrations and impaired glucose tolerance. At this stage, no differences were noted in the resting heart rate, blood pressure, and tissue and plasma CGRP concentrations between n2-STZ and control rats. At the age of 18 months, diabetic rats of both sexes had blood glucose levels not differing from the age-matched control values. However, impaired glucose tolerance associated with decreased insulin response to the glucose challenge and elevated pancreatic insulin concentrations suggested defective insulin secretion in rats after neonatal STZ administration. Although systolic pressure did not differ between the control and n2-STZ rats, diastolic pressure was significantly higher in diabetic animals of both sexes. CGRP concentrations were significantly lower in the plasma, arteries and heart compartments of diabetic rats, reaching 40-70% of the control values. Interestingly, female diabetic hearts displayed significantly lower peptide concentrations than diabetic males. With regard to the well-documented cardioprotective and vasodilatory effects of CGRP, changes in CGRP synthesis and release may be associated with the development of cardiovascular complications of type 2 diabetes, i.e. decreased protection of the myocardium against ischemic injury and increased total peripheral resistance. In addition, profound deficit in myocardial CGRP levels revealed in female diabetic rats might be implicated in the loss of "female advantage" in the prevalence of coronary heart disease frequently described in diabetic patients.

Chapter VI - Epilepsy remains a major medical problem. Despite widespread use of both classic and newer pharmacologic agents that target ion channels, amino acid transmission or receptors, there are many examples of therapeutic failures. The study of the neuropharmacologic actions of opioids in CNS has yielded new insights into epileptogenesis and epilepsy treatment. In animals, the dynorphin system in the hippocampus is known to regulate excitability and epilepsy vulnerability has been associated with a dynorphin promoter region polymorphism, or low dynorphin expression genotype, in man. Encephalitic temporal lobe viruses, such as herpes and Borna Disease virus in rats, cause seizures by reproducing this neurochemical marker of epilepsy, namely low dynorphin tone, and restoration of dynorphin tone with a kappa opioid receptor agonist reverses their seizures. This work has renewed interest in the role of dynorphin dysregulation in the pathogenesis of epilepsy and kappa opioids in anticonvulsant protection, but leaves open to question a full understanding of kappa opioid mechanisms of action and side effect profiles. In particular, recent human and animal studies indicate that activation of kappa opioid receptors also elicits dysphoric and hallucinatory responses, the same responses reported to accompany cessation of seizures in temporal lobe epilepsy. Altogether, the evidence is consistent with dynorphin system and kappa opioid receptors being key factors in the regulation of excitability in hippocampus, and in shaping the clinical presentation of the psychiatric disorders in postictal phenomena.

Chapter VII - Hypothalamo-pituitary-adrenal axis (HPA) is a crucial component of the homeostasis. During chronic stress the regulation of the HPA seems to be relying on a pronounced role of vasopressin. Our hypothesis is that during prolong stress regiment in the absence of vasopressin the chronic stress symptoms will be diminished. Therefore we studied stress-related variables after 14 and 28 days of daily restraint in littermate heterozygous and vasopressin-deficient (di/di) Brattleboro rats.

Stressed heterozygote rats showed retarded growth, decreased thymus and spleen weights, and hypertrophied adrenals. After 2 weeks these changes were equal in di/di and di/+ rats with some reduction in di/di after prolongation of the restraint to 4 week. In chronically stressed animals the POMC mRNA in the anterior lobe of the pituitary and the slightly, but not significantly elevated resting plasma corticosterone level were equal in di/+ and di/di rats after 2 weeks restraint. Prolongation of the stress regime for 4 weeks leads to significantly smaller elevation of CRH and vasopressin mRNA in the nucleus paraventricularis hypothalami, diminished enhancement of POMC mRNA in the anterior lobe of the pituitary and reduced rise in basal plasma corticosterone levels in di/di rats. In addition, the acute response to the first, 15th and 29th restraint was also smaller (but not abolished) in vasopressin-deficient as compared with heterozygous rats. The difference between ACTH and corticosterone responses (small decrease in ACTH but no habituation in corticosterone) hypothesised that the sensitivity of the adrenal gland to ACTH was changed but from superfusion experiments it seems that after 4 weeks the sensitivity was first decreased rather than increased with no significant effect of vasopressin deficiency.

The results from 14 days repeated restraint suggested that vasopressin is not indispensable for activating the HPA system by chronic stress although the prolongation of the stress lead to different conclusion. These later results supported the view that during long term repeated restraint the role of vasopressin becomes more important (but not exclusive) in the maintenance of the resting HPA axis activity. However, its effects become evident after prolonged stress exposure only.

Chapter VIII - Affective disorder and schizophrenia are associated with dysregulation of the hypothalamus-pituitary-adrenal (HPA) axis, which leads to repeated episodes of hypercortisolemia. Hypothalamic paraventricular neurons are believed to trigger these processes by aberrant generation and/or release of corticotrophin releasing hormone (CRH), vasopressin and oxytocin. Classically, this persisting overactivation of the HPA axis in neuropsychiatric disorders is thought to be maintained by a disturbed glucocorticoid-mediated feedback action. However, the discovery of strongly reduced numbers of nitric oxide synthase containing paraventricular (PVN) and suprachiasmatic (SCN) neurons as well as of fewer beta-endorphin generating arcuate nucleus neurons in post mortem brains of depressive and schizophrenic patients point to other possible mechanisms by which neuropeptide-synthesizing hypothalamic neurons may become disinhibited in affective disorder and schizophrenia, namely by disturbed cross-talk with local hypothalamic factors. We propose a model which suggests that at least four different factors may contribute to enhanced expression of paraventricular neuropeptides: reduced generation of nitric oxide (NO), reduced generation and intrahypothalamic release of beta-endorphin, disturbed neuronal control of the PVN by SCN neurons and, last but not least, reduced GABAergic inhibition of PVN neurons.

Chapter IX - Neurophysins are polypeptides with molecular weights of about 10 000 and are synthesized within hypothalamic nuclei (mainly magnocellular supraoptic and paraventricular). Different forms of pituitary and circulating neurophysins have been described, but it is generally accepted in human subjects that two types of immonureactive neurophysins exist: one associated with AVP and one with OT synthesis. In plasma and cerebrospinal fluid (CSF), neurophysins are more stable than their related biological active nonapeptides. The neurophysins assay may therefore be considered a better measure of long-term AVP and OT secretion activity than isolated AVP or OT determinations. The radioimmunoassay for neurophysins has been proposed and used as an index of neuropituitary function in different physiological and pathological conditions. OT and AVP are two chemically very similar neurohypophyseal neuropeptides which could be involved in mood disorders. A few studies have been published about neurophysins in psychiatric pathology. In bipolar depressed patients, CSF AVP- and OT-neurophysins levels seems to be higher than in unipolar depressed patients or schizophrenic patients. Moreover, manic patients exhibit significantly higher plasma AVP-neurophysins values than depressed and schizophrenic patients. Following a switch from a depressive to a manic state, an increase in plasma AVP-neurophysins concentration has been observed. Depressed patients strive for a reduced AVP secretion. AVP-neurophysins have been also involved in the pathophysiology of the cortisol postdexamethasone nonsuppression and in suicidal behavior. The purpose of this paper is to propose an overview of neurophysins in mood disorders.

Chapter X - The hypothalamo-neurohypophyseal system (HNS) secrets two physiologically important neuropeptides: vasopressin (VP) and oxytocin (OT), which regulate water secretion in kidney/blood pressure and milk ejection/uterus contraction, respectively. Both of these neuropeptides are synthesized by magnocellular neurons (MCNs) in the supraoptic nucleus (SON) and paraventricular nucleus of the hypothalamus and released from the posterior pituitary into the systemic circulation. Their synthesis and release are regulated by numerous synaptic input signals with the primary regulation of VP secretion coming from osmotic and cardiovascular signals. Cardiovascular signals are relayed through the A1 pathway to SON, where several neurotransmitters are released. Among them, a pair of co-released neurotransmitters, ATP and norepinephrine (NE), exerts unique regulation of VP and OT release. In vitro studies with ATP and phenylephrine (PE), an agonist to α-1 adrenergic receptor (α-1R), demonstrated that ATP and PE acted synergistically to stimulate VP and OT release from perifused HNS explants. Either of these agents alone only induced a transient and small VP or OT release. Combined application of these two agents transformed the response to a much larger and sustained one. This synergistic action on neuropeptide release also exists between another pair of neurotransmitters, angiotensin II and glutamate, which mediate osmotic signals to SON. Depletion of intracellular calcium stores by thapsigargin disrupted the synergistic response to ATP and PE, indicating involvement of calcium as a mediator. Since activation of P2X purinergic receptors (P2XR, ATP is an agonist) induces influx of calcium whereas activation of α-1Rs releases calcium from intracellular stores, it was anticipated that co-application of this pair of neurotransmitters might generate a different pattern of intracellular calcium, which served as the driving force for their synergistic action on neuropeptide release. Further studies employing a calcium imaging technique with calcium sensitive fluorescence dye, Fura-2 AM, revealed that either ATP or PE alone induced a large, rapid peak increase in intracellular calcium, but ATP and PE together induced a much larger sustained calcium increase than either agent alone. This larger sustained calcium response to combined ATP and PE treatment is not dependent on either pre-synaptic inputs, or continuous exposure to the agents. These results demonstrate that the pair of neurotransmitters acts post-synaptically to generate a different pattern of intracellular calcium, which in turn drives a synergistic neuropeptide release from posterior pituitary. The synergistic action on neuropeptide release may be physiologically important under circumstances such as severe blood loss induced hypo-volemia when large and sustained release of VP is required to maintain blood pressure until the volume loss is corrected.

Chapter XI - There is widespread agreement that the renin-angiotensin system (RAS) plays a pivotal role in the pathogenesis of insulin resistance and cardiovascular disease (CVD) in diabetes. Indeed, large clinical trials have demonstrated substantial benefit of the blockade of this system for end-organ protection. Thus the blockade of the RAS may be a promising strategy for the treatment of the patients with the metabolic syndrome, which consists of a constellation of coronary risk factors such as hypertension, insulin resistance, obesity and hyperlipidemia. Although several types of angiotensin II type 1 (AT_1) receptor blockers (ARBs) are commercially available for the treatment of patients with hypertension, we have recently found that telmisartan (Micardis®) could have the strongest binding affinity to AT_1 receptor. Telmisartan will be a promising cardiometabolic sartan due to its unique peroxisome proliferator-activated receptor-γ (PPAR- γ)-inducing properties as well. In this paper, we reviewed the role of angiotensin II-type 1 receptor system in cardiometabolic disorders, especially focusing on telmisartan. The authors discuss its potential therapeutic implication in these devastating disorders.

In: Neuropeptide Research Trends
Editor: Bernice A. Levine, pp. 1-3

ISBN: 978-1-60021-640-4

Expert Commentary A

Potential Utility of Agonists of Pigment Epithelium-Derived Factor (PEDF) Receptor for the Treatment of Diabetic Retinopathy

***Sho-ichi Yamagishi**[*] [a], **Kazuo Nakamura** [a], **Takanori Matsui** [a], and **Hiroyoshi Inoue** [b]*

[a] Department of Internal Medicine, [b] Radioisotope Institute for Basic and Clinical Medicine, Kurume University School of Medicine, Kurume 830-0011, Japan

Pigment epithelium-derived factor (PEDF) is a glycoprotein that belongs to the superfamily of serine protease inhibitors [1,2]. It was first purified from the conditioned media of human retinal pigment epithelial cells as a factor with potent neuronal differentiating activity [1,2]. Recently, PEDF has been shown to be a highly effective inhibitor of angiogenesis in cell culture and animal models. PEDF inhibits growth and migration of cultured endothelial cells (ECs), and it potently suppresses ischemia-induced retinal neovascularization [3-5]. Furthermore, we have recently found that PEDF inhibits advanced glycation end product (AGE)-induced vascular inflammation and retinal hyperpermeability by suppressing NADPH oxidase-mediated reactive oxygen species (ROS) generation [6-8]. In addition, PEDF acts on pericytes as well, another component of microvasculatures; it inhibits the AGE-induced ROS generation and subsequently blocks apoptotic cell death of retinal pericytes, the earliest histopathological hallmark of diabetic retinopathy [9]. PEDF levels in aqueous humor or vitreous are decreased in diabetic patients, especially with proliferative retinopathy, thus suggesting that a loss of PEDF in the eye may

[*] Reprints and correspondence to Sho-ichi Yamagishi MD, PhD, Department of Internal Medicine, Kurume University School of Medicine, Kurume 830-0011, Japan; Tel; +81-942-31-7580; Fax; +81-942-31-7707; E-mail; shoichi@med.kurume-u.ac.jp

contribute to the development and progression of diabetic retinopathy [10-12]. Taken together, these observations suggest that pharmacological up-regulation or substitution of PEDF may offer a promising strategy for halting the development and progression of diabetic retinopathy.

We have very recently found a single class of high affinity PEDF binding protein existed in microvascular ECs [6]. Moreover, antibody directed against a peptide of receptor binding region of PEDF neutralized the anti-permeability effects of PEDF in retinal vessels of diabetic serum-derived AGE-injected rats [6]. These findings indicate that ECs possess a system efficiently responding to PEDF, thus suggesting a possible involvement of EC surface receptor in the anti-permeability property of PEDF. Therefore, development of agonists for PEDF receptor in vascular wall cells may become a novel promising strategy for the treatment of diabetic retinopathy.

References

[1] Tombran-Tink J, Chader CG, Johnson LV. PEDF: Pigment epithelium-derived factor with potent neuronal differentiative activity. *Exp. Eye Res.* 1991;53:411-414.

[2] Tombran-Tink J, Barnstable CJ. PEDF: a multifaceted neurotropic factor. *Nat. Rev. Neurosci.* 2003;4:628-636.

[3] Dawson DW, Volpert OV, Gillis P, Crawford SE, Xu HJ, Benedict W, Bouck NP. Pigment epithelium-derived factor: A potent inhibitor of angiogenesis. *Science.* 1999;285:245-248.

[4] Duh EJ, Yang HS, Suzuma I, Miyagi M, Youngman E, Mori K, Katai M, Yan L, Suzuma K, West K, Davarya S, Tong P, Gehlbach P, Pearlman J, Crabb JW, Aiello LP, Campochiaro PA, Zack DJ. Pigment epithelium-derived factor suppresses ischemia-induced retinal neovascularization and VEGF-induced migration and growth. *Invest. Ophthalmol. Vis. Sci.* 2002;43:821-829.

[5] Abe R, Shimizu T, Yamagishi S, Shibaki A, Amano S, Inagaki Y, Watanabe H, Sugawara H, Nakamura H, Takeuchi M, Imaizumi T, Shimizu H. Overexpression of pigment epithelium-derived factor decreases angiogenesis and inhibits the growth of human malignant melanoma cells in vivo. *Am. J. Pathol.* 2004;164:1225-1232.

[6] Yamagishi S, Nakamura K, Matsui T, Inagaki Y, Takenaka K, Jinnouchi Y, Yoshida Y, Matsuura T, Narama I, Motomiya Y, Takeuchi M, Inoue H, Yoshimura A, Bucala R, Imaizumi T. Pigment epithelium-derived factor inhibits advanced glycation end product-induced retinal vascular hyperpermeability by blocking reactive oxygen species-mediated vascular endothelial growth factor expression. *J. Biol. Chem.* 2006;281:20213-20220.

[7] Yamagishi SI, Matsui T, Nakamura K, Takeuchi M, Imaizumi T. Pigment epithelium-derived factor (PEDF) prevents diabetes- or advanced glycation end products (AGE)-elicited retinal leukostasis. *Microvasc. Res.* 2006 (in press)

[8] Inagaki Y, Yamagishi S, Okamoto T, Takeuchi M, Amano S. Pigment epithelium-derived factor prevents advanced glycation end products-induced monocyte

chemoattractant protein-1 production in microvascular endothelial cells by suppressing intracellular reactive oxygen species generation. *Diabetologia.* 2003;46:284-287.

[9] Yamagishi S, Inagaki Y, Amano S, Okamoto T, Takeuchi M, Makita Z. Pigment epithelium-derived factor protects cultured retinal pericytes from advanced glycation end products-induced injury through its antioxidative properties. *Biochem. Biophys. Res. Commun.* 2002;296:877-882.

[10] Spranger J, Osterhoff M, Reimann M, Mohlig M, Ristow M, Francis MK, Cristofalo V, Hammes HP, Smith G, Boulton M, Pfeiffer FH. Loss of the antiangiogenic pigment epithelium-derived factor in patients with angiogenic eye disease. *Diabetes.* 2002;50:2641-2645.

[11] Ogata N, Tombran-Tink J, Nishikawa M, Nishimura T, Mitsuma Y, Sakamoto T, Matsumura M. Pigment epithelium-derived factor in the vitreous is low in diabetic retinopathy and high in rhegmatogenous retinal detachment. *Am. J. Ophthalmol.* 2001;132:378-382.

[12] Boehm BO, Lang G, Volpert O, jehle PM, Kurkhaus A, Rosinger S, Lang GK, Bouck N. Low content of the natural ocular anti-angiogenic agent pigment epithelium-derived factor (PEDF) in aqueous humor predicts progression of diabetic retinopathy. *Diabetologia.* 2003;46:394-400.

In: Neuropeptide Research Trends
Editor: Bernice A. Levine, pp. 5-12
ISBN: 978-1-60021-640-4

Expert Commentary B

Should the Endokinins Be Classified as Neuropeptides? An Analysis of Their Role in the Nervous System

Suzanne Newton and Nigel M. Page*
School of Life Sciences, Kingston University London, Penrhyn Road,
Kingston-upon-Thames, Surrey KT1 2EE, United Kingdom

Abstract

The tackykinin family are a group of bioactive neuropeptides. These originally only included substance P, neurokinin A and neurokinin B. Recent discoveries of new tachykinins, hemokinin-1, endokinin A and endokinin B have prompted us to consider their possible roles within the central nervous system. Here we discuss their differences and similarities to the original tachykinins, and whether there is any potential for their use to develop more targeted drugs to combat conditions such as neurodegenerative and psychiatric disorders where the tachykinins have been implicated.

Introduction

The tachykinin family consists of three traditional tachykinins, substance P (SP), neurokinin A (NKA) and neurokinin B (NKB), as well as the newly discovered endokinin A (EKA), endokinin B (EKB) and hemokinin-1 (HK-1) (Page, 2004). By definition the tachykinins are classed in the same family due to their common C-terminal sequence of Phe-X-Gly-Leu-Met-NH_2 (FXGLM-NH_2) where X is an aromatic or β-branched aliphatic amino acid (Page, 2006a). The tachykinins are encoded on three preprotachykinin (*TAC*) genes.

* Correspondence: School of Life Sciences; Kingston University London; Penrhyn Road; Kingston-upon-Thames; Surrey KT1 2EE; United Kingdom; Tel: +44 (0)20 8547 2000; FAX: +44 (0)20 8547 7562; E-mail: n.page@kingston.ac.uk

*TAC*1 encodes SP and NKA, *TAC*3 encodes NKB. The endokinins and HK-1 are encoded on the recently discovered *TAC*4 gene (Zhang *et al.,* 2000; Page *et al.*, 2003). The tachykinins mediate their effects through interaction with three transmembrane G-protein coupled tachykinin receptors, named NK_1, NK_2 and, NK_3 (Page *et al.* 2001). A truncated form of the NK_1 receptor has also been observed, where it is missing the C-terminal cytosolic region (Fong *et al.*, 1992; Mantyh *et al.*, 1996). Such an alteration has been shown to alter the receptor's ligand binding efficiency (Fong *et al.*, 1992; Meini *et al.*, 1994) and effector systems (Fong *et al.*, 1992). The C-terminal tail also contains a series of sites that mediate the phosphorylation and internalisation of the receptor following agonist binding and activation (Bohm *et al.*, 1997; Vigna, 1999). This would suggest an alteration in G-protein interaction; however a distinctive role for this subtype has yet to be determined (Page and Bell, 2002). Each member of the tackykinin family can ligand bind with each receptor, though with varying affinities (Page *et al.*, 2003). SP has the greatest affinity for NK_1 and NKA for NK_2 whereas NKB is the preferred ligand for NK_3 (Patacchini and Maggi, 2001). EKA and B along with HK-1 have been found to act as full agonists at all three receptors with the greatest preference for the NK_1 receptor (Page *et al.*, 2003; Berger and Paige, 2005). Moreover, each exhibits a remarkable selectivity and potency for the tachykinin NK_1 receptor similar to SP (Morteau *et al.*, 2001; Bellucci *et al.*, 2002; Kurtz *et al.*, 2002; Duffy *et al.*, 2003; Page *et al.*, 2003). More recently, SP has been implicated in depression, anxiety and some neurodegenerative disorders (Kramer *et al.,* 1998; Stout *et al.*, 2001). This begs the questions - what are the possible roles for these new tachykinins in the brain and could they explain some of the physiology's previously assigned to the traditional tachykinins?

Hemokinin-1

Expression analysis of mRNA has shown that HK-1 is broadly expressed throughout the brain (Duffy *et al.*, 2003). HK-1 expression has also been identified in astrocytoma cells (Page *et al.*, 2003; Berger and Paige, 2005), dendritic and microglial cells (Nelson *et al.*, 2004). Nevertheless, it should be noted that quantitative mRNA expression studies in the whole brain have consistently shown significantly lower mRNA expression levels of *TAC*4 expression when compared directly to *TAC*1. Duffy *et al.* (2003) showed 20 times lower expression of *TAC*4 and Page (2006a) demonstrated 15 times lower expression. This is important to note as many studies have shown HK-1 to have similar biological effects in the brain to SP (Duffy *et al.*, 2003; Berger and Paige, 2005; Fu *et al.*, 2005; Endo *et al.*, 2006; Fu *et al.*, 2006; Naono *et al.*, 2007). Moreover, do these studies demonstrate an important role for HK-1 in the brain when it is clearly expressed at significantly much lower levels than SP, or do these studies just reflect the similar pharmacological profiles of HK-1 and SP? Duffy *et al.* (2003) has performed the most comprehensive comparison of *TAC*1 and *TAC*4 mRNA expression in distinct regions of the mouse brain (spinal cord, cerebellum, trigeminus, brain stem, cortex, caudate, hippocampus, middle brain, hypothalamus and thalamus) and whilst the regional distribution of HK-1 and SP mRNA were similar, HK-1 expression was consistently present at lower levels than SP (5 – 200 fold), except in the cerebellum that had a 10 fold higher expression level of *TAC*4 than *TAC*1. These findings suggest the HK-1 could act as an important SP-like agonist within the central nervous system particularly in the cerebellum.

Administration of HK-1 *in vivo* produces similar effects to SP, inducing foot tapping and scratching in mice and gerbils similar to those observed following central administration of SP or the NK_1 receptor agonist, GR-73632 (Duffy *et al.*, 2003). These behavioural effects could be blocked by the selective NK_1 receptor antagonist, MK-869 (Duffy *et al.*, 2003). Again, in the mouse, HK-1 induced potent dose dependent analgesic effects following intracerebroventricular administration as observed by using the tail flick test (Fu *et al.*, 2005). A NK_1 receptor antagonist could fully antagonise the analgesic effect of HK-1, whilst the addition of the selective NK_2 receptor antagonist, SR48968 hardly effected the nociceptive effects of HK-1 (Fu *et al.*, 2005). From this study, Fu *et al.* (2005) conclude that this provided indirect evidence favouring a role for NK_1 supraspinal receptors in the inhibitory control of descending pain pathways. The authors have subsequently demonstrated that HK-1 can significantly potentiate the effects of morphine and suggest a synergistic analgesic relationship between HK-1 and the central opioid systems on the direct modulation of local nociceptive responses (Fu *et al.*, 2006). Similar pain-related effects have also been observed by Endo *et al.* (2006) when HK-1 was intrathecally administered causing scratching behaviour, a behaviour which could be inhibited by pre-treatment with an NK_1 receptor antagonist. In addition, SP was found to decrease the delay of the withdrawal of the hind paw when exposed to noxious heat stimulation 20 to 30 minutes after intrathecal administration. The authors found that HK-1 did not produce the same response, and so concluded that this could mean there is a specific HK-1 receptor (Endo *et al.,* 2006). Nevertheless, overwhelming evidence currently exists to demonstrate the remarkable selectivity and potency for HK-1 at the NK_1 receptor (Morteau *et al.*, 2001; Bellucci *et al.*, 2002; Camarda *et al.*, 2002; Kurtz *et al.*, 2002; Duffy *et al.*, 2003; Page *et al.*, 2003) and that the physiological effects of HK-1 mimic those of SP *in vivo* (Camarda *et al.*, 2002; Kurtz *et al.*, 2002; Duffy *et al.*, 2003; Page *et al.*, 2003; Metwali *et al.*, 2004). An alternative explanation could lend itself to the controversy over the potential sub-types and isoforms of the NK_1 receptor (Glowinski, 1995; Page, 2005). Subtypes of the NK_1 receptor so far described have included the "classic", "septide-sensitive", "new NK_1-sensitive" and those of at least two types of C-terminal tail truncated isoforms (reviewed by Beaujouan *et al.*, 2004; Page, 2005). What is the role of all of these different sub-types and isoforms? Is there also a "hemokinin-sensitive" sub-type or could one of these already documented forms represent the "hemokinin-sensitive" binding site? Does HK-1 bind with an even greater affinity to the truncated form of the NK_1 receptor? As the truncated NK_1 receptor lacks most of the cytosolic C-terminal, it could be suggested that this allows different signalling pathways to be activated, as well as differential rates of receptor desensitisation and internalisation.

Binding affinities for HK-1 to the NK_1 receptor have been shown to be similar to that of SP, indicating that NK_1 is the preferred receptor of HK-1 (Berger and Paige, 2005). Berger and Paige (2005) also demonstrated within astrocytoma cells that nano molar concentrations of HK-1 stimulated calcium ion mobilisation as well as inducing an increase in cytokine expression, including IL-1β, IL-6, LIF and GM-CSF. No increase in expression was observed for the immunosuppressant cytokine TGF-β. Nano molar concentrations of SP also induced this response. The release of these cytokines may influence tumour angiogenesis, progression and proliferation as seen in astrocytomas. Antagonists specifically developed for HK-1 could suppress tumour growth allowing for smaller treatment doses to kill the tumour quickly, and more successfully. HK-1, as it has similar functions to SP within the brain means that a target specific approach for one brain area could pose a problem. However, an understanding of the manner in which HK-1 can bind to each of the potential sub-types and isoforms of the NK_1 receptor may prove vital in designing specific drugs to differentiate the actions of SP and HK-1.

The Endokinins

EKA and B are different from the traditional tachykinins as they lack the N-terminal dibasic residue cleavage site which SP, NKA and NKB possess. Instead they possess a potential monbasic cleavage site (a single lysine) (Kurtz *et al.*, 2002). This has the potential to generate elongated tachykinin peptides that could be 47 or 41 amino acid residues in length, respectively (Page *et al.*, 2003) and also those which could be much shorter if monobasic cleavage were to occur (Kurtz *et al.*, 2002). The extensions seen in these new tachykinins could alter their receptor interactions, which could lead to separate signalling pathways being activated. Indeed, Page *et al.* (2003) has shown equipotent affinity of the longer variants of EKA/B comparable to those of SP, however, these were approximately 6.6 times higher affinity than those of the much shorter EKA/B proposed to occur by monobasic cleavage. It should be stressed at the current time that it is unclear in which conformations EKA and EKB exist. A specific antibody for the EK peptides has not yet been developed and a further complication is that some established and commercially available antibodies to SP have 100% cross-reactivity with the common C-terminal decapeptide of the EK peptides (Page, 2004, Graham *et al.*, 2004). It is highly possible that earlier peptide analysis of SP in the brain may have also included the detection of EK and HK-1.

Despite this, the C terminal decapeptide sequence of EKA and EKB are closely related to SP's, where they share the last six amino acid residues. This high degree of homology would suggest that these endokinins would have similar physiological effects to SP *in vivo* as this sequence is crucial to the interaction of the tachykinins with the NK_1 receptor (Page, 2005). Many studies have proved this is the case (Page *et al.*, 2003; Yoshioka *et al.*, 2006). Recently it has been demonstrated that mRNA encoding EKB is expressed in the spinal cord and brain (Page *et al.*, 2003; 2004), and the binding of the common C terminal decapeptide of EKA and EKB to the NK_1 receptor has similar pharmacology to SP (Page *et al.*, 2003). Yoshioka *et al.* (2006) looked at the common C terminal decapeptide in EKA and EKB to see if it triggered the same pain processing responses of scratching and hyperalgesia in rats as seen with SP. They found that it did, and it also worked through the NK_1 receptor. They also concluded that their rat model was reliable in predicting similar effects in humans, as the amino acid sequence of SP is conserved in both species, and even though rats do not express EKA or EKB, they displayed similar hemodynamic effects to SP in rats. This therefore means that EKA and EKB could potentially contribute to pain processing and vasodilation in humans (Yoshioka *et al.*, 2006). In another study, Naono *et al.* (2007) administered the C-terminal decapeptide region of EKA/B intrathecally into rats. Doses of both SP and EKA/B (at 10^{-3}M) evoked scratching behaviour and when each peptide was administered for a second time after 15 minutes there was remarkable desensitising to the scratching behaviour. Interestingly, in the rats' administration of HK-1 only produced very weak cross desensitising leading to the speculation that in the rat there may be another HK-1 preferred receptor (Naono *et al.*, 2007). However, there is still along way to go before we understand the precise conformations of the HK-1 and EK peptides and the potential ways that these could interact with the different binding sites on the tachykinin receptors. The conformations of the final protein, as well as the extension of the peptide compared to SP are most likely to cause altered receptor interactions. For example, it will be fascinating to see whether SP antagonists which may have a preference to one particular sub-type or isoform of the NK_1 receptor will provide methods to differentiate between the actions of SP and the EK peptides.

*TAC*4 mRNA expression levels in the human central nervous system remain largely uncharacterised. Nonetheless, the tachykinins have been a family of peptides traditionally classified as neuropeptides. This has come to be challenged recently, and even the placenta a tissue devoid of nervous system innervation has been found to express levels of NKB mRNA at 2.6 fold higher than those in the brain (Page *et al.*, 2006). The endokinins have been found to be predominantly expressed in the periphery leading to the proposal that HK-1 and EKA/B are the peripheral SP-like endocrine/paracrine agonists where SP is not expressed (Page *et al.*, 2003). Nevertheless, the translation and processing of the tachykinins to active peptides in the periphery has not been fully determined. A study aimed at quantifying the amount of HK-1-like immunoactivity (using a SP assay) in mouse B cells came to the conclusion that there was only limited HK-1 peptide production (Nelson and Bost, 2005). In the case of the placenta even though this tissue was found to express 2.6 fold higher levels of *TAC*3 mRNA in comparison to peptide levels found in the brain the placenta was found to possess very little NKB peptide (Page, 2006a). In contrast, the (neuronal) rat brain appeared efficient at fully processing NKB to its decapeptide (Page, 2006a). It is apparent in the later case that the brain has the necessary machinery to fully process a biologically active tachykinin. From this point of view even though HK-1 and EKA/B are more abundantly expressed in the periphery are they actually present as active peptides?

Summary

Very little research has been undertaken to look at the effects of the new tachykinins on the brain. The research to date shows SP homology in the physiological responses they cause. What needs to be established is how HK-1 and EKA/B bind to their receptors and whether their differential sequence and lengths compared to that of SP causes differing signalling pathways to be induced or alternative receptor binding conformations. Also, the locations and metabolic states which cause expression and processing within the central nervous system need to be established. Potentially, EKA/B and HK-1 have significant roles to play within the central nervous system, and this could lead to significant implications for the treatment of many tachykinin-linked diseases in the future such as neurodegenerative and psychiatric disorders.

References

Beaujouan, J.C.; Torrens, Y.; Saffroy, M.; Kemel, M.L. and Glowinski. J. (2004). A 25 year adventure in the field of tachykinins. *Peptides*, 25, 339-357.

Bellucci, F.; Carini, F.; Catalani, C.; Cucchi, P.; Lecci, A.; Meini, S.; Patacchini, R.; Quartara, L.; Ricci, R.; Tramontana, M.; Giuliani, S. and Maggi, C.A. (2002). Pharmacological profile of the novel mammalian tachykinin, hemokinin 1. *British Journal of Pharmacology*, 135, 266-274

Berger, A. and Paige, C.J. (2005). Hemokinin-1 has substance P like behaviour in U-251 MG astrocytoma cells. A pharmacological and functional study. *Journal of Neuroimmunology*, 164, 48-56.

Bohm, S. K.; Khitin, L. M.; Smeekens, S. P.; Grady, E.F.; Payan, D. G. and Bunnett, N. W. (1997). Identification of potential tyrosine-containing endocytic motifs in the carboxyl-tail and seventh transmembrane domain of the neurokinin 1 receptor. *Journal of Biological Chemistry*, 272, 2363-2372.

Camarda, V.; Rizzi, A.; Calo, G.; Guerrini, R.; Salvadori, S. and Regoli, D. (2002). Pharmacological profile of hemokinin 1: a novel member of the tachykinin family. *Life Science* 71, 363-370.

Duffy, R.A.; Hedrick, T.A.; Randolph, G.; Morgan, C.A.; Cohen-Williams, M.E.; Vassileva, G.; Lachowicz, T.E.; Laverty, M.; Maguire, M.; Sham, L.S.; Gustafson, E. and Varty, G.B. (2003). Centrally administered hemokinin1 (HK-1), a neurokinin NK1 receptor agonist produces substance P-like behavioural effects in mice and gerbils. *Neuropharmacology*, 45, 242-250.

Endo, D.; Ikeda, T.; Ishida, Y.; Yoshioka, D. and Nishimon, T. (2006). Effect of intrathecal administration of hemokinin 1 on the withdrawal response to noxious thermal stimulation of the rat hind paw. *Neuroscience Letters*, 392, 114-117.

Fong, T. M.; Anderson, S. A.; Yu, H.; Huang, R. R.; Strader, C. D. (1992) Differential activation of intracellular effector by two isoforms of human neurokinin-1 receptor. *Molecular Pharmacology*, 41, 24-30.

Fu, C.Y.; Kong, Z.Q.; Wang, K.R.; Yang, Q.; Zhai, K.; Chen, Q. and Wang, R. (2005) Effects and mechanisms of supraspinal administration of rat/mouse hemokinin-1, a mammalian tachykinin peptide, on nociception in mice. *Brain Research*, 1056, 51-58.

Fu, C.Y.; Yang, Q.; Wang, K.R.; Kong, Z.Q.; Chen, Q. and Wang, R. (2006) Rat/mouse hemokinin-1, a mammalian tachykinin peptide, markedly potentiates the antinociceptive effects of morphine administered at the peripheral and supraspinal level. *Behaviour Brain Research*, 170, 293-301.

Glowinski, J. (1995) The 'septide-sensitive' tachykinin receptor: still an enigma. *Trends in Pharmacological Science*, 16, 365-367.

Graham, G.J.; Stevens, J.M.; Page, N.M.; Grant, A.D.; Brain, S.D.; Lowry, P.J.; Gibbins, J.M. (2004) Tachykinins regulate the function of platelets. *Blood*, 104, 1058-1065.

Konick, Y. and Henry, J.L. (1991). Substance P-mediated slow excitatory postsynaptic potential elicited in dorsal horn neurons, *Proceedings of the Natural Academy of Sciences of the United States of America*, 88, 11344-11348.

Kramer, M.S.; Cutler, N.; Feighner, J.; Shrivastava, R.; Carman, J.; Sramek, J.J.; Reines, S.A.; Liu, G.; Snavely, D.; Wyatt-Knowles, E.; Hale, J.J.; Mills, S.G.; MacCoss, M.; Swain, C.J.; Harrison, T.; Hill, R.G.; Hefti, F.; Scolnick, E.M.; Cascieri, M.A.; Chicchi, G.G.; Sadowski, S.; Williams, A.R.; Hewson, L.; Smith, D.; Carlson, E.J.; Hargreaves, R.J. and Rupniak, N.M. (1998). Distinct mechanism for antidepressant activity by blockade of central substance P receptors. *Science,* 281, 1640-1645.

Kurtz, M.M.; Wang, R.; Clements, M.K.; Cascieri, M.A.; Austin, C.P.; Cunningham, B.R.; Chicchi, G.G. and Liu Q. (2002) Identification, localization and receptor characterization of novel mammalian substance P-like peptides. *Gene*, 296, 205-112.

Mantyh, P. W.; Rogers, S. D.; Ghilardi, J. R.; Maggio, J. E.; Mantyh, C. R. and Vigna, S. R. (1996) Differential expression of two isoforms of the neurokinin-1 (substance P) receptor in vivo. *Brain Research*, 719, 8-13.

Meini, S.; Patacchini, R.; Maggi, C. A. (1994). Tachykinin NK1 receptor subtypes in the rat urinary bladder. *British Journal of Pharmacology*, 111, 739-746.

Metwali, A.; Blum, A.M.; Elliott, D.E.; Setiawan, T. and Weinstock, J.V. (2004) Cutting edge: hemokinin has substance P-like function and expression in inflammation. *Journal of Immunology*, 172, 6528-6532.

Morteau, O.; Lu, B.; Gerard, C. and Gerard, N.P. (2001) Hemokinin 1 is a full agonist at the substance P receptor. *Nature Immunology*, 2, 1088-1088.

Naono, R.; Yoshioka, D.; Ikeda, T; Nakayama, T. and Nishimori, T. (2007) The common carboxyl-terminal region of novel tachykinin peptides contributes to induce desensitization in scratching behavior of rats. *Brain Research Bulletin*, 71, 461-465.

Nelson, DA.; Marriott, I. and Bost, K.L. (2004) Expression of hemokinin 1 mRNA by murine dendritic cells. *Journal of Neuroimmunology,* 155, 94-102.

Nelson, D.A. and Bost, K.L. (2005) Quantification of hemokinin-1 peptide production and secretion from mouse B cells. *Cellular Immunology,* 237, 115-122.

Page, N.M. (2006a) Characterisation of the gene structures, precursor processing and pharmacology of the endokinin peptides. *Vascular Pharmacology*, 45, 200-208.

Page, N.M. (2006b) Brain Tachykinins, Handbook of biologically active peptides, Elsevier (Academic press, Ed. Kastin, A.J.), London, U.K., 763-769.

Page, N.M.; Dakour, J. and Morrish, D.W. (2006) Gene regulation of neurokinin B and its receptor NK3 in late pregnancy and pre-eclampsia. *Molecular Human Reproduction*, 12, 427-333.

Page, N.M.; Bell, N.J.; Gardiner, S.M.; Manyonda, I.T.; Brayley, K.L.; Strange, P.G. and Lowry, P.J. (2003). Characterisation of endokinins: human tachykinins with cardiovascular activity. *Proceedings of the Natural Academy of Sciences of the United States of America*, 100, 6245-6250.

Page, N.M. and Bell, N.J. (2002). The human tachykinin NK_1 (short form) and the tachykinin NK_4 receptor; a reappraisal. *European Journal of Pharmacology*, 437, 27-30.

Page, N.M.; Woods, R.J. and Lowry P.J. (2001) A regulatory role for neurokinin B in the placental physiology and pre-eclampsia. *Regulatory Peptides*, 98, 97-104.

Patacchini, R. and Maggi, C.A. (2001) Peripheral tachykinin receptors as new targets for drugs. *European Journal of Pharmacology*, 429, 13-21.

Severini, C.; Imprata, G.; Falconieri-Erspanier, G.; Salvadori, S. and Erspanier, V. (2002), The Tachykinin Peptide Family, *Pharmacological Reviews*, 54, 285-322.

Stout, S.C.; Owens, M.J. and Nemeroff, C.B. (2001) Neurokinin(1) receptor antagonists as potential antidepressants. *Annual Review of Pharmacology and Toxicology* 41, 877-906.

Vigna, S. R. (1999) Phosphorylation and desensitization of neurokinin-1 receptor expressed in epithelial cells. *Journal of Neurochemistry*, 73, 1925-1932.

Yoshioka, D.; Takebuchi, F.; Nishimori, T.; Naono, R.; Ikeda, T. and Nakayama, T. (2006), Intrathecal administration of the common carboxyl-terminal decapeptide in endokinin A and endokinin B evokes scratching behaviour and thermal hyperalgesia in the rat. *Neuroscience Letters*, 410, 193-197.

Zhang, Y.; Lu, L.; Furlong, C.; Wu, G.E. and Paige, C.J. (2000) Hemokinin is a hematopoietic-specific tachykinin that regulates B lymphoiesis. *Nature Immunology*, 1, 392-397

In: Neuropeptide Research Trends
Editor: Bernice A. Levine, pp. 13-63
ISBN: 978-1-60021-640-4

Chapter I

Neurotensin: Physiological Properties and Role in Cardiovascular Regulation

Oleg Osadchii[1,2], Angela Woodiwiss[2] and Gavin Norton[2]
1 Normal Physiology Department, Kuban Medical Academy, Krasnodar, Russia;
2 Cardiovascular Pathophysiology and Genomics Research Unit, School of Physiology, University of the Witwatersrand, Medical School, Parktown, Johannesburg, South Africa

Abstract

Neurotensin is a tridecapeptide originally extracted from bovine hypothalamus by Carraway and Leeman (1973), and later found in a variety of peripheral tissues with the highest levels determined in the jejuno-ileal portion of the gastrointestinal tract. Neurotensin possess a broad spectrum of physiological activity, and could act as a neurotransmitter and neuromodulator in the central nervous system and as a local hormone in gastrointestinal tract. Importantly, there is increasing evidence to suggest that neurotensin plays a pivotal role in regulation of the cardiovascular system. Indeed, neurotensin-containing neural fibers are found in close contact with atrial and ventricular cardiomyocytes, sinoatrial and atrioventricular nodal cells, neurons of intracardiac ganglia, as well as coronary vessels in humans and various animal species. The density of neurotensin-immunoreactive myocardial innervation is reduced in cardiac disease. Neurotensin produces a variety of cardiovascular actions including potent effects on blood pressure, heart rate, myocardial contractility, coronary vascular tone, venous smooth muscle tone as well as regional blood flow in the gastrointestinal tract, skin and adipose tissue. Neurotensin also modifies the electrophysiological properties of the heart and exerts a protective effect in vagally-induced cardiac arrhythmias. Moreover, neurotensin could trigger cardiovascular reflexes by stimulating primary visceral afferents synaptically connected with preganglionic sympathetic neurons in the spinal cord. Neurotensin-induced cardiovascular effects are strongly influenced by species-dependent differences, and rapidly subjected to tachyphylaxis. Structural determinants of biological activity of neurotensin reside primarily in the C-terminal portion of its molecule. Indeed, C-terminal neurotensin hexapeptide fragment, neurotensin (8-13), is shown to be equipotent to intact neurotensin in eliciting cardiovascular effects, as well as in binding to the purified neurotensin receptor. In contrast, short N-terminal neurotensin fragments, neurotensin (1-6), (1-8) and (1-10), do not possess cardiovascular activity, and exhibit very low affinity for the neurotensin receptor. Neurotensin-induced cardiovascular responses are mediated via direct activation of neurotensin receptors, as well as through stimulation of release of various endogenous biologically active substances such as histamine, serotonin, catecholamines and prostaglandins. To date, three subtypes of neurotensin receptors (NTS_1, NTS_2 and NTS_3) have been identified and found to be expressed in the brain and various peripheral tissues including the myocardium. NTS_1, a G-protein-coupled receptor with seven putative transmembrane domains, represents a high-affinity ($K_d \sim 0.1$-0.3 nM) neurotensin binding site coupled to phospholipase C-inositoltrisphosphate transduction pathway. The NTS_1 receptor seems to play a pivotal role in the mechanism of neurotensin-induced cardiovascular responses, whereas a putative significance of NTS_2 and NTS_3 receptor subtypes remains to be determined.

Introduction

Neurotensin (NT), a tridecapeptide with a broad spectrum of physiological activity, was originally extracted from the bovine hypothalamus [24] and later found in the bovine [108] and human [73] small intestine. Since the discovery of NT in 1973, rapid progress has been achieved in understanding various aspects related to NT function in the central nervous system and peripheral organs, as well as exploring the nature of the mechanisms involved in the physiological effects of NT in health and disease. The main focus of research in this area over the last 30 years has been on the mechanisms of central NT effects such as antinociception, hypothermia and various behavioural actions [45, 185, 233], as well as elucidating the role of NT in hormonal regulation of the digestive system [53] and proliferation of normal and cancerous cells [231]. The potential therapeutic relevance of these findings for the diagnosis and treatment of schizophrenia, cancer and obesity as well as for pain suppression has been recently reviewed [111, 233]. However, there has been much less attention given to the experimental findings demonstrating potent and diverse effects of NT on the cardiovascular system. In the present chapter we review the current knowledge related to the role of NT in various aspects of cardiovascular regulation, with a particular focus on the mechanisms mediating NT effects upon myocardial and vascular tissue. For systematic purposes, a detailed overview of NT effects upon the cardiovascular system will be preceded by a brief summary of the data related to the structural properties as well as the pharmacokinetics and metabolism of the NT molecule. We will also briefly discuss the results of some studies designed to evaluate concentrations of endogenous NT in blood plasma and its distribution profile in peripheral tissues and organs.

Neurotensin: Processing from a Precursor Molecule

NT is synthesized as a part of prohormone which also contains neuromedin N, a NT-like hexapeptide [235]. The NT/neuromedin N precursor is found in both the brain and various peripheral tissues including myocardium [34, 110] as well as in the human circulation [51], but the highest concentrations are measured in the jejuno-ileum part of the gastrointestinal tract [34, 110]. The NT/neuromedin N precursor is a polypeptide of 170 amino acids, with a structure that is 76% homologous in bovine, dogs and rats [107, 110, 207]. NT and neuromedin N are contained as adjacent sequences within the C-terminal of the precursor, and liberated in a 1:1 molar ratio following precursor processing by proprotein convertases [110]. The presence of the NT/neuromedin N precursor in various tissues accounts for a rapid (over seconds) and significant (by 25-200-fold) elevation of immunoreactive NT levels following the action of endogenous peptidases. In particular, large amounts of biologically active NT-related peptides are generated following the exposure of rat and chicken gastric tissue to aqueous acid, an effect accounted for by the proteolytic activity of pepsin-related enzymes (pepsin, cathepsin and renin) [29]. The presence of large (16-70 kDa) molecular weight protein substrates which can liberate immunoreactive NT upon pepsin treatment has also been shown to occur in a variety of feline tissues [32], as well as blood plasma of several mammalian species [31, 140, 220].

Scheme 1. NT structure in humans and various animal species (adapted from [207] with modifications)

	1	2	3	4	5	6	7	8	9	10	11	12	13
Human, dog, rat, pig, bovine	pGlu –	Leu –	Tyr –	Glu –	Asn –	Lys –	Pro –	Arg –	Arg –	Pro –	Tyr –	Ile –	Leu
Guinea-pig	----	----	----	----	----	----	*Ser*	----	----	----	----	----	----
Possum	----	----	*His*	*Val*	----	----	*Ala*	----	----	----	----	----	---
Chicken	----	----	*His*	*Val*	----	----	*Ala*	----	----	----	----	----	---
Alligator, python	----	----	*His*	*Val*	----	----	*Ala*	----	----	----	----	----	---
Frog	----	*Ala*	*His*	*Ile*	*Ser*	----	*Ala*	----	----	----	----	----	---
Toad	----	*Ala*	*Ile*	*Val*	*Ser*	----	*Ala*	----	----	----	----	----	---

The interrupted lines indicate the presence of the same amino acid residue in NT structure. The underlined amino acid sequence represents a C-terminal portion of NT molecule responsible for NT receptor activation.

A biologically active NT derived from the precursor molecule is a 13 amino acid neuropeptide whith a C-terminal portion that contains structural determinants primarily responsible for NT receptor activation (Scheme 1). The amino acid sequence of NT is similar in most mammalian species (Scheme 1). In the guinea-pig, however, Pro^7 is substituted for Ser in the NT molecule. In birds, reptiles and amphibian species the structure of NT is different from mammals due to the substitution of three-to-five amino acid residues in the N-terminal part of the NT molecule (Scheme 1). However, the amino acid sequence of the C-terminal portion of the NT molecule responsible for the physiological activity of this neuropeptide is highly conserved across species.

Pharmacokinetics and Metabolism

Exogenous NT is rapidly cleared from circulation upon intravenous administration as evidenced by significant peripheral arteriovenous extraction. The metabolic clearance rate and half-life time of NT determined in normal human subjects and two animal species are shown in table 1. Importantly, the C-terminal fragments of NT are metabolized significantly faster than the N-terminal fragments [5, 91, 121, 163, 218, 220].

Table 1. Parameters of NT pharmacokinetics

Species	Antibodies used	Clearance, ml/кg/min	Half-life, min	References
Human	N-terminal directed antisera	16±1	3.8±0.2	[19]
	C-terminal directed antisera	88±25	< 0.5	[219]
	N-terminal directed antisera	9.9±0.8	13.9±1.5	
	C-terminal directed antisera	36 (21-54)	1.7 (0.7-2.8)	[91]
	N-terminal directed antisera	11 (6.7-21.7)	8.3 (4.7-13.8)	
	C-terminal directed antisera	ND	1.4	[121]
	N-terminal directed antisera	ND	6.0	
Sheep	C-terminal directed antisera	25.8±3.2	2.1±0.3	[218]
	N-terminal directed antisera	12.8±1.4	6.3±0.5	
Rat	C-terminal directed antisera	ND*	0.55	[5, 220]
	N-terminal directed antisera	26.0±4.0	5.0	

ND, not determined. *No significant increase in C-terminal NT immunoreactivity in blood plasma was detected following intravenous infusion of intact NT.

The half-life of NT, as determined after incubating fresh human blood plasma with NT, is 226 min, which is significantly longer than the half-life found upon systemic infusions of intact NT in vivo (~1.4 min) [122]. Likewise, no N-terminal NT fragments are detected following a short-term (1-10 min) incubation of synthetic NT with rat blood [5]. These findings suggest that NT catabolism in vivo is mainly provided by tissue enzymes rather than peptidases present in blood plasma. In conscious sheep, the kidney, brain and gut are the major organs involved in the catabolism of intravenously injected synthetic NT [218]. In humans, the liver and kidney are the two visceral organs primarily involved in NT catabolism [91, 219]. The NT molecule is catabolised by angiotensin-converting enzyme, neutral endopeptidase 24.11 (enkephalinase), endopeptidase 24.15 and endopeptidase 24.16 which cleave Arg^8-Arg^9, Pro^{10}-Tyr^{11} and Tyr^{11}-Ile^{12} bonds [13, 41, 110, 135, 136, 222]. The cleavage products detected following degradation of the intact NT are represented by free tyrosine, C-terminal dipeptide Ile^{12}-Leu^{13} as well as NT fragments such as NT_{1-7}, NT_{1-8}, NT_{1-10}, NT_{1-11} and NT_{11-13}. The overall rate of degradation of intact NT in a culture of human endothelial cells is 80 nM per 1 mg of protein per hour [144].

Blood Plasma Concentrations

Picomolar concentrations of endogenous NT are present in blood plasma (table 2) as well as cerebrospinal fluid [79, 216, 243] and human milk [50]. In rats, about 85% of the total plasma NT-like immunoreactivity is represented by N-terminal NT fragments such as NT_{1-8} (73%) and NT_{1-11} (11%), whereas only 15% is represented by intact NT [220]. In human blood plasma, the concentrations of N-terminal NT fragments are about 5-fold higher as compared to intact NT [163, 219].

Table 2. Plasma NT concentrations in humans and various animal species

Species	NT concentration, pM/l	References
Human		
healthy adults	14-65	[17-19, 26, 31, 33, 55, 74, 91, 97, 105, 121, 125, 130, 131, 163, 204, 217, 219]
healthy children of 1-14 years old	24	[16]
infants	29-48	[125]
pregnant women	29	[130]
Dog	2-40	[6, 26, 132, 139, 199]
Cat	25-30	[31, 61, 62]
Rabbit	37	[26]
Rat	20-48	[25, 26, 31, 200, 220, 230]
Bovine	15-25	[25, 26]
Frog		
Xenopus laevis	0.05-1.6	[28]
Rana catesbeiana	0.5	
Rana pipiens	0.03-0.53	

No age- or gender-related differences have been found in fasting plasma NT levels in adult humans [163]. However, NT concentrations are reported to be higher in newborns as compared to adults, and full-termed infants have higher plasma NT levels as compared to preterm infants at birth [125].

NT concentrations in blood plasma are increased upon food intake. In humans, food intake elicits a biphasic elevation of NT concentrations which peaks at 20-30 and 90 min after the ingestion of a large meal, and enhanced blood plasma NT concentrations are maintained at least over 3 hours after ingestion [17, 55, 131, 204, 217]. Interestingly, the elevation of plasma NT concentration following food intake is faster and achieves a higher peak value in patients subjected to total or partial gastrectomy, an effect related to accelerated gastroduodenal passage of chyme [208]. Food intake-induced elevations of the total NT-like immunoreactivity in blood plasma is largely attributed to a significant (up to 10-fold) increase in concentrations of the stable N-terminal NT fragments such as NT_{1-8} and NT_{1-11} [74].

The increase in NT concentrations in blood plasma upon food intake is due to nutrient-induced stimulation of NT secretion by mucosal endocrine-like N-cells localised within the epithelial layer of small intestine [81, 167, 227]. Indeed, elevated NT concentrations in portal blood are found following perfusion of the isolated small intestine with solutions containing sodium oleate, bile acids or peptones in dogs [60], rats [49, 230] and cats [34]. A similar effect is observed following intraduodenal infusion of oleic acid in humans [105]. The most potent stimulant of NT secretion is fat, with a less significant effect produced by glucose [18, 49, 130]. NT secretion, however, is not affected by the intake of amino acids, coffee and alcohol [131]. During fasting, plasma NT levels are decreased below threshold of chromatographic detection.

Apart from a local stimulatory action of nutrients, NT secretion by N-cells is stimulated by some gastrointestinal hormones released into the portal circulation following food intake. In the isolated, vascularly perfused rat small intestine, a potent stimulation of NT secretion is induced by bombesin, calcitonin gene-related peptide, gastrin-releasing peptide, substance P and leu-enkephalin [48, 86, 200], whereas somatostatin exerts an inhibitory effect [54]. There is some evidence to suggest that the secretory activity of N-cells is stimulated by vagal nerve fibers. In particular, both basal and meal-stimulated NT concentrations in human blood plasma are reduced following the infusion of atropine, a muscarinic cholinergic receptor blocker, to healthy volunteers [55]. Consistently, NT secretion by rat N-cells is significantly increased by cholinergic agonists [86].

Increased basal plasma NT levels have been described in various types of pathology such as Parkinson's disease [214], diabetes mellitus [130], dumping syndrome in gasterectomized patients [18], and menopausal hot flashes in women [113]. Furthermore, plasma NT concentrations are markedly increased in patients with NT-secreting neuroendocrine tumors (neurotensinomas) localised in the liver, pancreas, stomach or prostate [33, 97]. Alternatively, an elevation of endogenous NT concentrations in blood plasma could result from a reduced degradation by renal and liver peptidases. In particular, elevated concentrations of NT are found in the systemic circulation in patients with liver cirrhosis [91] and chronic renal failure [219]. In rats subjected to bilateral nephrectomy, basal plasma NT concentrations have been reported to be 3-fold higher as compared to unoperated control rats, an effect associated with significantly reduced metabolic clearance rate of NT [220]. The clinical significance of these findings remains to be determined.

Tissue and Organ Distribution Profile

In rats, only about 10% of the total amount of NT-immunoreactive material is present in the brain, whereas 90% is distributed throughout a variety of peripheral tissues, with the highest amount (85%) found in the intestine [25]. In the brain, the highest concentrations of NT are found in the hypothalamus (60 pM/g) and pituitary gland (29 pM/g), and the lowest in the cerebellum (0.8 pM/g) [25].

In the rat gastrointestinal tract, the highest (50 pM/g) NT concentrations are found in the jejunum and ileum, whereas much lower (1-8 pM/g) concentrations are detected in the esophagus, stomach, duodenum and large intestine [25]. NT-immunoreactive cells (N-cells) have been found in the ileal mucosa of 13 animal species [81], being more abundant in non-mammalian vertebrates as compared to mammals [179]. In rats, the density of N-cells is about 150 cells per mm^2 of mucosal surface in terminal ileum [179]. These cells have triangular or slightly elongated shape, are located in the epithelium of the villi and crypts, and their apical pole reaches the gut lumen and constitutes part of the brush border ("open type" cells). NT is stored in round electron-dense secretory granules with an average diameter of about 260-290 nM, which are located in the basal portion of the N-cells [167, 179, 227].

In peripheral structures of the autonomic nervous system, NT is mainly localised in presynaptic terminals of preganglionic sympathetic nerves originating in the thoracic spinal cord [126, 182]. In cats, NT-containing sympathetic preganglionic neurons are mostly present in segments T4-T7 [115, 116]. The axons of these neurons project to the stellate ganglion, L7 ganglion of sympathetic chain, and coeliac ganglion [23, 126, 128, 129]. In the stellate ganglion, NT-containing nerve terminals are widespread and closely associated both with cardiac and non-cardiac sympathetic postganglionic neurons [23]. In guinea-pigs, a considerable number of NT-immunoreactive neural fibers are found in para- and prevertebral sympathetic ganglia [182]. NT-like immunoreactivity, however, is not found in perikarya of the principal ganglionic cells. In preganglionic sympathetic fibers, NT is subjected to anterograde axonal transport as evidenced by accumulation of NT-like immunoreactivity proximal to the ligation [126-128] as well as depletion of NT-like immunoreactivity following chronic decentralization of sympathetic ganglia [35].

With respect to endocrine glands, the highest NT levels are found in the adrenal glands where it is stored in the chromaffin granules present in a subpopulation of norepinephrine-containing cells [43, 65, 168, 187, 201, 229]. In cats, about 60-70% of norepinephrine-containing cells exhibit positive NT-like immunoreactivity. The highest adrenal NT concentrations were found in cat (64 pM/g), as compared to lower levels noted in bovine (6 pM/g), rat, guinea-pig and rabbit (1.3-2.1 pM/g) adrenal tissue [65]. In humans, the NT concentration in adrenal tissue is ~0.7 pM/g [201]. Importantly, NT could act as a neurotransmitter/neuromodulator regulating secretory function of chromaffin cells. Indeed, NT-immunoreactive fibers innervate norepinephrine-containing adrenal chromaffin cells in the cat [165], hamster [164] and frog [221]. Moreover, a stimulatory effect of NT on the release of catecholamines has been shown to occur in cultured bovine adrenal chromaffin cells [20].

High (50 Hz) frequency stimulation of the splanchnic nerve in cats induces a 3.6-fold increase of NT concentration in blood plasma collected from the adrenal vein, an effect antagonized by hexamethonium [62, 201]. Likewise, NT release from adrenal medulla has been demonstrated in stress models designed to stimulate the sympathoadrenal axis such as shock induced by splanchnic artery occlusion [62] and hemorrhagic hypovolemia [61]. In isolated, retrogradely perfused cat adrenal glands, NT secretion is stimulated by acetylcholine, nicotine or high potassium concentrations [43]. Both in vivo and vitro studies have shown that NT release from adrenal tissue occurs concurrently with catecholamines, enkephalins, neuropeptide Y and somatostatin [43, 61, 62, 201]. Taken together, these findings suggest that NT could act as an adrenal hormone involved in physiological responses to stress stimuli.

Cardiovascular Innervation

The presence of NT-immunoreactive neural fibers in the cardiovascular system of a variety of mammalian and non-mammalian species has been demonstrated in many studies (table 3).

Following the quantitative assessment of the total content of NT-like immunoreactive material in the heart of various species, the highest myocardial NT concentrations are found in the guinea-pig (3.1 pM/g) and hen (3.0 pM/g), followed by rat (2.2 pM/g), cat (1.0 pM/g), dog (0.8 pM/g) and human (0.8 pM/g) [201]. With regards to comparison of NT content in myocardium versus non-cardiac tissues, NT concentrations in rat atria and ventricles are found to be 1.8 and 2.3 pM/g, respectively, which is comparable to NT levels in rat liver, lung and oesophagus, but higher than NT concentrations in renal, duodenal and splenic tissue [67].

NT-Immunoreactive Cardiac Innervation

Using immunohistochemistry and radioimmunoassay, NT-containing fibers were found in close contact with atrial and ventricular cardiomyocytes as well as neurons of intracardiac ganglia localised in the vicinity of the sinoatrial node [59, 180, 181, 183, 184]. The NT-immunoreactive synaptic inputs to myocardial ganglionic cell bodies presumably mediate excitatory effects as evidenced by an increase in activity of right atrial neurons in the dog heart in situ following a local intra-arterial NT infusion [4]. In the rat [36], guinea-pig [181, 183] and rabbit [150] myocardium, NT levels are higher in the atria as compared to the ventricles, and higher NT concentrations are found in right atrium and ventricle as compared to the left side of the heart [149, 150]. A dense network of NT-containing fibers is found in the sinoatrial and atrioventricular node [181, 183, 184, 240] as well as in the bundle of His, where the density of NT-immunoreactive innervation is about 10-fold higher as compared to right atrial tissue [63]. In experiments conducted by our own group, NT was noted to be present at picomolar concentrations in both the rat ventricular myocardium and in samples obtained from the coronary effluent collected from isolated, perfused heart preparations from the rat [157]. These findings suggest that NT is stored in and could be released from neural terminals localised in rat myocardial tissue.

Table 3. Distribution profile of NT-immunoreactive nerve fibers in myocardial tissue as determined by immunohistochemistry and radioimmunoassay

Species	Localization of NT-like immunoreactivity	References
Human	Cardiac ventricles, aorta, pulmonary vein and coronary vessels. No NT-immunoreactive fibers are found in atrial tissue and pulmonary artery.	[150, 180, 184]
Monkey	Sinoatrial node, intracardiac ganglia and coronary vessels. Coronary arteries are more densely innervated than veins.	[240]
Dog	Sinoatrial node, atrioventricular node, and coronary vasculature. In less amounts, NT-immunoreactive nerve fibers are found in close contact with atrial and ventricular cardiomyocytes, as well as in intracardiac ganglia.	[59, 240, 241]
Rabbit	Atrial tissue, aorta, pulmonary artery and vein. Lower NT levels are found in ventricular tissue.	[150]
Cat	Sinoatrial node, intracardiac ganglia and coronary vessels. Coronary arteries are more densely innervated than veins.	[240]
Rat	The highest NT levels are found in pulmonary vein, followed by right atrium and ventricle. In lower amounts, NT is present in aorta, left cardiac chambers and pulmonary artery.	[36, 149, 150, 241]
Guinea-pig	Sinoatrial node, atrioventricular node, the bundle of His, adventitia of the ascending aorta, aortic arch and pulmonary trunk. In coronary vasculature, arterial side is more densely innervated than veins. NT-containing nerve fibers are also found in intracardiac ganglia.	[63, 180, 181, 183, 240, 241]
Chicken	NT level in heart homogenates is found to be 3.0 ± 0.7 pM/g. NT-immunoreactive fibers are found to supply ascending aorta, coronary arteries and intracardiac ganglia.	[184, 201]
Quail	Small coronary arteries as well as atrial and ventricular cardiomyocytes.	[184]
Frog	NT level in heart homogenates is found to be 1.4 ± 0.1 pM/g. NT-immunoreactive fibers are associated with aortic trunk, coronary vasculature, atrial and ventricular cardiomyocytes.	[28, 67, 184]
Lizard	Adventitia of ascending aorta and coronary vessels, paravascular plexus, atrial and ventricular cardiomyocytes.	[184]
Australian lungfish, *Neoceratodus forsteri*	Sinus venosus, atrial myocytes, atrial ganglion cells.	[90]
Teleost bony fish, *Cottus scorpius*	Ventricular cardiomyocytes.	[184]

The origin of NT-immunoreactive cardiac innervation remains uncertain. The somata of NT-containing neurons whose axons project to cardiac structures seem to have extracardiac localization as no NT-like immunoreactivity is found in perikarya of intrinsic cardiac neurons [59, 78, 181]. In the rat myocardium, NT-immunoreactive innervation is not affected by pretreatment with capsaicin, a neurotoxin targeting primary afferent sensory fibers, or 6-

hydroxydopamine, an agent that destroys postganglionic sympathetic fibers [67]. Therefore, it is possible that NT is predominantly contained in cardiac preganglionic parasympathetic fibers [67]. However, to our knowledge, no studies assessing the effect of colchicine treatment and antibodies against choline acetyltransferase have been done to ascertain whether NT-immunoreactive fibers are related to cardiac parasympathetic innervation. Furthermore, there is no evidence to demonstrate the presence of NT-immunoreactivity in perikarya of cardiac preganglionic neurons in the dorsal motor nucleus of the vagus nerve [88]. Thus, the origin of a putative parasympathetic NT-immunoreactive innervation of the heart is uncertain.

NT-like immunoreactivity has not been found either in sympathetic or parasympathetic ganglionic cell bodies [126, 182, 184]. However, Weihe et al. have hypothesized that a fraction of ganglionic cells could contain a NT precursor polypeptide, a molecule which does not react with NT-specific antisera [241]. The NT-precursor may be delivered to presynaptic terminals of cardiac nerves by axonal transport and subjected there to enzymatic cleavage to produce the biologically active NT [241]. In support of this argument, high concentrations of NT precursor are found in cat myocardial tissue [34]. Nevertheless, further studies are required to assess this hypothesis.

NT-immunoreactive innervation is altered in cardiac disease. In rat model of monocrotaline-induced cardiac hypertrophy, endogenous NT levels are significantly decreased in all cardiac chambers, especially on the right side, and both cardiac and circulating NT concentrations are reduced with the transition from cardiac hypertrophy to heart failure [36].

NT-Immunoreactive Vascular Innervation

Numerous evenly distributed NT-containing fibers are detected in peri- and paravascular plexuses innervating coronary vessels [59, 181, 183, 240]. These fibers are found in all segments of coronary vasculature, but the coronary arteries are more densely innervated than the veins. Furthermore, NT-immunoreactive innervation of coronary vessels is more abundant in atria as compared to ventricles [184]. Numerous NT-containing neural fibers are also found in the adventitia of the ascending aorta, aortic arch, and pulmonary vein, whereas only scarce NT-immunoreactive innervation is detected in superior and inferior vena cava [149, 150, 181, 184].

NT-immunoreactive innervation has been found in large parietal and splachnic vessels in the guinea-pig and dog [184], with the density of innervation generally being higher in arteries as compared to veins. In large arteries, NT-containing neural fibers are mainly found at the adventitia-media border. NT-immunoreactive innervation is particularly pronounced in cortical arteries of the kidney, renal and intrapancreatic arteries, as well as in the arteries of the skin. However, very few NT-immunoreactive fibers are found in cerebral vessels [184]. Amongst venous vessels, only the portal vein exhibits relatively dense innervation by NT-immunoreactive fibers [184].

Vascular Effects of NT

Blood Pressure

The modulation of vascular smooth muscle tone is one of the typical physiological effects produced by NT, and this action was reported in the first publication describing the extraction procedure and biological activity of this neuropeptide [24]. Indeed, the name "NT" was originally derived from the potent vascular effects produced by this regulatory peptide extracted from the neural tissue. However, NT effect on systemic blood pressure is largely species-specific. In the rat [39, 44, 58, 138, 177, 190], dog [15, 52, 202] and rabbit [99], intravenous infusion of NT at nanomolar concentrations produces marked hypotensive effect lasting at least 40-60 min. The NT-induced hypotensive effect is rapidly subjected to tachyphylaxis thus enabling only one effective NT concentration to be tested in each animal. On the other hand, NT produces dose-dependent pressor effects in guinea-pigs [100, 143, 196, 197] and sheep [190]. In anesthetized rats [71, 101, 146, 148, 177, 191] and guinea-pigs [100, 143, 190], NT also elicits biphasic and triphasic variations (e.g. initial decrease followed by a transient increase and a further long-lasting decrease) of blood pressure. In humans, at physiological concentrations representing postprandial plasma NT levels, NT has no effect on blood pressure [18, 121, 124, 205, 219, 234], but produces a moderate pressor response following administration at a pharmacological ($\sim 10^{-5}$ M) dose [156].

Rat strain-related differences in NT effects upon blood pressure are noted in conscious animals. In particular, intravenous infusion of NT induces a marked fall in blood pressure in Sprague-Dawley rats [226] while producing pressor responses in Long Evans rats [10]. In anesthetized rats, the pattern of NT-induced changes in blood pressure could be influenced by the type of anesthetic employed. For instance, NT induces a uniform, prolonged hypotensive effect in urethane-anesthetized rats, whereas it elicits a biphasic response in rats anesthetized with sodium thiobutabarbitone, a decrease in blood pressure being preceded by a marked but transient pressor effect [39].

Hypotensive effects of NT in rats are abolished by H_1-histamine receptor antagonists or agents depleting histamine stores in mast cells such as compound 48/80 [39, 71, 146, 177, 190], as well as by SR 48692, a NT receptor antagonist [71, 212]. Consistently, NT exerts a direct, NT receptor-mediated stimulatory action on histamine release from isolated rat mast cells [14, 27, 30, 114, 118, 206, 215], an effect which will be discussed in more detail below. However, stimulation of endogenous histamine release seems to be only one of as a number of mechanisms mediating NT-induced vasodepressor responses. Indeed, NT-induced hypotension could be accounted for, at least in part, by an inhibitory NT effect on norepinephrine release from rat perivascular nerve fibers [232]. Furthermore, hypotensive effects of NT in anesthetized rats are strongly attenuated following acute bilateral nephrectomy as well as captopril or saralasin pretreatment [101, 190] indicating that NT could decrease the activity of renin-angiotensin system.

In anesthetized guinea-pigs, pressor effects of NT are completely abolished by ganglion blocking agents, reserpine or a mixture of α- and β-adrenoreceptor antagonists, thus indicating the involvement of a catecholamine-dependent mechanism [100]. Likewise, the hypertensive phase of a biphasic depressor-pressor NT effect is significantly reduced by adrenalectomy or α-adrenoreceptor antagonist administration in anesthetized ganglion-blocked rats [191]. There is some evidence to suggest that NT-induced sympathetic activation represents an indirect effect mediated by stimulation of substance P and/or calcitonin gene-related peptide release from capsaicin-sensitive sensory neurons coupled to spinal preganglionic sympathetic neurons [9, 10]. Indeed, sympathetically-mediated pressor effects produced by intravenous NT administration in anesthetized guinea-pigs [9, 197] and conscious, unrestrained Long Evans rats [10] are abolished by chronic pretreatment with capsaicin, a neurotoxin that depletes substance P and calcitonin gene-related peptide stores in primary sensory neurons. Consistently, NT has been shown to promote stimulus-induced release of substance P from the myenteric plexus of the guinea-pig small intestine [92]. Importantly, NT-induced pressor responses in guinea-pigs are inhibited by SR 48692, a NT receptor antagonist [143], suggesting that the functional NT receptor is located on capsaicin-sensitive sensory neurons involved in NT-induced sympathetic activation. The functional significance of these effects for reflex regulation of cardiovascular system will be discussed below.

It can be hypothesized that the potent stimulatory effects of NT upon release of various vasoactive endogenous neuromodulators could mask its own direct effect on vascular smooth muscle tone. However, NT has no effects on isolated rat [44] and rabbit [198] aortic strips, as well as the rabbit pulmonary artery [145] in in vitro studies. Likewise, only weak endothelium-dependent relaxation of the dog isolated carotid artery was elicited by NT at concentrations about two orders of magnitude higher than those of tachykinins, reference vasoactive peptides [46]. Furthermore, NT-induced vasodepressor responses in anesthetized rats were not affected by prior administration of L-NAME, an inhibitor of nitric oxide synthase [71]. Taken together, these findings suggest that a direct effect of NT on vascular smooth muscle tone does not significantly contribute to changes in blood pressure evoked by systemic NT administration.

Coronary Vascular Tone

NT induces coronary vasoconstriction in rats [171, 172, 176, 192], but evokes a dilation of coronary arteries in guinea-pigs [8] and dogs [15, 52]. On the other hand, NT has no effect on coronary vascular tone in the isolated, perfused rabbit heart [171]. In isolated, electrically-driven guinea-pig heart preparation, NT-induced coronary vasodilator effects were shown to be larger at high basal coronary perfusion pressures [8, 52]. The vasodilator effects were abolished by atropine but markedly potentiated by neostigmine, a cholinesterase inhibitor, indicating that NT actions on coronary arteries are mediated via stimulation of presynaptic acetylcholine release [8].

In the dog, NT has no effect on coronary blood flow upon systemic administration [6], but produces coronary vasodilation upon intracoronary injection, an effect significantly potentiated by captopril, an inhibitor of angiotensin-converting enzyme [15]. NT-induced coronary vasodilatation in anesthetized open-chest dogs is abolished by indomethacin thus suggesting that this effect is mediated by prostanoid products produced by cyclooxygenase [52]. Moreover, NT-induced effects on coronary vascular tone is abolished following gradual reduction of coronary perfusion pressure, an effect presumably related to preactivation of the prostaglandin system in the ischemic myocardium [52].

NT-induced coronary vasoconstriction in isolated, perfused rat heart preparation is abolished by the NT receptor antagonists, [D-Trp^{11}]-NT and [Tyr(Me)^{11}]-NT, but not by antagonists of angiotensin II, serotonin, H_1-histamine and α-adrenergic receptors, suggesting a direct, NT receptor-mediated effect on coronary vascular tone [176]. Interestingly, this effect is potentiated in hearts pretreated with compound 48/80 or dexamethasone [192] indicating that NT receptor-mediated coronary vasoconctrictor responses could be partially antagonized by concomitant NT-induced stimulation of histamine release from cardiac mast cells.

The hypotensive and coronary vasoconstrictor effects of NT in rats are markedly enhanced following 24-72 hours fasting [172], an effect related to reduction of plasma NT concentrations which presumably contributes to up-regulation of NT receptors in the cardiovascular system.

Gastrointestinal Blood Flow

NT reduces gastric mucosal blood flow in human subjects [56]. On the other hand, NT induces a significant dose-dependent increase in intestinal blood flow in rats [148], cats [77] and dogs [6, 112, 202], an effect related to reduced intestinal vascular resistance. In the dog, NT-induced increases in regional intestinal blood flow are observed in the muscular layer of the duodenum, jejunum, ileum and colon while mucosal blood flow is unchanged [6]. The intestinal vasodilation following systemic NT administration is induced at concentrations below those eliciting hypotension, suggesting that locally released NT could potentially act as a mediator of postprandial intestinal hyperemia. This function, however, seems not to be preserved in ischemic intestinal tissue. Indeed, NT aggravates the intestinal microcirculatory disturbances produced by local ischemia-reperfusion in rats [87].

NT increases pancreatic blood flow in the dog [112]. Consistently, NT-immunoreactive neural fibers are found in close contact with intrapancreatic vasculature in the dog [184].

Cerebral Blood Flow

NT, at pharmacological concentrations (~10^{-5} M), produces a significant contractile effect on isolated human cerebral arteries [242]. In this regard, the efficacy of NT is higher as compared to some reference vasoconstrictor agents such as norepinephrine, serotonin, arginine vasopressin and angiotensin II. On the other hand, perivascular infusion of NT over a wide range of concentrations has no effect on the smooth muscle tone of feline cerebral arteries in situ [238]. Likewise, intravenous NT infusion does not affect regional cerebral blood flow in anesthetized dogs [6].

Regional Blood Flow in other Areas

NT infusion produces cutaneous vasodilation in anesthetized rats [24], an effect mediated via stimulation of histamine release from skin mast cells [42]. NT also moderately increases perfusion medium flow in an intact, in situ perfused rat adrenal gland [89], but has no effect on regional blood flow in adrenal glands and kidney following intravenous administration in anesthetized dogs [6]. On the other hand, NT induces a delayed vasoconstriction in denervated subcutaneous adipose tissue in dogs [202, 203] and abdominal adipose tissue in humans [124], suggesting a physiological role of NT in the regulation of postprandial uptake of nutrients by adipose tissue.

NT induces a vasoconstrictor effect in the isolated, perfused rat hindquarter, an action antagonized by methysergide, a serotonin receptor blocker [102]. Consistently, NT elicits a prolonged vasoconstrictor effect in renal, superior mesenteric and hindquarter vascular beds upon systemic administration to conscious rats [10]. NT-induced vasoconstrictor responses are significantly reduced by α-adrenoreceptor blocker indicating an interacton between NT and the sympathetic nervous system.

Venous Smooth Muscle Tone

NT induces dose-dependent increases in basal tension, spontaneous contraction rate and peak phasic contraction of the longitudinal strips of the isolated rat portal vein [80, 188, 238]. In this regard, NT is more potent than substance P, bradykinin and norepinephrine, but less potent than angiotensin II [80]. NT-induced venoconstrictor effects are markedly inhibited by indomethacin, suggesting an involvement of prostaglandin-dependent mechanisms [188]. Consistently, NT induces prostacyclin release from endothelial cells of the human umbilical vein, as well as elevation of circulating 6-keto-prostaglandin $F_{1\alpha}$ levels in anesthetized rats [212]. Both NT-induced venoconstrictor responses [188, 189] and an increase in plasma levels of prostaglandin $F_{1\alpha}$ [212] are abolished by the selective NT receptor antagonists, [D-Trp^{11}]-NT, [Tyr(Me)^{11}]-NT and SR 48692. NT, however, has no detectable effect on the smooth muscle tone of the guinea-pig portal vein [238].

Cardiac Effects

Basal Heart Rate

NT induces a significant tachycardia in rats [10, 169, 190, 225], guinea-pigs [7, 100, 143, 169, 196, 197], cats [12, 152] and sheep [190]. In healthy human volunteers, NT has no effect on basal heart rate upon intravenous infusion at physiological (picomolar) concentrations [19, 121, 124, 205, 219], but produces a tachycardia following bolus injection at a pharmacological ($\sim 10^{-5}$ M) dose in patients with cardiac disease [156]. NT does not modify the rate of spontaneous contraction in isolated rabbit heart preparations [169, 202]. Furthermore, no consistent evidence is available to show reproducible chronotropic responses to NT in dogs. Indeed, although NT has no effect on basal heart rate upon systemic administration [202], it produces a modest bradycardia after intracoronary injection [15, 52] or marginal heart rate acceleration following injection into the intact sinus node artery in anesthetized, open chest dogs [186].

Interestingly, NT has been shown to modify basal heart rate in some non-mammalian animal species. In particular, an increase in heart rate follows systemic NT administration in the estuarine crocodile [98]. On the other hand, NT produces an atropine-dependent bradycardia in free-swimming, unanesthetized Australian lungfish [90]. NT has no effect upon basal heart rate in anesthetized hens [47].

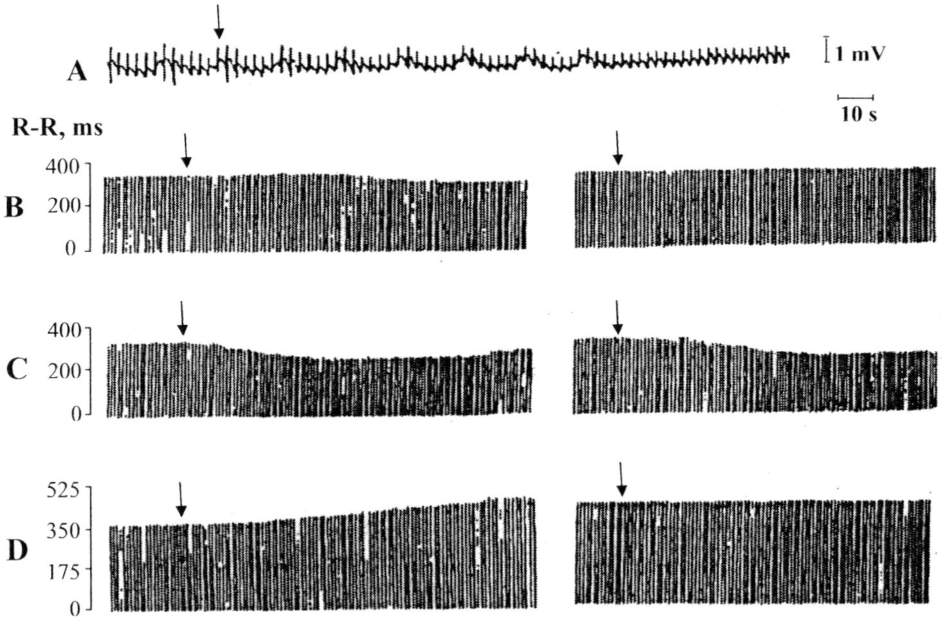

Figure 1. Modifications of the cardiac electrogram (panel A) and changes of the cardiac cycle length (R-R interval of the cardiac electrogram, ms) (panels B-D) following systemic administration of neurotensin and epinephrine in anesthetized cats. Neurotensin (4.0 x 10^{-8} M) and epinephrine (0.5 x 10^{-8} M) were administered as intravenous bolus injections. Arrows in all fragments indicate the moment of injection. Panel B: positive chronotropic response to neurotensin (left fragment) is subjected to rapid tachyphylaxis as evidenced by the lack of shortening of the cardiac cycle following the second neurotensin injection (right fragment). Panel C: positive chronotropic responses following the first (left fragment) and second (right fragment) epinephrine administration are very similar indicating the lack of tachyphylaxis. Panel D: neurotensin-induced effect is blunted (right fragment) after prior intravenous administration of propranolol (1.0 mg/kg), a β-adrenoreceptor blocker (left fragment). Note a reduction in the voltage of R waves on cardiac electrogram trace following neurotensin injection (panel A), and lengthening of the cardiac cycle following propranolol administration (panel D, left fragment). The time interval between two consequtive injections of neurotensin (panel B) or epinephrine (panel C), as well as administrations of propranolol and neurotensin (panel D) was 15 min. A comparison of the mean values of variables characterising the dynamics of neurotensin- and epinephrine-induced chronotropic responses is given in table 4.

In guinea-pigs, heart rate acceleration induced by systemic NT administration in vivo [100] as well as in isolated, spontaneously beating heart [7] is not abolished by ganglion blocking agents or antagonists of β-adrenoreceptors, H_2-histamine and 5-HT receptors, thus indicating a direct effect of NT on the sinoatrial node. The magnitude of NT-induced positive chronotropic responses tends to be larger in heart preparations with lower basal rates of contraction [7]. On the other hand, NT-induced tachycardia in rats [10, 191], cats [152] (figure 1) and sheep [190] is abolished by β-adrenoreceptor blockers, reserpine pretreatment or adrenalectomy thus indicating that chronotropic responses to NT are mediated by the release of adrenal catecholamines. Indeed, plasma epinephrine levels are increased by ~4-fold following systemic NT administration in anesthetized rats, an effect suppressed by adrenalectomy [147].

Catecholamine-dependent tachycardia induced by NT in cats could be mediated by facilitation of sympathetic ganglionic transmission [11, 12, 23]. Indeed, NT is released by tonic firing of preganglionic sympathetic axons projecting to the stellate ganglion [128, 129]. Furthermore, a profound heart rate acceleration can be induced by close intra-arterial injection of NT into the decentralized stellate ganglion in anesthetized spinal cats [12]. NT-induced tachycardia is abolished by propranolol, a β-adrenoreceptor antagonist, as well as by section of the inferior cardiac nerve thus indicating that the NT effect is related to excitation of sympathetic postganglionic neurons [12].

Chronotropic responses elicited by equimolar concentrations of NT and epinephrine upon systemic administration in anesthetized cats are of a similar magnitude (table 4). However, both latency and duration of NT-induced chronotropic responses are longer as compared to epinephrine-induced effects (table 4). Moreover, in contrast to epinephrine-induced cardioaccelearation, the chronotropic responses to NT rapidly undergo tachyphylaxis (figure 1) lasting at least for two hours after systemic NT administration.

Table 4. Parameters of the positive chronotropic responses induced by NT and epinephrine in anesthetized cats

Agent used	Basal cardiac cycle length, ms	Peak shortening of the cardiac cycle, ms	Latency of chronotropic response, s	Time to peak of chronotropic response, s	Duration of chronotropic response, s
Neurotensin, (n=7)	372.5±10.6	315.0±9.8*	25.4±5.1**	11.9±1.5	193.6±23.1**
Epinephrine, (n=8)	368.8±18.1	293.8±13.5*	7.8±2.2	9.5±2.2	98.9±12.6

Both agents were infused intravenously at equimolar concentrations (~10^{-8} M) in anesthetized cats maintained on artificial lung ventilation. Cardiac cycle length (R-R interval of the cardiac electrogram) was recorded on a beat-to-beat basis. *P<0.01 vs. baseline cardiac cycle length; **P<0.01 vs. the same value after epinephrine infusion.

Cardiac Electrophysiology

In patients with cardiac disease, NT administered intravenously at a pharmacological concentration (~10^{-5} M) produces a variety of effects on electrophysiological properties of the cardiac conduction system [156]. In particular, NT reduces cardiac cycle (R-R interval

ECG) duration, shortens the effective refractory period of the atrioventricular junction, as well as increases the value of Wenckebach's point following rapid atrial pacing (figure 2). Taken together, these findings indicate that NT exerts a positive chronotropic response associated with increases in myocardial excitability and atrioventricular conduction in the human heart, which is similar to the cardiac effects induced by sympathetic stimulation. Sinus node recovery time was slightly reduced upon NT administration, an effect largely attributed to a marked NT-induced shortening of the cardiac cycle. Indeed, an evaluation of the corrected sinus node recovery time (pre-automatic pause) revealed that a restoration of sinus node automaticity upon cessation of the rapid atrial pacing was significantly slowed following NT administration (figure 2, panel F). Further clinical studies are needed to examine whether a sustained elevation of endogenous plasma NT levels (e.g. in patients with NT-secreting neuroendocrine tumors) could potentially contribute to sinus node dysfunction.

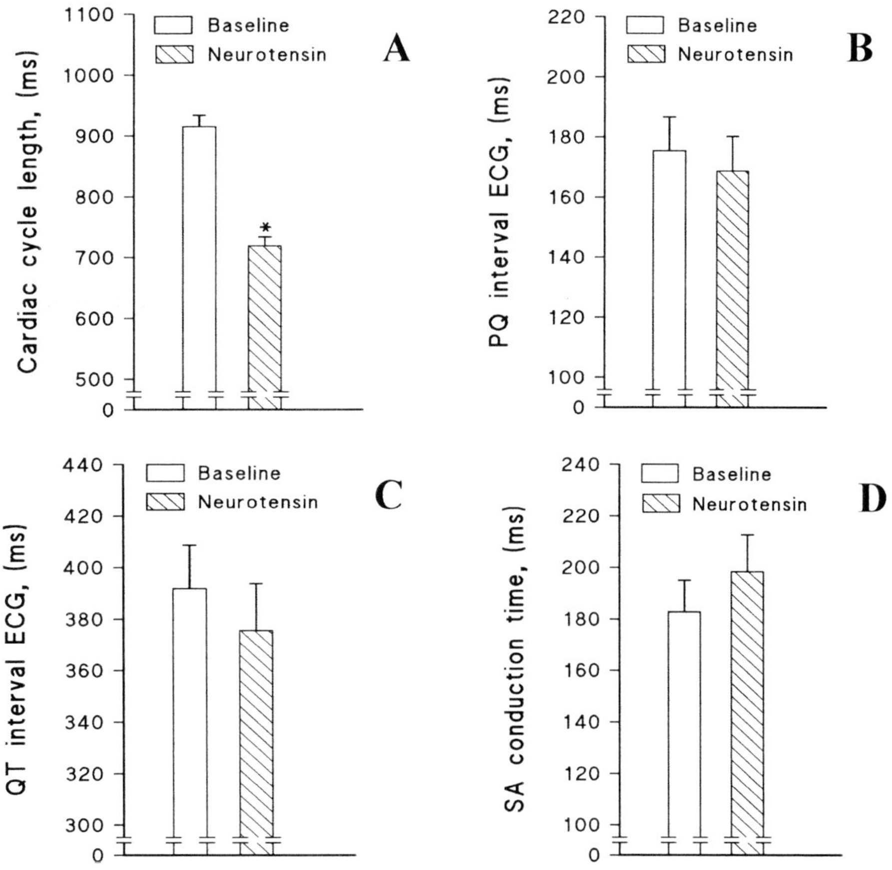

Figure 2 Continued.

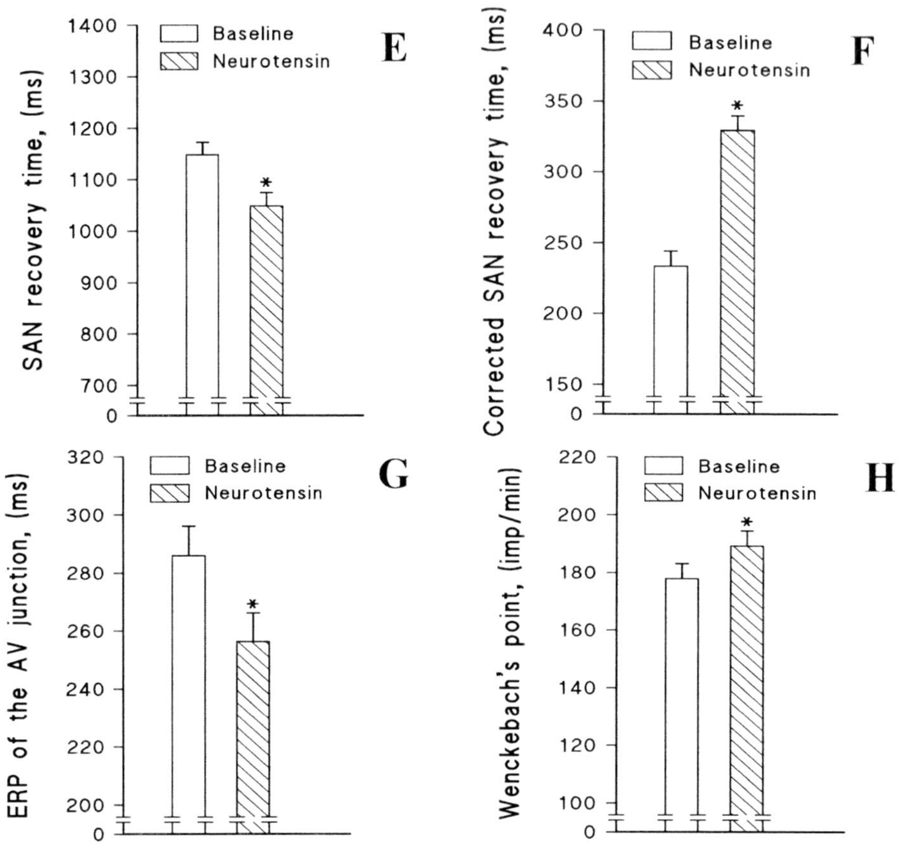

Figure 2. Effects of neurotensin on electrophysiological properties of the human myocardium. Electrophysiological parameters of the cardiac conduction system were determined in patients with cardiac disease (n=40) by programmed transesophageal atrial pacing. In recumbent patients, a bipolar electrode probe introduced intranasally into the esophagus towards the level of the left atrium was used to stimulate the myocardium with a series of rectangular pulses (10 ms, 12.8±2.3 mA). Neurotensin (~10^{-5} M) was administered as an intravenous bolus injection. ECG, electrocardiogram; SA, sinoatrial; SAN, sinoatrial node; ERP, effective refractory period; AV, atrioventricular. The corrected SAN time (panel F) was determined by subtracting the cardiac cycle length from the corresponding value of the SAN recovery time. *P<0.05 as compared to baseline value.

Autonomic Regulation of the Heart

NT-induced stimulation of sinoatrial node automaticity contributes to reduction of the magnitude of cardioinhibitory responses induced by concomitant activation of parasympathetic nervous system. In anesthetized cats, NT reduces the magnitude of the lengthening of the cardiac cycle evoked by vagal stimulation [151-153, 155] and abolishes vagally-induced cardiac arrhythmias such as second degree atrioventricular block and isorhythmic atrioventricular dissociation [154] (figure 3). A relatively long latency of NT-induced antiarrhythmic effects (second degree atrioventricular block: 40.7±4.6 s; atrioventricular dissociation: 58.7±10.3 s) indirectly suggests that NT action is mediated via stimulation of the release of some endogenous substance with a cardiostimulatory profile of action. Interestingly, the modulatory influence of NT upon vagal chronotropic and arrhythmogenic effects is abolished both by [D-Trp^{11}]-NT, a competitive NT receptor antagonist, and by propranolol, a β-adrenoreceptor blocker. Thus, NT-induced reduction of the vagal cardioinhibitory effects in cats seems to be largely related to the cardiostimulatory action of endogenous catecholamines presumably released by NT upon systemic administration.

NT effects upon autonomic regulation of the heart are species-specific. Indeed, in contrast to potent modulation of parasympathetically-mediated cardiac responses in cats, NT has no effect on the negative chronotropic responses elicited by vagal stimulation in the dog [106]. Furthermore, NT does not modify the magnitude of maximal heart rate deceleration induced by parasympathetic activation following the Valsalva maneuver in human subjects [156].

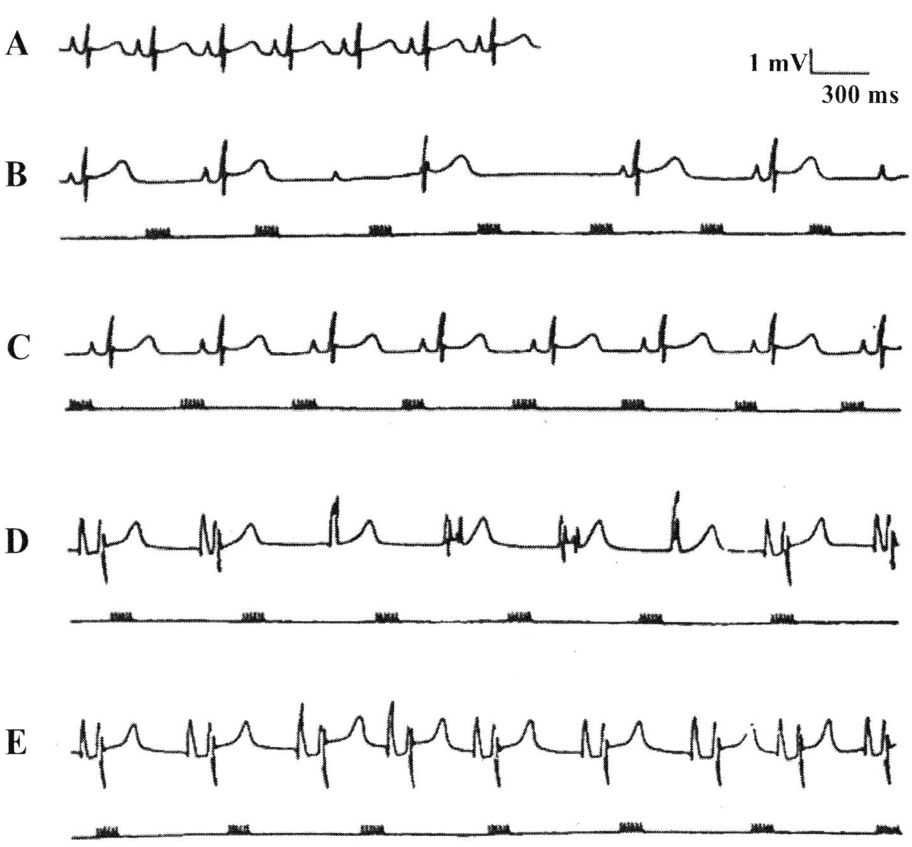

Figure 3. The impact of neurotensin on vagally-induced cardiac arrhythmias in anesthetized cats. Cardiac electrogram was recorded at baseline (panel A) and following stimulation of the right vagus nerve with bursts of six pulses (40 Hz, 2 ms, 2.5-4.0 V) (panels B-E) in anesthetized cats (n=14). Vagally-induced second degree atrioventricular (AV) block (panel B) and isorhythmic AV dissociation (panel D) were abolished following intravenous bolus neurotensin administration (4.0 x 10^{-8} M) (panels C and E, respectively).

Myocardial Contractility

NT at nanomolar concentrations increases contractility of isolated rat and guinea-pig atria [9, 143, 169, 173, 225]. On the basis of a comparison of pEC_{50} values, NT has been shown to be 20-30-fold more potent than epinephrine at eliciting atrial contractile responses, although inotropic efficacy of NT was lower than that of epinephrine. In rat atria, NT also potentiates norepinephrine-induced inotropic responses [225]. In isolated guinea-pig atria, positive inotropic and chronotropic effects of NT are preserved following blockade of β-adrenoreceptors, but significantly reduced by chronic pretreatment with capsaicin [9, 173] as well as after prior administration of SR 48692, a nonpeptide NT receptor antagonist [143]. Therefore, cardiostimulatory responses induced by NT in this experimental model are ascribed to NT receptor-mediated stimulation of substance P and/or calcitonin gene-related peptide release from cardiac capsaicin-sensitive afferent neurons [9]. Interestingly, contractile effects of NT in isolated guinea-pig atria are also reduced by prior somatostatin infusion [170], an effect presumably related to a reduced availability of the intracellular Ca^{2+} pool mobilized by NT to elicit the inotropic response.

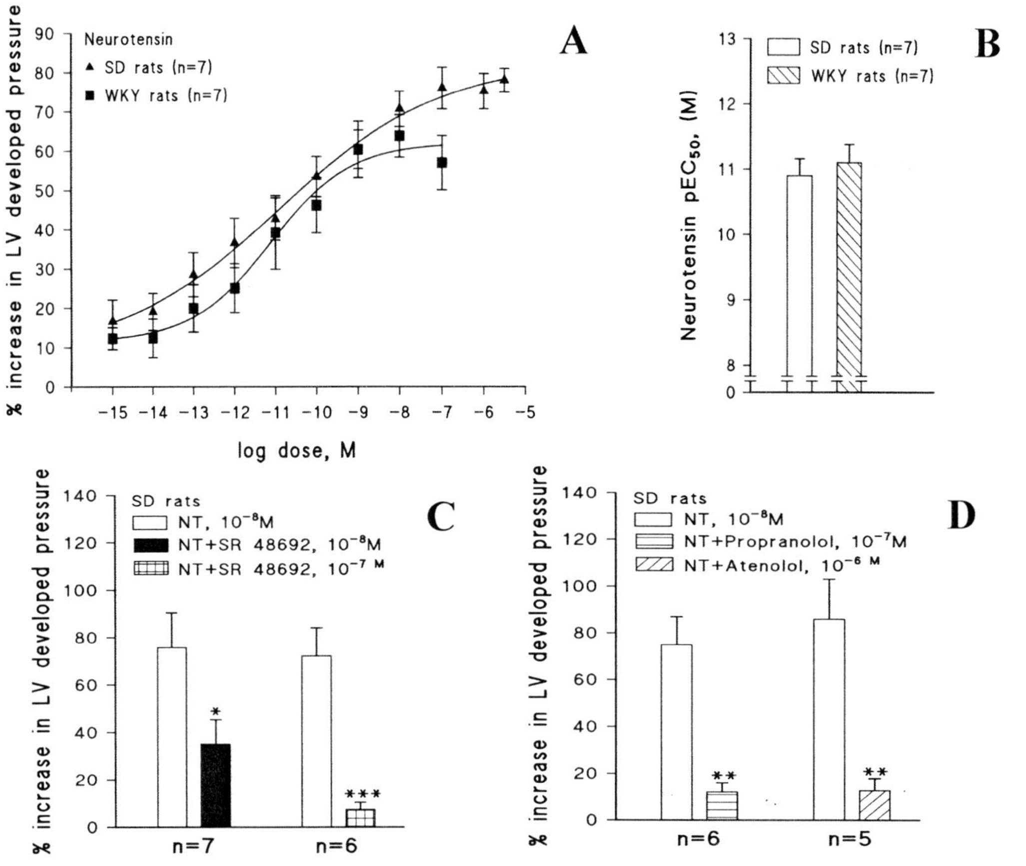

Figure 4. Neurotensin-induced positive inotropic responses in rat ventricular myocardium. % increase in left ventricular (LV) developed pressure was determined in isolated, perfused heart preparations of male Sprague-Dawley (SD) and Wistar-Kyoto (WKY) rats exposed to a wide range of neurotensin (NT) concentrations. Cumulative dose-response curves (panel A) and pEC_{50} values (panel B) are shown.

Neurotensin (10^{-8} M)-induced inotropic responses were dose-dependently inhibited by SR 48692 (panel C), a specific neurotensin receptor antagonist, as well as by β-adrenoreceptor blockers (panel D), propranolol and atenolol. SR 48692 and β-adrenoreceptor blockers were infused for 10 min prior to NT administration. $^{*}P<0.05$, $^{**}P<0.01$, $^{***}P<0.001$ vs. basal value.

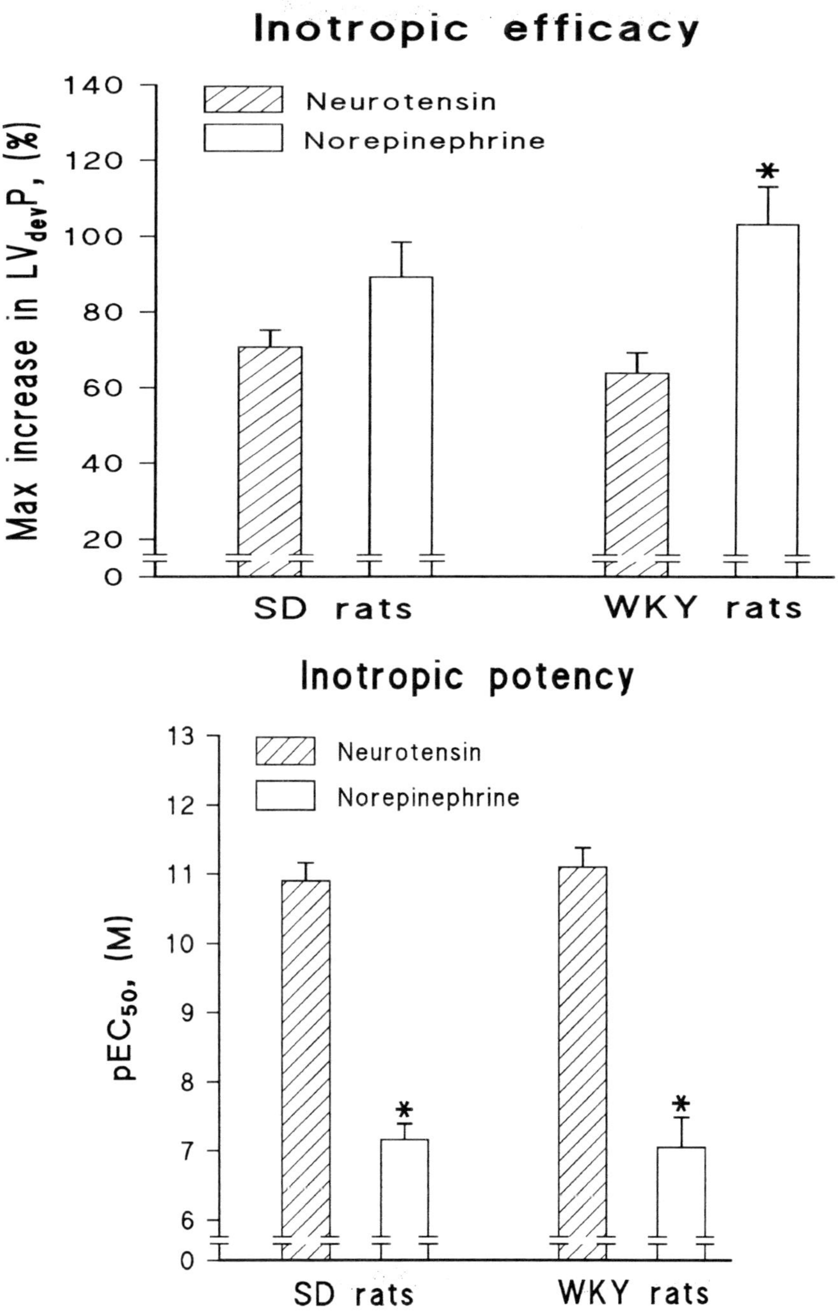

Figure 5. A comparison of neurotensin and norepinephrine inotropic efficacy (upper panel) and potency (lower panel) in the rat ventricular myocardium. The maximal (Max) increase in left ventricular developed pressure ($LV_{dev}P$) was elicited by neurotensin and norepinephrine at concentrations of 10^{-8} M and 10^{-5} M, respectively. Sample size: Sprague-Dawley (SD) rats: neurotensin (n=7), norepinephrine (n=8); Wistar-Kyoto (WKY) rats: neurotensin (n=7), norepinephrine (n=9). * $P<0.05$ vs. neurotensin value.

NT exerts a potent contractile effect on the isolated rat ventricular myocardium (figure 4, panels A-B). On the basis of a comparison of pEC_{50} values, NT is ~four orders of magnitude more potent as an inotrope than norepinephrine, an endogenous sympathetic neurotransmitter (figure 5, lower panel). With regards to the efficacy of inotropic action, the maximal NT-induced contractile effects are equivalent in magnitude to those of norepinephrine in Sprague-Dawley (SD) rats, but lower than norepinephrine-induced responses in Wistar-Kyoto (WKY) rats (figure 5, upper panel). NT-induced inotropic responses in the rat left ventricle are largely attributed to the C-terminal portion of its molecule. Indeed, the effect of the C-terminal NT hexapeptide fragment, NT_{8-13}, on myocardial contraction is very similar to that of intact NT, whereas the N-terminal NT hexapeptide fragment, NT_{1-6}, has no effect on the isolated rat left ventricle [157].

NT-induced contractile effects on the rat left ventricle are dose-dependently inhibited by SR 48692, a specific NT receptor antagonist (figure 4, panel C). Interestingly, NT-induced contractile effects are also abolished by β-adrenoreceptor blockers (figure 4, panel D), but are unaffected by antagonists of serotonin, histamine or angiotensin II receptors [157]. These findings suggest a pivotal role of sympathetic activation in NT-induced positive inotropic responses. Indeed, a significant elevation of norepinephrine concentrations in the coronary effluent is observed when isolated, perfused rat heart preparations are exposed to NT (figure 6, panel A). The magnitude of NT-induced stimulatory effect upon myocardial norepinephrine release is higher in SD as compared to WKY rats (figure 6, panel A). Importantly, a strong positive correlation is noted between NT-induced stimulation of myocardial norepinephrine release and contractile responses to this neuropeptide (figure 6, panel B). Furthermore, like NT-induced ventricular contractile responses (figure 4, panel C), NT effects upon myocardial norepinephrine release are inhibited by SR 48692, a specific NT receptor antagonist (figure 6, panel A). Taken together, these findings indicate that NT-induced contractile responses in isolated rat ventricular tissue are largely mediated via presynaptic stimulation of norepinephrine release from myocardial adrenergic terminals. A complete abolishment of NT-induced inotropic responses by β-adrenoreceptor blockers (figure 4, panel D) suggests that NT produces no direct contractile effect on the rat ventricular myocardium.

NT effects on the ventricular myocardium are modified in cardiac disease [158, 159]. In particular, positive inotropic responses to NT are reduced over wide range of concentrations following sustained sympathetic activation produced by chronic administration of isoproterenol, a β-adrenoreceptor agonist, to rats (figure 7, panels A and C). Furthermore, the inotropic potency of NT as assessed by pEC_{50} values is decreased in rats with a cardiomyopathy induced by chronic (3 months) isoproterenol administration (figure 7, panel D). A downregulation of NT-induced contractile responses following chronic β-adrenoreceptor activation is attributed to a blunted stimulatory effect of NT upon myocardial norepinephrine release in isoproterenol-treated rats (figure 7, panel E).

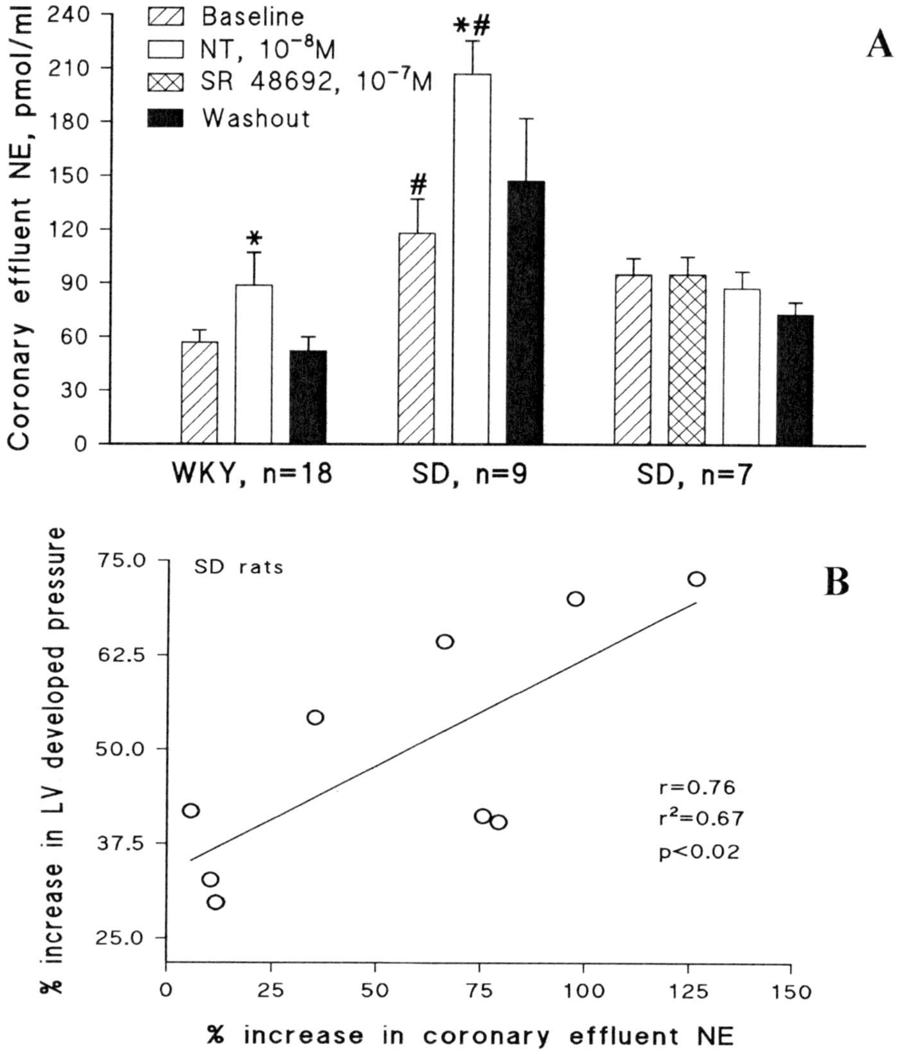

Figure 6. Stimulatory effects of neurotensin (NT) on myocardial norepinephrine (NE) release in Wistar-Kyoto (WKY) and Sprague-Dawley (SD) rats. Norepinephrine concentrations were determined by high-performance liquid chromatography in samples of coronary effluent collected from the isolated, perfused heart preparations. Neurotensin (10^{-8} M) effects on myocardial norepinephrine release were determined at baseline and after prior 10 min infusion of SR 48692, a specific neurotensin receptor antagonist (panel A). A positive correlation was found between neurotensin (10^{-8} M)-induced stimulation of myocardial norepinephrine release and neurotensin-induced left ventricular (LV) contractile responses (panel B). Panel A: *$P<0.05$ vs. basal value; #$P<0.05$ vs. corresponding value in WKY rats.

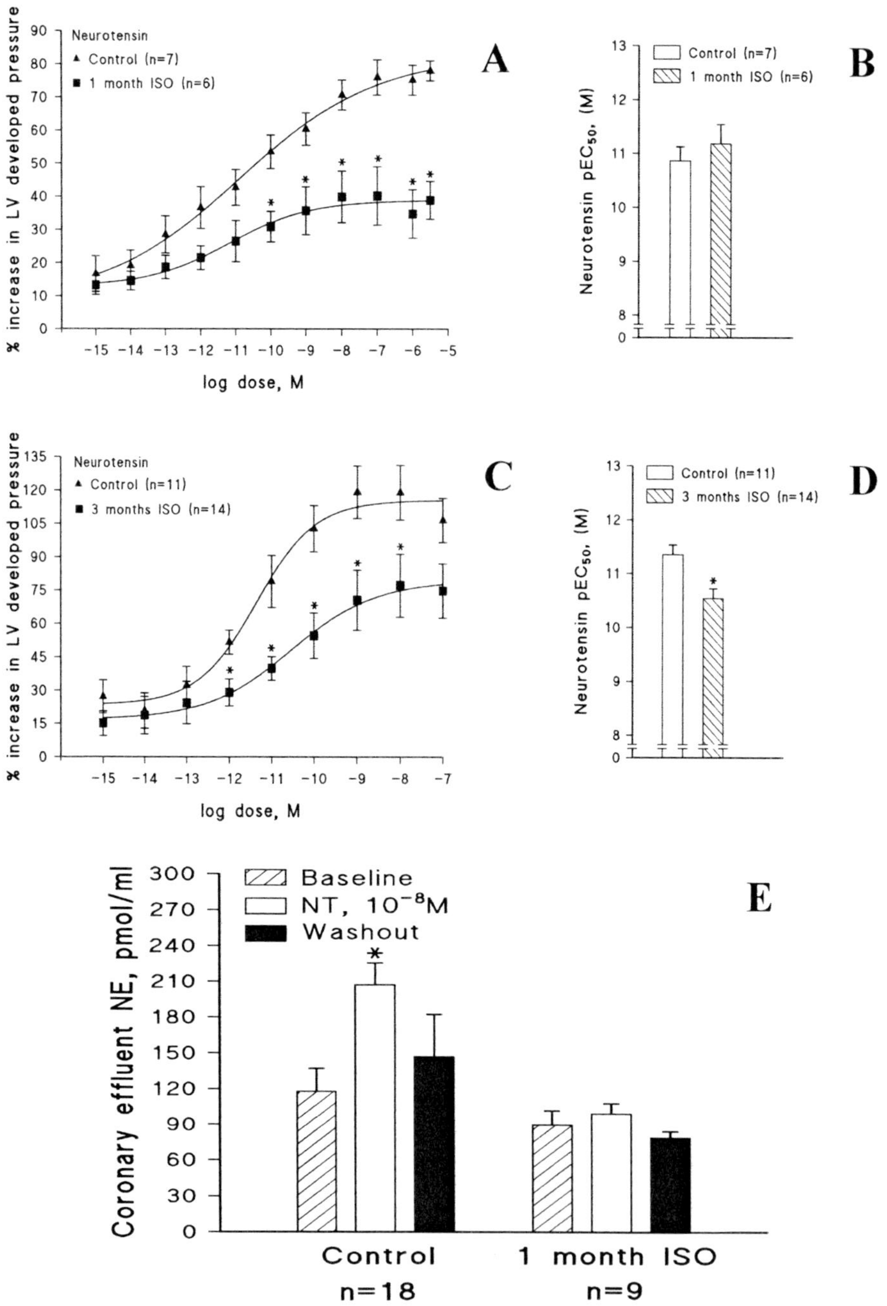

Figure 7. Left ventricular (LV) inotropic responses to neurotensin (NT) (panels A-D) and neurotensin-induced stimulation of myocardial norepinephrine (NE) release (panel E) in control and isoproterenol (ISO)-treated Sprague-Dawley rats. Isoproterenol, a β-adrenoreceptor agonist, was administered daily intraperitoneally at a dose of 0.05 mg/kg over 1 month (panels A-B) or 3 months (panels C-D) to male Sprague-Dawley rats. Control rats were injected with a similar (~0.2 ml) volume of saline vehicle. Isolated, perfused heart preparations were used to assess contractile responses to neurotensin. Cumulative dose-response curves (panels A, C) and pEC_{50} values (panels B, D) are shown. Neurotensin (10^{-8} M)-induced stimulation of myocardial norepinephrine release (panel E) was assessed by high-performance liquid chromatography measurements of norepinephrine concentrations in samples of coronary effluent. Panels A, C and D: $^*P<0.05$ vs. corresponding value in control rats; panel E: $^*P<0.05$ vs. baseline value.

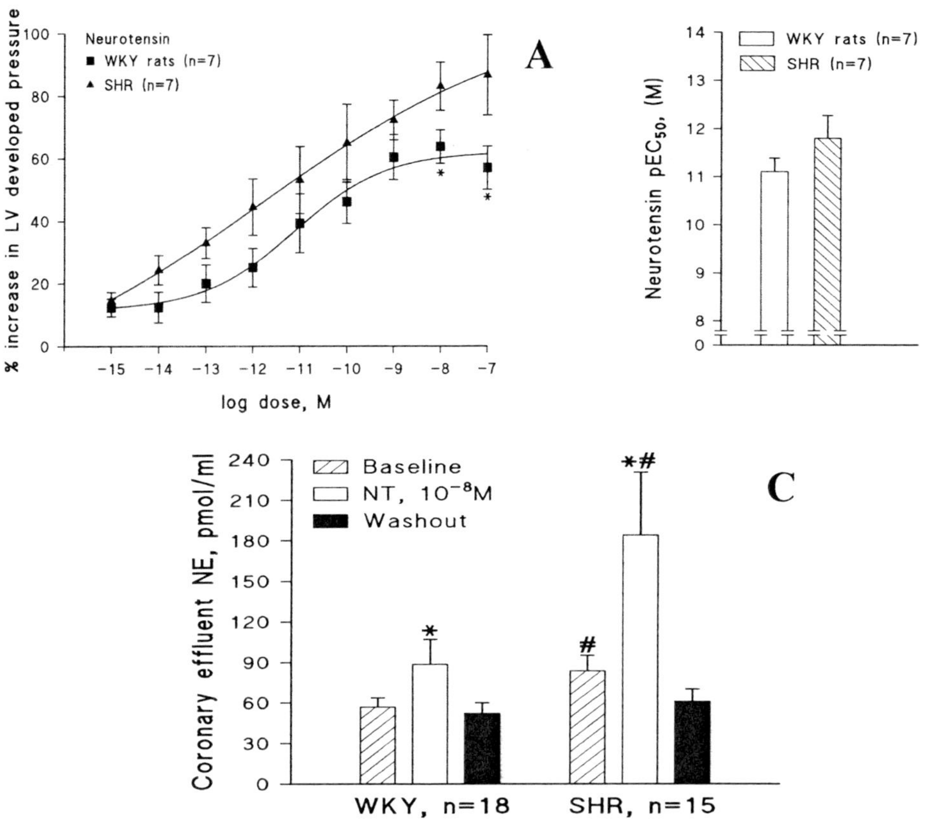

Figure 8. Left ventricular (LV) inotropic responses to neurotensin (NT) (panels A-B) and neurotensin-induced stimulation of myocardial norepinephrine (NE) release (panel C) in Wistar-Kyoto (WKY) and spontaneously hypertensive rats (SHR). Isolated, perfused heart preparations of young (6-8 months age) WKY rats and SHR were used to assess contractile responses to neurotensin. Cumulative dose-response curves (panel A) and pEC_{50} values (panel B) are shown. Neurotensin (10^{-8} M)-induced stimulation of myocardial norepinephrine release (panel C) was assessed by high-performance liquid chromatography measurements of norepinephrine concentrations in samples of coronary effluent. Panel A: *P<0.05 vs. corresponding value in SHR rats. Panel C: *P<0.05 vs. baseline value; # P<0.05 vs. corresponding value in WKY rats.

A different pattern of changes in myocardial contractile responses to NT is noted in compensated pressure-overload cardiac hypertrophy [158]. Specifically, the maximal ventricular inotropic responses to NT are increased in young spontaneously hypertensive rats as compared to normotensive control Wistar-Kyoto rats (figure 8, panel A). This change is associated with significant enhancement of NT-induced stimulation of the myocardial norepinephrine release (figure 8, panel C). The putative significance of these changes for the progression of the cardiac disease remains to be determined.

As with many other physiological effects of NT, its action on myocardial contractility is species-specific. In particular, in contrast to rat and guinea-pig, NT has no contractile effects on the dog [6, 52, 187], rabbit [169] and frog [225] atrial and ventricular myocardium.

Cardiovascular Reflexes

NT-immunoreactive fibers are involved in sensory innervation of the cardiovascular system. In particular, NT-immunoreactive nerve endings are found in the wall of guinea-pig carotid sinus where they are mostly distributed in adventitia and tunica media [68, 117]. These neural terminals have sensory origin as they degenerate after transection of the carotid sinus nerve. Moreover, retrograde tracing has revealed the presence of NT-like immunoreactivity in primary afferent neurons of the petrosal ganglion which projects to the carotid sinus [117]. In the rat brain stem, NT-containing cells are found in dorsomedial region of the nucleus of the tractus solitarius, which receives numerous synaptic inputs from afferent fibers of aortic nerves [88, 95]. Furthermore, NT-containing neural fibers are found in close contact with cardiac vagal preganglionic neurons in the lateral portion of the dorsal vagal nucleus at the level of the obex [88]. In rats, intracerebroventricular administration of NT elicits significant pressor [226] or depressor [190] responses presumably mediated via activation of specific NT binding sites found both in the nucleus of the tractus solitarius and dorsal vagal nucleus [104].

Collectively, these findings suggest an important role played by NT in cardiovascular reflexes originating in reponse to stimulation of arterial baroreceptors.

There is evidence to suggest that NT is also involved in cardiovascular reflexes originating in reponse to stimulation of primary afferents localised in some visceral organs. In particular, topical application of NT at picomolar concentrations on the epicardial surface of the left ventricle [194] or to serousal surface of abdominal organs (ileum, stomach or right hepatic lobe) [195] induces transient dose-dependent increases in blood pressure and heart rate in anesthetized guinea-pigs. These hemodynamic responses are significantly attenuated by ganglion blockers or α-adrenoreceptor antagonists thus indicating that these NT-induced effects are mediated via an increased sympathetic outflow. Furthermore, both NT-induced pressor effects and tachycardia are markedly suppressed by chronic pretreatment of guinea-pigs with capsaicin, suggesting an excitatory effect of NT upon sensory, substance P and calcitonin gene-related peptide-containing neurons [194, 195]. Importantly, NT-induced cardiovascular responses are also prevented by prior topical application of lidocaine, a local anesthetic, to the surface of the left ventricle or abdominal organs. Taken together, these findings suggest that NT could trigger cardiovascular reflexes by stimulating visceral capsaicin-sensitive primary afferents which activate preganglionic sympathetic neurons at the spinal cord via a substance P and/or a calcitonin gene-related peptide-dependent mechanism. Once activated, these neurons transmit impulses to para- and prevertebral ganglionic cells innervating the heart and vessels which ultimately results in a pressor effect and a tachycardia.

The afferent limb of these reflexes is provided by sympathetic rather than parasympathetic fibers as NT-induced cardiovascular responses are preserved after acute bilateral cervical vagotomy [194-196]. Importantly, cardiovascular responses following NT-induced activation of visceral afferents are accompanied by transient twitch-like contractions of the guinea-pig abdominal wall, an effect inhibited by pancuronium and morphine but potentiated by naloxone [196, 197]. Furthermore, both hemodynamic and abdominal motor responses elicited by NT are abolished by acute spinalization. Thus, NT-induced pressor effects and tachycardia could represent a cardiovascular component of the complex spinal nociceptive reflex originating in response to distension of hollow visceral organs, transient episodes of intestinal ischemia, or any other visceral nociceptive stimulus [196, 197].

Structural Determinants of Cardiovascular Activity of NT

The cardiovascular effects of NT are not modified by removal of the N-terminal sequence $pGlu^1$-Leu^2-Tyr^3-Glu^4-Asn^5-Lys^6-Pro^7 thus indicating that the major determinants of biological activity of this neuropeptide reside primarily in the C-terminal portion of its molecule. Indeed, the C-terminal NT hexapeptide, $NT_{8\text{-}13}$, has been shown to be equivalent to intact NT in eliciting vasodepressor responses [44], coronary vasoconstrictor effects [176], contractile effects on the isolated portal vein [189], as well as positive inotropic responses in the isolated atrial [173] and ventricular [157] myocardium. A shorter C-terminal NT fragment, $NT_{9\text{-}13}$, has only about 2% of intact NT potency but it still able to elicit the same maximum response at high concentrations, whereas the $NT_{10\text{-}13}$ tetrapeptide is completely inactive. Therefore, the C-terminal sequence Arg^9-Pro^{10}-Tyr^{11}-Ile^{12}-Leu^{13}-OH seems to be the minimum structure required for displaying cardiovascular activity of NT [173, 174, 176, 189].

These findings are consistent with NT structure-activity relationships determined in receptor binding studies. In particular, the C-terminal NT pentapeptide, $NT_{9\text{-}13}$, is the shortest C-terminal fragment exhibiting properties of a full agonist in binding with neuronal and extraneuronal NT receptors, whereby it possesses about 3% of the potency of intact NT [1, 69, 109, 133]. The addition of Arg^8 to this sequence yields a hexapeptide fragment, $NT_{8\text{-}13}$, which is as potent as intact NT in binding to NT receptors expressed in various cell lines [1, 37, 109, 211], or purified from newborn human [247] and mouse [133] brain tissue. On the other hand, the N-terminal NT fragments, $NT_{1\text{-}6}$, $NT_{1\text{-}8}$ and $NT_{1\text{-}10}$, do not possess cardiovascular activity [157, 173, 174, 176, 189] and exhibit a very low affinity ($IC_{50} > 10$ μM) for the NT receptor in binding studies [247].

Of the C-terminal portion of NT molecule, amino acids Ile^{12} and Leu^{13} contribute mainly to the affinity of NT for its receptor, whereas the chemical groups responsible for NT receptor activation reside in the sequence Arg^9-Pro^{10}-Tyr^{11} [174]. Indeed, NT fragments missing Ile^{12} and Leu^{13} ($NT_{1\text{-}12}$ and $NT_{1\text{-}11}$), although being less potent than intact NT, are able to elicit the same maximal contractile responses in isolated guinea-pig atria [173]. On the other hand, substitutions in positions 9, 10 and 11 yield NT analogues exhibiting a very low potency in eliciting cardiovascular effects [173, 174, 176, 189]. Amongst C-terminal amino acids, Tyr^{11} is particularly important for NT receptor activation [173, 176]. Indeed, NT analogues where Tyr^{11} is replaced with D-Tyr, D-Phe, Phe, Ala or Leu are three-four orders of magnitude less potent than the intact NT in eliciting positive inotropic responses in isolated guinea-pig atria [173, 174], coronary vasoconstrictor effect in isolated, perfused rat hearts [176] and contractile effects in the isolated rat portal vein [189]. Likewise, in contrast to intact NT, a NT analogue in which Tyr^{11} is replaced with D-Trp ([D-Trp^{11}]-NT), has no contractile effect on isolated rat ventricular myocardium [157], as well as basal heart rate and vagally-induced cardioinhibitory effects in anesthetized cats [151-155]. Consistently, [Phe^{11}]-NT and [D-Tyr^{11}]-NT analogues are several orders of magnitude less potent than intact NT in binding to NT receptors expressed in human umbilical vein endothelial cells [211]. These findings raise the possibility that the presence in position 11 of an aromatic residue with the L-configuration is important for NT binding with its receptor. Indeed, replacing Tyr^{11} with Trp, an aromatic residue with a longer side chain, yields a NT analogue, [Trp^{11}]-NT, which is equivalent to intact NT in eliciting cardiovascular responses in rats [176, 189].

Interestingly, NT analogues in which Tyr^{11} is substituted for Tyr(Me)^{11} or D-Trp^{11} exhibit properties of competitive NT receptor antagonists at concentrations below 10^{-6} M while behaving like weak agonists at higher concentrations [176, 189]. NT receptor antagonist effects produced by these compounds have been shown in a number of studies (table 5). Importantly, [D-Trp^{11}]-NT and Tyr(Me)^{11}-NT antagonize coronary vasoconstrictor responses induced by native NT in isolated perfused rat hearts [175, 176] as well as its contractile effects on isolated rat portal vein [189], but has no effect on NT-induced positive inotropic responses in isolated guinea-pig atria [175]. Taken together, these findings raise the possibility that receptors underlying the influence of NT on the vascular smooth muscle tone are pharmacologically distinct from those mediating myocardial effects of NT, at least in rats.

Mechanisms of NT-Induced Cardiovascular Effects

NT Receptors

NT Receptor Subtypes and Specific Antagonists

To date, three subtypes of NT receptors (NTS_1, NTS_2 and NTS_3) have been identified [235]. All three NT receptor subtypes recognize the same C-terminal NT sequence, Arg^9-Pro^{10}-Tyr^{11}-Ile^{12}-Leu^{13}-OH. NTS_1 and NTS_2 are G-protein-coupled receptors with seven putative transmembrane domains, whereas the NTS_3 receptor is an intracellular NT binding site structurally identical to gp95/sortilin, a protein of ~95 kDa with a single transmembrane domain [166].

To study putative physiological roles of NT receptors, two selective nonpeptide NT receptor antagonists, SR 48692 and SR 142948A, have been synthesized by Sanofi Research [70, 72]. Both compounds at nanomolar concentrations competetively inhibit ^{125}I-labeled NT binding to NT receptors of neural and extraneural localization as well as NT-induced changes in intracellular levels of the second messengers [70, 72, 94, 160, 211, 213]. While SR 48692 has a higher affinity for NTS_1 as compared to NTS_2 receptor subtypes, SR 142948A equally recognizes both NT receptor subtypes and therefore exhibits a wider spectrum of pharmacological activity than SR 48692 [72]. Furthermore, in receptor binding studies SR 142948A exhibits about a 60-fold higher affinity for the human NTS_1 receptor as compared to SR 48692 [213]. Importantly, in contrast to [D-Trp^{11}]-NT and Tyr(Me)^{11}-NT, NT analogues possessing properties of NT receptor antagonists, SR 48692 and SR 142948A do not exhibit an intrinsic activity even at concentrations about 100-fold higher than those required for 50% inhibition of NT receptor binding [71, 72, 143, 213].

Table 5. Inhibition of NT-induced cardiovascular effects by selective NT receptor antagonists

Species	Preparation used/In vivo experiment	NT effect	NT receptor antagonist used	NT receptor antagonist dose	Reference
Rat	Isolated portal vein	Venoconstrictor effect	[D-Trp^{11}]-NT	4-11 x 10^{-7} M	[188]
			[Tyr(Me)^{11}]-NT	10^{-7}-10^{-6} M	[189]
	Isolated, perfused heart	Coronary vasoconstrictor effect	[D-Trp^{11}]-NT	10^{-7}-10^{-6} M	[175]
			[Tyr(Me)^{11}]-NT	10^{-7}-10^{-6} M	[176]
		Positive ventricular inotropic responses	SR 48692	10^{-8}-10^{-7} M	[157-159]
	In vivo experiment	Hypotensive effect	SR 48692	0.1-10 mg/kg orally, 1 hour before NT	[71, 212]
			SR 142948A	0.01-1.0 mg/kg orally, 1 hour before NT	[213]
		Increase in cutaneous vascular permeability	SR 48692	5.0 pM, intradermally	[138, 213]
			SR 142948A	0.01-1.0 nM, intradermally	[213]
Guinea -pig	Isolated atria	Positive chronotropic and inotropic effects	SR 48692	IC_{50}~ 10^{-9} M	[143]
	In vivo experiment	Pressor effect	SR 48692	6-400 μg/kg, i.v.	[143]
Cat	In vivo experiment	Tachycardia	[D-Trp^{11}]-NT	4.0 x 10^{-8} M, i.v.	[152]
		Reduction of vagally-induced bradycardia	[D-Trp^{11}]-NT	4.0 x 10^{-8} M, i.v.	[151, 153, 155]
		Protective effect in vagally-induced cardiac arrhythmias	D-Trp^{11}]-NT	4.0 x 10^{-8} M, i.v.	[154]
Human	Endothelial cells of umbilical vein	Stimulation of phosphoinositide turnover, elevation of intracellular Ca^{2+} level, and stimulation of prostacyclin release	SR 48692 SR 142948A	IC_{50} ~ 41-86 nM IC_{50} ~ 17-24 nM	[211-213]

Although expression of NTS_3 receptor mRNA has been found in mouse [85], rat [123] and human [166] myocardial tissue, the intracellular localization of this NT receptor subtype limits the chances of it being involved in NT-induced cardiac effects. Hence the functional significance of the other two NT receptor subtypes, NTS_1 and NTS_2, only will be discussed.

NTS_1 Receptor

The NTS_1 receptor, purified from human [247] and mouse [133] brain, is a polypeptide with an apparent molecular weight of 100 kDa that reperesents a high-affinity (K_d~0.1-0.3 nM) NT binding site coupled to phospholipase C via pertussis toxin-insensitive G-proteins [2, 22]. The NTS_1 receptor which was cloned for the first time by Tanaka et al. (1990) from the rat brain, has a structure that consists of 424 amino acids. A polypeptide chain of the human NTS_1 receptor cloned from the colonic adenocarcinoma cell line HT 29 includes 418 amino acids and exhibits a high degree of structural identity (84%) with the rat NTS_1 receptor [236]. NTS_1 receptor subtype mRNA expression has been found in rat myocardial tissue [228] as well as human umbilical vein and aortic endothelial cells [211]. Furthermore, a specific binding of ^{125}I-labeled NT has been demonstrated in human umbilical vein (K_d ~ 0.23 nM, B_{max} ~ 5500 sites/cell) and aortic endothelial cell monolayers (K_d ~ 0.6 nM, B_{max} ~ 32000 sites/cell) [211, 213].

Activation of the NTS_1 receptor expressed in various cells exerts a strong stimulatory influence on phosphatidylinositol turnover as evidenced by a transient elevation of intracellular inositol phosphate levels with a subsequent stimulation of Ca^{2+} release from intracellular stores (table 6). With regards to the cardiovascular system, there is a strong positive correlation between the dissociation constant of NT in binding experiments and EC_{50} values for stimulation of $^{45}Ca^{2+}$ efflux in human endothelial cells [211]. Furthermore, NT-induced Ca^{2+} mobilization in human umbilical vein endothelial cells is abolished by SR 48692 and SR 142948A, selective NT receptor antagonists [213], suggesting a specific NT receptor-mediated effect.

Activation of the NTS_1 receptor expressed in mouse neuroblastoma N1E115 cells produces a transient increase in intracellular cGMP levels [1, 3, 64, 160]. However, no changes in intracellular cGMP levels were noted following activation of NTS_1 receptors expressed in fibroblasts [37], colonic adenocarcinoma HT29 cells [2, 160] and Chinese hamster ovary cells [160] thus indicating that this NT receptor subtype is functionally coupled to guanylate cyclase only in neural tissue. Likewise, no consistent evidence is available to show the impact of NTS_1 receptor activation on adenylate cyclase activity. Indeed, an increase [160, 244], a decrease [1] as well as no change [2, 37, 66, 239] in intracellular cAMP levels has been found following NTS_1 receptor activation in different studies. Collectively, these findings suggest that NTS_1 receptor-mediated stimulation of phosphatidylinositol turnover rather than a change in intracellular levels of cyclic nucleotides is involved in the cardiovascular effects of NT.

Table 6. Stimulation of phosphoinositide hydrolysis and intracellular Ca^{2+} mobilization following activation of NTS_1 receptor subtype

Species	Cell/tissue expressing NT receptor	NT dose	Reference
Human	Colonic adenocarcinoma HT29 cells	EC_{50} 3.5±0.5 nM	[2, 22, 160]
	Pancreatic cancer cells	10^{-12}-10^{-9} M	[93, 94]
	Umbilical vein and aorthic endothelial cells	3-100 nM	[211, 212]
Mouse	Neuroblastoma N1E115 cells	EC_{50} 0.46-0.9 nM	[3, 64, 96, 160, 223]
Rat	Brain slices	EC_{50} 224±18 nM	[66]
	Brain cortex neurons	10^{-6}-10^{-9} M	[210]
	Serosal mast cells	10^{-7} M	[14]
	Fibroblasts	EC_{50} 4.84 nM	[37]
	Pheochromacetoma cells	10^{-7} M	[83]
	CHO cells	EC_{50} 3.0 nM	[82, 84, 160, 239]

NT effects on blood pressure, heart rate, myocardial contractility and phosphoinositide turnover in endothelial cells, are dose-dependently inhibited by SR 48692 (table 5), a NT receptor antagonist exhibiting a significantly higher affinity for the NTS_1 (IC_{50} ~ 5.6 nM) as compared to the NTS_2 (IC_{50} ~ 300 nM) receptor subtype [70]. These findings suggest that NT-induced cardiovascular responses are primarily mediated via an activation of the NTS_1 receptor subtype. This notion is also supported by results of studies designed to compare the relative potency of different NT receptor antagonists in abolishing NT-induced cardiovascular responses [213]. In particular, a higher affinity of SR 142948A as compared to SR 48692 for the NTS_1 receptor accounts for a significantly higher potency of SR 142948A in antagonizing NT effects on blood pressure and cutaneous vascular permeability in anesthetized rats, as well as Ca^{2+} levels and prostacyclin release from isolated human umbilical vein endothelial cells [213].

NTS_2 Receptor

The NTS_2 receptor subtype represents a low-affinity (K_d~3-5 nM) NT binding site sensitive to levocabastine, a histamine H_1 receptor antagonist, which competitively inhibit NT binding to NTS_2 [235]. A polypeptide chain of the brain NTS_2 receptor comprises 417 (mouse), 416 (rat) or 410 (human) amino acids (~46 kDa) and shows a high degree of structural homology to NTS_1 [40, 134, 237]. Expression studies has revealed the presence of low levels of NTS_2 receptor mRNA in rat [40] and human [237], but not in mice [134] myocardial tissue. Interestingly, expression of the NTS_2 receptor subtype in neonatal rat cardiomyocytes is enhanced upon long-term exposure to leukaemia inhibitory factor, a cytokine which belongs to the interleukin-6 family and which is involved in cardiac remodeling and the development of heart failure [57]. However, the functional significance of the myocardial NTS_2 receptor remains uncertain. Indeed, although NT specifically binds to recombinant human [237] and mouse [21] NTS_2 receptors expressed in CHO or HEK 293 cell lines, it does not elicit a physiological response as assessed by measurements of intracellular levels of second messengers. Furthermore, only small elevations of intracellular Ca^{2+} levels are elicited by very high (1 μM) NT concentrations upon activation of the rat NTS_2 receptor expressed in CHO cells [245]. In contrast, nonpeptide NTS_1 receptor antagonists, SR 48692 and SR 142948A, behave as potent agonists of the NTS_2 receptor, stimulating polyphosphoinositide hydrolysis and increasing intracellular Ca^{2+} levels [237, 245]. Surprisingly, these effects are fully antagonized by NT suggesting that some yet unidentified endogenous compound rather than NT could act as a natural agonist for the NTS_2 receptor [237, 245]. In the light of these findings, it is very unlikely that the NTS_2 receptor subtype is involved in NT-induced cardiovascular responses.

Stimulation of Endogenous Neurotransmitter/Neuromodulator Release

Histamine

NT could act as a modulator producing a potent stimulatory effect on the release of a variety of endogenous biologically active substances (table 7). Only those modulatory effects of NT that could contribute to the cardiovascular activity of this neuropeptide will be briefly discussed.

A number of cardiovascular effects produced by NT are mediated via stimulation of histamine release from mast cells. In particular, the histamine-releasing effect of NT accounts for vasodepressor responses, peripheral vasodilatation, increased capillary permeability and plasma extravasation following intravenous NT infusion in anesthetized rats [24, 27, 38, 42, 71, 77, 148, 137].

The stimulatory effect of NT on histamine release has been demonstrated using different experimental models. In anesthetized rats, intravenous NT infusion induces a transient, dose-dependent increase in plasma histamine levels which is abolished by pretreatment with compound 48/80, a mast cell degranulating agent [27, 147]. Furthermore, NT-induced stimulation of histamine release has been shown in the isolated, perfused rat heart [192, 193], hindquarter [102, 103], skin slices [42] as well as isolated mast cells [14, 27, 30, 114, 118, 206, 215]. On a molar basis, NT is 5-6 times as active as compound 48/80 in increasing histamine concentrations in coronary effluent collected from the isolated, spontaneously beating rat heart [192]. NT-induced stimulation of histamine release is independent of heart rate and coronary perfusion pressure, but is suppressed in Ca^{2+}-free perfusion media [193]. There is some evidence to suggest that endogenous NT is involved in stress-induced stimulation of histamine release from cardiac mast cells [161]. Indeed, cardiac mast cell degranulation induced by acute immobilization stress in rats is completely abolished by pretreatment with SR 48692, a NT receptor antagonist [161].

Table 7. Stimulatory effects of NT upon release of various endogenous biologically active substances from peripheral tissues

Endogenous substance released by NT	Species	Preparation used	NT dose	References
Histamine	Rat	Isolated, perfused heart	EC_{50} 0.34 nM	[192, 193]
		Isolated, perfused hindquarter	10^{-8}-10^{-6} g/ml	[103]
		Isolated, perfused skin slices	10^{-13}-10^{-6} M	[42]
		Blood samples collected from carotid artery	1 nM, i.v.	[27, 137, 147]
		Peritoneal mast cells	10^{-9}-10^{-5} M	[14, 27, 30, 118, 137, 206]
		Pleural mast cells	EC_{50} 3 x 10^{-8} M	[14, 114, 137]
	Human	Jejunal mast cells	10^{-9}-10^{-6} M	[215]
Norepinephrine	Rat	Perfused mesenteric arteries	10^{-6}-10^{-7} M	[232]
		Isolated, perfused heart	10^{-8} M	[157-159]
		Pheochromacetoma cells	EC_{50} 59 μM	[162]
Epinephrine	Rat	Samples of blood plasma	10^{-9} M, i.v.	[147]
Serotonin	Rat	Isolated, perfused hindquarter	5-15 x 10^{-7} /ml	[102]
		Peritoneal mast cells	3 μM	[30]
Acetylcholine	Guinea-pig	Myenteric plexus-longitudinal muscle strips of small intestine	3 pM-3 nM	[178, 246]
			10^{-10}-10^{-7} M	[141]
Prostaglandins	Rat	Samples of blood plasma	2-3 nM/kg, i.v.	[212]
	Human	Umbilical vein endothelial cells	EC_{50} 14±1 nM	[212]
Substance P	Guinea-pig	Inferior mesenteric ganglion	5.0 x 10^{-6} M	[224]
		Myenteric plexus-longitudinal muscle strips of small intestine	60 nM	[92]
γ-aminobutyric acid	Guinea-pig	Strips of ileum	10^{-10}-10^{-7} M	[141]
Somatostatin	Rat	Samples of portal venous blood	3 μg/100 g BW, i.v.	[209]
			0.24-0.72 pM/kg/hr , i.v.	[76]
			7.2 nM/kg/hr, i.v.	[75]

In isolated serosal rat mast cells, the maximal amount of histamine released by NT constitutes about 20% of the total histamine content [114, 206]. However, mast cell degranulating capacity of NT is significantly reduced in Ca^{2+}- or glucose-free perfusion media [27, 30]. There is a strong evidence to suggest that the stimulatory effect of NT upon histamine release is mediated via activation of a mast cell NT receptor. Indeed, NT-induced elevation of plasma histamine levels [137], stimulation of histamine release from isolated rat mast cells [137, 206], as well as increases in cutaneous vascular permeability [137, 213] are blocked by the selective NT receptor antagonists, SR 48692, SR 142948A and [D-Trp^{11}]-NT. Consistently, a specific and reversible ^{125}I-NT binding (K_d=154 nM) has been found in rat mast cells [119]. These findings are supported by results of structure-activity studies demonstrating that the histamine releasing effect of NT requires a preserved integrity of the C-terminal portion of the NT molecule [103, 206]. Indeed, a replacement of Arg^8, Arg^9 or Tyr^{11} with various amino acids yields NT analogues exhibiting a very low potency at eliciting histamine release in the isolated, perfused rat hindquarter [103]. On the other hand, in rat mast cell binding studies, substitutions of N-terminal amino acids with their D-isomers has no significant impact on NT binding, whereas modifications performed within the C-terminal portion of NT, residues Pro^{10} through Leu^{13}, considerably reduces NT binding capacity [120].

C-terminal NT hexa- and pentapeptide fragments, NT_{8-13} and NT_{9-13}, are found to possess about 70% of the relative affinity for mast cell NT receptors as compared to intact NT [120].

Mast cell NT receptors are coupled with phospholipase C via pertussis toxin-sensitive G-proteins [14]. In isolated rat serosal mast cells, NT-induced stimulation of histamine release is accompanied by a rapid and transient increase in intracellular free Ca^{2+} and inositol trisphosphate levels [14]. These effects were suppressed by pretreatment with U-73122, an inhibitor of phospholipase C, as well as by pertussis toxin.

Serotonin

Serotonin is a mast cell mediator released in parallel fashion with histamine upon NT exposure. In isolated peritoneal rat mast cells, NT induces a prompt (within seconds) and dose-dependent concomitant liberation of histamine and serotonin, with both secretory effects being Ca^{2+}- and energy-dependent [30]. In isolated, perfused rat hindquarters, NT-induced stimulation of serotonin release is associated with a marked increase in vascular resistance [102]. Both serotonin releasing effect of NT and NT-induced vasoconstriction are inhibited by compound 48/80.

Catecholamines

NT induces a transient increase in plasma epinephrine concentrations following intravenous administration in anesthetized rats, an effect attenuated by adrenalectomy [147]. However, NT does not affect plasma norepinephrine concentrations. On the other hand, NT exerts a potent stimulatory effect on myocardial norepinephrine release in isolated, perfused rat heart preparations (figure 6, panel A) [157-159].

Prostaglandins

NT induces a dose-dependent prostacyclin release from human umbilical vein endothelial cells, an effect associated with a transient elevation of intracellular Ca^{2+} levels [212].

Furthermore, intravenous infusion of NT ~doubles circulating levels of 6-keto-prostaglandin $F_{1\alpha}$ in anesthetized rats, an effect accompanied by a significant fall in blood pressure [212]. All of these effects are abolished by nanomolar concentrations of SR 48692 and SR 142948A, nonpeptide NT receptor antagonists, suggesting a direct, high-affinity NT receptor-mediated stimulatory effect upon release of endogenous prostaglandins [212, 213]. A prostaglandin-dependent mechanism is also involved in NT-induced contractile effects on the isolated rat portal vein [188] as well as the NT-induced vasodilatory effects on the coronary vasculature in anesthetized dogs [52].

References

[1] Amar, S; Mazella, J; Checler, F; Kitabgi, P; Vincent, JP. (1985). Regulation of cyclic GMP levels by neurotensin in neuroblastoma clone N1E115. *Biochem. Biophys. Res. Commun., V.129,* 117-125.

[2] Amar, S; Kitabgi, P; Vincent, JP. (1986). Activation of phosphatidylinositol turnover by neurotensin receptors in the human colonic adenocarcinoma cell line HT29. *FEBS Lett., V.201,* 31-36.

[3] Amar, S; Kitabgi, P; Vincent, JP. (1987). Stimulation of inositol phosphate production by neurotensin in neuroblastoma N1E-115 cells: implication of GTP-binding proteins and relationship with the cyclic GMP response. *J. Neurochem., V.49,* 999-1006.

[4] Armour, JA (1996). Comparative effects of endothelin and neurotensin on intrinsic cardiac neurons in situ. *Peptides, V.17,* 1047-1052.

[5] Aronin, N; Carraway, RE; Ferris, CF; Hammer, RA; Leeman, SE. (1982). The stability and metabolism of intravenously administered neurotensin in the rat. *Peptides, V.3,* 637-642.

[6] Baca, I; Mittmann, U; Feurle, GE; Haas, M; Muller, Th. (1981). Effect of neurotensin on regional intestinal blood flow in the dog. *Res. Exp. Med. (Berl), V.179,* 53-58.

[7] Bachelard, H; St-Pierre, S; Rioux, F. (1985). The chronotropic action of neurotensin in the guinea-pig isolated heart. *Peptides, V.6,* 841-845.

[8] Bachelard, H; St-Pierre, S; Rioux, F. (1986). The coronary vasodilator effect of neurotensin in the guinea-pig isolated heart. *Peptides, V.7,* 431-435.

[9] Bachelard, H; St-Pierre, S; Rioux F. (1987). Participation of capsaicin-sensitive neurons in the cardiovascular effects of neurotensin in guinea-pigs. *Peptides, V.8,* 1079-1087.

[10] Bachelard, H; Gardiner, SM; Kemp, PA; Bennett, T. (1992). Mechanisms contributing to the regional haemodynamic effects of neurotensin in conscious, unrestrained *Long Evans* rats. *Br. J. Pharmacol., V.105,* 191-201.

[11] Bachoo, M; Ciriello, J; Polosa, C. (1987). Effect of preganglionic stimulation on neuropeptide-like immunoreactivity in the stellate ganglion of the cat. *Brain Res., V.400,* 377-382.

[12] Bachoo, M; Polosa, C. (1988). Cardioacceleration produced by close intra-arterial injection of neurotensin into the stellate ganglion of the cat. *Can. J. Physiol. Pharmacol., V.66,* 408-412.

[13] Barelli, H; Fox-Threlkeld JET; Dive, V; Daniel, EE; Vincent, JP; Checler, F. (1994). Role of endopeptidase 3.4.24.16 in the catabolism of neurotensin, in vivo, in the vascularly perfused dog ileum. *Br. J. Pharmacol., V.112,* 127-132.

[14] Barrocas, AM; Cochrane, DE; Carraway, RE; Feldberg, RS. (1999). Neurotensin stimulation of mast cell secretion is receptor-mediated, pertussis-toxin sensitive and requires activation of phospholipase C. *Immunopharmacol., V.41,* 131-137.

[15] Bauer, B; Neubauer, S; Spindler, M; Durr, R; Becker, HH; Ertl, G.(1995). Interference of angiotensin-converting enzyme inhibition with vasoactive peptides in the coronary circulation of dogs. *J. Cardiovasc. Pharmacol., V.25,* 756-762.

[16] Besterman, HS; Bloom, SR; Sarson, DL; Blackburn, AM; Johnston, DI; Patel, HR; Stewart, JS; Modigliani, R; Guerin, S; Mallinson, CN. (1978). Gut-hormone profile in coeliac disease. *Lancet, V.1,* 785-788.

[17] Blackburn, AM; Bloom, SR. (1979). A radioimmunoassay for neurotensin in human plasma. *J. Endocrinol., V.83,* 175-181.

[18] Blackburn, AM; Christofides, ND; Ghatei, MA; Sarson, DL; Ebeid, FH; Ralphs DN; Bloom, SR. (1980). Elevation of plasma neurotensin in the dumping syndrome. *Clin. Sci., V.59,* 237-243.

[19] Blackburn, AM; Fletcher, DR; Adrian, TE; Bloom, SR. (1980). Neurotensin infusion in man: pharmacokinetics and effect on gastrointestinal and pituitary hormones. *J. Clin. Endocrinol. Metab., V.51,* 1257-1261.

[20] Bommer, M; Herz, A. (1989). Neurotensin affects metabolism of opioid peptides, catecholamines and inositol phospholipids in bovine chromaffin cells. *Life Sci., V.44,* 327-335.

[21] Botto, JM; Chabry, J; Sarret, P; Vincent, JP; Mazella J. (1998). Stable expression of the mouse levocabastine-sensitive neurotensin receptor in HEK 293 cell line: binding properties, photoaffinity labeling, and internalization mechanism. *Biochem. Biophys. Res. Commun., V.243,* 585-590.

[22] Bozou, JC; Rochet, N; Magnaldo, I; Vincent, JP. (1989). Neurotensin stimulates inositol trisphosphate-mediated calcium mobilization but not protein kinase C activation in HT29 cells. *Biochem. J., V.264,* 871-878.

[23] Carbo, R; Zetina, ME; Corkidi, G; Morales, MA. (1997). Topographic relationship of neurotensin-containing axon terminals with cardiac and non-cardiac principal ganglion cells in the stellate ganglia of the cat. *Synapse, V.25,* 277-284.

[24] Carraway, R; Leeman, SE. (1973). The isolation of a new hypotensive peptide, neurotensin, from bovine hypothalami. *J. Biol. Chem., V.248,* 6854-6861.

[25] Carraway, R; Leeman, SE. (1976). Characterization of radioimmunoassayable neurotensin in the rat. *J. Biol. Chem., V.251,* 7045-7052.

[26] Carraway, R; Hammer, RA; Leeman, SE. (1980). Neurotensin in plasma: immunochemical and chromatographic character of acid/acetone-soluble material. *Endocrinology, V.107,* 400-406.

[27] Carraway, R; Cochrane, DE; Lansman, JB; Leeman, SE; Paterson, BM; Welch, H.J. (1982). Neurotensin stimulates exocytotic histamine secretion from rat mast cells and elevates plasma histamine levels. *J. Physiol., V.323,* 403-414.

[28] Carraway, R; Ruane, SE; Feurle, GE; Taylor, S. (1982). Amphibian neurotensin is not xenopsin: dual presence of neurotensin-like and xenopsin-like peptides in various amphibian. *Endocrinology, V.110,* 1094-1101.

[29] Carraway, RE. (1984). Rapid proteolytic generation of neurotensin-related peptides and biologic activity during extraction of rat and chicken gastric tissues. *J. Biol. Chem., V.259,* 10328-10334.

[30] Carraway, RE; Cochrane, DE; Granier, C; Kitabgi, P; Leeman, S; Singer, EA. (1984). Parallel secretion of endogenous 5-hydroxytryptamine and histamine from mast cells stimulated by vasoactive peptides and compound 48/80. *Br. J. Pharmacol., V.81,* 227-229.

[31] Carraway, RE; Mitra, SP; Ferris, CF. (1986). Pepsin treatment of mammalian plasma generates immunoreactive and biologically active neurotensin-related peptides in micromolar concentrations. *Endocrinology, V.119,* 1519-1526.

[32] Carraway, RE; Mitra SP. (1987).Precursor forms of neurotensin in cat: processing with pepsin yields NT-(3-13) and NT-(4-13). *Regul. Peptides, V.18,* 139-154.

[33] Carraway, RE; Mitra, SP; Feurle, GE; Hacki, W.H. (1988). Presence of neurotensin and neuromedin N within a common precursor from a human pancreatic neuroendocrine tumor. *J. Clin Endocrinol. Metab., V.66,* 1323-1328.

[34] Carraway, RE; Mitra, SP; Spaulding, G. (1992). Posttranslational processing of the neurotensin/neuromedin N precursor. *Ann. N.Y. Acad. Sci., V.668,* 1-16.

[35] Caverson, MM; Bachoo, M; Ciriello, J; Polosa, C. (1989). Effect of preganglionic stimulation or chronic decentralization on neurotensin-like immunoreactivity in sympathetic ganglia of the cat. *Brain Res., V.482,* 365-370.

[36] Ceconi, C; Condorelli, E; Quinzanini, M; Rodella, A; Ferrari, R; Harris P. (1989). Noradrenaline, atrial natriuretic peptide, bombesin and neurotensin in myocardium and blood of rats in congestive cardiac failure. *Cardiovasc. Res., V.23,* 674-682.

[37] Chabry, J; Labbe-Jullie, C; Gully, D; Patrick, K; Vincent, JP; Mazella, J. (1994). Stable expression of the cloned rat brain neurotensin receptor into fibroblasts: binding properties, photoaffinity labeling, transduction mechanisms, and internalization. *J. Neurochem., V.63,* 19-27.

[38] Chahl, L.A. (1979). The effect of putative peptide neurotransmitters on cutaneous vascular permeability in the rat. *Naunyn-Schmiedeberg's Arch. Pharmacol., V.309,* 159-163.

[39] Chahl, LA; Walker, SB. (1981). Responses of the rat cardiovascular system to substance P, neurotensin and bombesin. *Life Sci., V.29,* 2009-2015.

[40] Chalon, P; Vita, N; Kaghad, M; Guillemot, M; Bonnin, J; Delpech, B; Le Fur, G; Ferrara, P; Caput, D. (1996). Molecular cloning of a levocabastine-sensitive neurotensin binding site. *FEBS Lett., V.386,* 91-94.

[41] Checler, F; Kostolansca, B; Fox, J. (1988). In vivo inactivation of neurotensin in dog ileum: major involvement of endopeptidase 24.11. *J. Pharmacol. Exp. Ther., V.244,* 1040-1044.

[42] Cochrane, DE; Emigh, C; Levine, G; Carraway, RE; Leeman, SE. (1982). Neurotensin alters cutaneous vascular permeability and stimulates histamine release from isolated skin. *Ann. N.Y. Acad. Sci., V.400,* 396-397.

[43] Corder, R; Mason, DF; Perrett, D; Lowry, PJ; Clement-Jones, V; Linton EA; Besser GM; Rees, LH. (1982). Simultaneous release of neurotensin, somatostatin, enkephalins and catecholamines from perfused cat adrenal glands. *Neuropeptides, V.3,* 9-17.

[44] Di Paola, ED; Richelson E. (1990). Cardiovascular effects of neurotensin and some analogues on rats. *Eur. J. Pharmacol., V.175,* 279-283.

[45] Dobner, PR; Deutch, AY; Fadel, J. (2003). Neurotensin: dual roles in psychostimulant and antipsychotic drug responses. *Life Sci., V.73,* 801-811.

[46] D'Orleans-Juste, P; Dion, S; Mizrahi, J; Regoli, D. (1985). Effects of peptides and non-peptides on isolated arterial smooth muscles: role of endothelium. *Eur. J. Pharmacol., V.114,* 9-21.

[47] Duke, GE; Carraway, RE; Raven, JA. (1993). Cardiovascular effects of neurotensin in anesthetized white leghorn hens. *Poult. Sci., V.72,* 1606-1610.

[48] Dumoulin, V; Dakka, T; Plaisancie, P; Chayvialle, JA; Cuber, JC. (1995). Regulation of glucagons-like peptide-1-(7-36) amide, peptide YY, and neurotensin secretion by neurotransmitters and gut hormones in the isolated vascularly perfused rat ileum. *Endocrinology, V.136,* 5182-5188.

[49] Dumoulin, V; Moro, F; Barcelo, A; Dakka, T; Cuber, JC. (1998). Peptide YY, glucagon-like peptide-1, and neurotensin responses to luminal factors in the isolated vascularly perfused rat ileum. *Endocrinology, V.139,* 3780-3786.

[50] Ekman, R; Ivarsson, S; Jansson, L. (1985). Bombesin, neurotensin and pro-gamma-melanotropin immunoreactants in human milk. *Regul. Peptides, V.10,* 99-105.

[51] Ernst, A; Hellmich, S; Bergmann, A. (2006). Proneurotensin 1-117, a stable neurotensin precursor fragment identified in human circulation. *Peptides, V.27,* 1787-1793.

[52] Ertl, G; Bauer, B; Becker H. (1993). Effects of neurotensin and neuropeptide Y on coronary circulation and myocardial function in dogs. *Am. J. Physiol., V.264,* H1062-H1068.

[53] Evers, BM. (2002). Endocrine gene neurotensin: molecular mechanisms and a model of intestinal differentiation. *World J. Surg., V.26,* 799-805.

[54] Ferris, CF; Parker, MC; Armstrong, MJ; Leeman, SE. (1985). Inhibition of neurotensin release by a cyclic hexapeptide analog of somatostatin. *Peptides, V.6,* 945-948.

[55] Fletcher, DR; Shulkes, A; Bladin, PH; Booth, D; Hardy, KJ. (1983). Cholinergic inhibition of meal stimulated plasma neurotensin-like immunoreactivity in man. *Life Sci., V.33,* 863-869.

[56] Fletcher, DR; Shulkes, A; Hardy, KJ. (1985). The effect of neurotensin and secretin on gastric acid secretion and mucosal blood flow in man. *Regul. Peptides, V.11,* 217-226.

[57] Florholmen, G; Andersson, KB; Yndestad, A; Austbo, B; Henriksen, UL; Christensen G. (2004). Leukaemia inhibitory factor alters expression of genes involved in rat cardiomyocyte energy metabolism. *Acta Physiol. Scand., V.180,* 133-142.

[58] Folkers, K; Chang, D; Humphries, J; Carraway, R; Leeman, SE; Bowers CY. (1976). Synthesis and activities of neurotensin, and its acid and amide analogs: possible natural occurrence of [Gln^4]-neurotensin. *Proc. Natl. Acad. Sci. USA, V.73,* 3833-3837.

[59] Forssmann, WG; Reinecke, M; Weihe, E. (1982). Cardiac innervation. In: Bloom SR, Polak JM, Lindenlaub E. *Systemic role of regulatory peptides*. Symposia Medica Hoechst 18. 329-349.

[60] Fujimura, M; Khalil, T; Sakamoto, T; Greeley, GH; Salter, MG; Townsend, CM; Thompson, JC. (1989). Release of neurotensin by selective perfusion of the jejunum with oleic acid in dogs. *Gastroenterology, V.96,* 1502-1505.

[61] Gaumann, DM; Yaksh, TL. (1988). Effects of hemorrhage and opiate antagonists on adrenal release of neuropeptides in cats. *Peptides, V.9,* 393-405.

[62] Gaumann, DM; Yaksh, TL; Tyce, GM; Stoddard, SL. (1989). Adrenal vein catecholamines and neuropeptides during splanchnic nerve stimulation in cats. *Peptides, V.10,* 587-592.

[63] Gerstheimer, FP; Simon, T; Kolb, J; Hopker, W; Metz, J. (1988). Computer-assisted morphometric study of the innervation of the guinea-pig heart. *Histochemistry, V.88,* 545-551.

[64] Gilbert, JA; McCormick, DJ; Pfenning, A; Kanba, KS; Enloe, LJ; Moore, A; Richelson, E. (1989). Neurotensin (8-13): comparison of novel analogs for stimulation of cyclic GMP formation in neuroblastoma clone N1E-115 and receptor binding to human brain and intact N1E-115 cells. *Biochem. Pharmacol., V.38,* 3377-3382.

[65] Goedert, M; Reynolds, GP; Emson PC. (1983). Neurotensin in the adrenal medulla. *Neurosci. Lett., V.35,* 155-160.

[66] Goedert, M; Pinnock, RD; Downes, CP; Mantyh, PW; Emson, PC. (1984). Neurotensin stimulates inositol phospholipid hydrolysis in rat brain slices. *Brain Res., V.323,* 193-197.

[67] Goedert, M; Sturmey, N; Williams, BJ; Emson, PC. (1984). The comparative distribution of xenopsin- and neurotensin-like immunoreactivity in *Xenopus laevis* and rat tissues. *Brain Res., V.308,* 273-280.

[68] Gorgas, K; Reinecke, M; Weihe, E; Forssman, WG. (1983). Neurotensin and substance P immunoreactive nerve endings in the guinea-pig carotid sinus and their ultrastructural counterparts. *Anat. Embryol., V.167,* 347-354.

[69] Granier, C; Van Rietschoten, J; Kitabgi, P; Poustis, C; Freychet, P. (1982). Synthesis and characterization of neurotensin analogues for structure/activity relationship studies. Acetyl-neurotensin-(8-13) is the shortest analogue with full binding and pharmacological Activities. *Eur. J. Biochem., V.124,* 117-125.

[70] Gully, D; Canton, M; Boigegrain, R; Jeanjean, F; Molimard, JC; Poncelet, M; Gueudet, C; Heaulme, M; Leyris, R; Brouard, A; Pelaprat, D; Labbe-Jullie, C; Mazella, J; Soubrie, P; Maffrand, JP; Rostene, W; Kitabgi, P; Le Fur, G. (1993). Biochemical and pharmacological profile of a potent and selective nonpeptide antagonist of the neurotensin receptor. *Proc. Natl. Acad. Sci. USA, V.90,* 65-69.

[71] Gully, D; Lespy, L; Canton, M; Rostene, W; Kitabgi, P; Le Fur, G; Maffrand JP. (1996). Effect of the neurotensin receptor antagonist SR 48692 on rat blood pressure modulation by neurotensin. *Life Sci., V.58,* 665-674.

[72] Gully, D; Labeeuw, B; Boigegrain, R., Oury-Donat, F; Bachy, A; Poncelet, M; Steinberg, R; Suaud-Chagny, MF; Santucci, V; Vita, N; Pecceu, F; Labbe-Jullie, C; Kitabgi, P; Soubrie, P; Le Fur, G; Maffrand, JP. (1997). Biochemical and

pharmacological activities of SR 142948A, a new potent neurotensin receptor antagonist. *J. Pharmacol. Exp. Ther., V.280,* 802-812.

[73] Hammer, RA; Leeman, SE; Carraway, R; Williams, RH. (1980). Isolation of human intestinal neurotensin. *J. Biol. Chem., V.255,* 2476-2480.

[74] Hammer, RA; Carraway, RE; Leeman, SE. (1982). Elevation of plasma neurotensin-like immunoreactivity after a meal. *J. Clin. Invest., V.70,* 74-81.

[75] Hammer, RA; Fernandez, C; Ertan, A; Arimura, A. (1991). Anesthetic dependence of the inhibitory effect of neurotensin on pentagastrin-stimulated acid secretion in rats. A possible role for somatostatin. *Life Sci., V.48,* 333-339.

[76] Hammer, RA; Ochoa, A; Fernandez, C; Ertan, A; Arimura, A. (1992). Somatostatin as a mediator of the effect of neurotensin on pentagastrin-stimulated acid secretion in rats. *Peptides, V.13,* 1175-1179.

[77] Harper, SL; Barrowman, JA; Kvietys, PR; Granger, DN. (1984). Effect of neurotensin on intestinal capillary permeability and blood flow. *Am. J. Physiol., V.247,* G161-G166.

[78] Hassall, CJS; Burnstock, G. (1986). Intrinsic neurons and associated cells of the guinea-pig heart in culture. *Brain Res., V.364,* 102-113.

[79] Hedner, J; Hedner, T; Lundell, KH; Bissette, G; O'Connor, L; Nemeroff, CB. (1989). Cerebrospinal fluid concentrations of neurotensin and corticotrophin-releasing factor in pediatric patients. *Biol. Neonate, V.55,* 260-267.

[80] Helle, KB; Serck-Hanssen, G; Jorgensen, G; Knudsen, R. (1980). Neurotensin-induced contractions in venous smooth muscle. *J. Auton. Nerv. Syst., V.2,* 143-155.

[81] Helmstaedter, V; Taugner, C; Feurle, GE; Forssman, WG. (1977). Localization of neurotensin-immunoreactive cells in the small intestine of man and various mammals. *Histochemistry, V.53,* 35-41.

[82] Hermans, E; Maloteaux, JM; Octave, JN. (1992). Phospholipase C activation by neurotensin and neuromedin N in Chinese hamster ovary cells expressing the rat neurotensin receptor. *Mol. Brain Res., V.15,* 332-338.

[83] Hermans, E; Gailly, P; Octave, JN; Maloteaux, JM. (1994). Rapid desensitization of agonist-induced calcium mobilization in transfected PC12 cells expressing the rat neurotensin receptor. *Biochem. Biophys. Res. Commun., V.198,* 400-407.

[84] Hermans, E; Gailly, P; Gillis, JM; Octave, JN; Maloteaux, JM. (1995). Lack of rapid desensitization of calcium responses in transfected CHO cells expressing the rat neurotensin receptor despite agonist-induced internalization. *J. Neurochem., V.64,* 2518-2525.

[85] Hermey, G; Riedel, B; Hampe, W; Schaller, HC; Hermans-Borgmeyer, I. (1999). Identification and characterization of SorCS, a third member of a novel receptor family. *Biochem. Biophys. Res. Commun., V.266,* 347-351.

[86] Herrmann, C; Cuber, JC; Abello, J; Dakka, T; Bernard, C; Chayvialle, JA. (1991). Release of ileal neurotensin in the rat by neurotransmitters and neuropeptides. *Regul. Peptides, V.32,* 181-192.

[87] Heuser, M; Gralla, O; Pfaar, O; Nustede, R; Grone, HJ; Post, S. (2000). Differential effects of neurotensin and cholecystokinin on intestinal microcirculation after ischemia-reperfusion. *Langenbeck's Arch. Surg., V.385,* 357-362.

[88] Higgins, GA; Hoffman, GE; Wray, S; Schwaber, JS. (1984). Distribution of neurotensin-immunoreactivity within baroreceptive portions of the nucleus of the tractus solitarius and the dorsal vagal nucleus of the rat. *J. Comp. Neurol., V.226,* 155-164.

[89] Hinson, JP; Cameron, LA; Purbrick, A; Kapas, S. (1994). The role of neuropeptides in the regulation of adrenal vascular tone: effects of vasoactive intestinal polypeptide, substance P, neuropeptide Y, neurotensin, met-enkephalin and leu-enkephalin on perfusion medium flow rate in the intact perfused rat adrenal. *Regul. Peptides, V.51,* 55-61.

[90] Holmgren, S; Fritsche, R; Karila, P; Gibbins, I; Axelsson, M; Franklin, C; Grigg, G; Nilsson, S. (1994). Neuropeptides in the Australian lungfish *Neoceratodus forsteri*: effects in vivo and presence in autonomic nerves. *Am. J. Physiol., V.266,* R1568-R1577.

[91] Holst Pedersen, J; Andersen, HO; Olsen, PS; Henriksen, JH. (1989). Pharmacokinetics and metabolism of neurotensin in man. *J. Clin. Endocrinol. Metab., V.68,* 294-300.

[92] Holzer, P. (1984). Characterization of the stimulus-induced release of immunoreactive substance P from the myenteric plexus of the guinea-pig small intestine. *Brain. Res., V.297,* 127-136.

[93] Ishizuka, J; Townsend, CM; Thompson, JC. (1993). Neurotensin regulates growth of human pancreatic cancer. *Ann. Surg., V.217,* 439-446.

[94] Iwase, K; Evers, M; Hellmich, M; Kim, HJ; Higashide, S; Gully, D; Thompson, JC; Townsend, CM. (1997). Inhibition of neurotensin-induced pancreatic carcinoma growth by a nonpeptide neurotensin receptor antagonist, SR 48692. *Cancer, V.79,* 787-793.

[95] Jennes, L; Stumpf, WE; Kalivas, PW. (1982). Neurotensin: topographical distribution in rat brain by immunohistochemistry. *J. Comp. Neurol., V.210,* 211-224.

[96] Kanba, KS; Richelson, E. (1987). Comparison of the stimulation of inositol phospholipid hydrolysis and of cyclic GMP formation by neurotensin, some of its analogs, and neuromedin N in neuroblastoma clone N1E-115. *Biochem. Pharmacol., V.36,* 869-874.

[97] Kapuscinski, M; Shulkes, A; Read, D; Hardy, KJ. (1990). Expression of neurotensin in endocrine tumors. *J. Clin. Endocrinol. Metab., V.70,* 100-106.

[98] Karila, P; Axelsson, M; Franklin, CE; Fritsche, R; Gibbins, IL; Nilsson, S; Holmgren S. (1995). Neuropeptide immunoreactivity and co-existence in cardiovascular nerves and autonomic ganglia of the estuarine crocodile, *Crocodylus porosus*, and cardiovascular effects of neuropeptides. *Regul. Peptides, V.58,* 25-39.

[99] Kataoka, K; Taniguchi, A; Shimizu, H; Soda, K; Okuno, S; Yajima, H; Kitagawa. K. (1978). Biological activity of neurotensin and its C-terminal partial sequences. *Brain Res. Bull., V.3,* 555-557.

[100] Kerouac, R; Rioux, F; St-Pierre, S. (1981). Mechanism of neurotensin-induced pressor effect and tachycardia in guinea-pigs. *Life Sci., V.28,* 2477-2487.

[101] Kerouac, R; St-Pierre, S; Manning, M; Rioux, F. (1983). Partial blockade of neurotensin-induced hypotension in rats by nephrectomy, captopril and saralasin. Possible mechanisms. *Neuropeptides, V.3,* 295-307.

[102] Kerouac, R; St-Pierre, S; Rioux, F. (1984). Participation of mast cell 5-hydroxytryptamine in the vasoconstrictor effect of neurotensin in the rat perfused hindquarter. *Life Sci., V.34,* 947-959.

[103] Kerouac, R; St-Pierre, S; Rioux, F. (1984). Histamine release by neurotensin in the rat hindquarter: structure-activity studies. *Peptides, V.5,* 695-699.

[104] Kessler, JP; Beaudet, A. (1989). Association of neurotensin binding sites with sensory and visceromotor components of the vagus nerve. *J. Neurosci., V.9,* 466-472.

[105] Kihl, B; Rokaeus, A; Rosell, S; Olbe, L. (1981). Fat inhibition of gastric acid secretion in man and plasma concentrations of neurotensin-like immunoreactivity. *Scand. J. Gastroenterol., V.16,* 513-526.

[106] Kilborn, MJ; Potter, EK; McCloskey, DI. (1986). Effects of periods of conditioning stimulation and of neuropeptides on vagal action at the heart. *J. Auton. Nerv. Syst., V.17,* 131-142.

[107] Kislauskis, E; Bullock, B; McNeil, S; Dobner, PR. (1988). The rat gene encoding neurotensin and neuromedin N. *J. Biol. Chem., V.263,* 4963-4968.

[108] Kitabgi, P; Carraway, R; Leeman, SE. (1976). Isolation of a tridecapeptide from bovine intestinal tissue and its partial characterization as neurotensin. *J. Biol. Chem., V.251,* 7053-7058.

[109] Kitabgi, P; Poustis, C; Granier, C; Van Rietschoten, J; Rivier, J; Morgat, JL; Freychet, P. (1980). Neurotensin binding to extraneural and neural receptors: comparison with biological activity and structure-activity relationships. *Mol. Pharmacol., V.18,* 11-19.

[110] Kitabgi, P; De Nadai, F; Rovere, C; Bidard, JN. (1992). Biosynthesis, maturation, release and degradation of neurotensin and neuromedin N. *Ann. N.Y. Acad. Sci., V.668,* 30-42.

[111] Kitabgi, P. (2002). Targeting neurotensin receptors with agonists and antagonists for therapeutic purposes. *Curr. Opin. Drug Discov. Devel., V.5,* 764-776.

[112] Konturek, SJ; Jaworek, J; Cieszkowski, M; Pawlik, W; Kania, J; Bloom, SR. (1983).Comparison of effects of neurotensin and fat on pancreatic stimulation in dogs. *Am. J. Physiol., V.244,* G590-G598.

[113] Kronenberg, F; Carraway, RE. (1985). Changes in neurotensin-like immunoreactivity during menopausal hot flashes. *J. Clin. Endocrinol. Metab., V.60,* 1081-1086.

[114] Kruger, PG; Aas, P; Onarheim, J; Helle, KB. (1982). Neurotensin-induced release of histamine from rat mast cells in vitro. *Acta Physiol. Scand., V.114,* 467-469.

[115] Krukoff, TL; Ciriello, J; Calaresu, FR. (1985). Segmental distribution of peptide-like immunoreactivity in cell bodies of the thoracolumbar sympathetic nuclei of the cat. *J. Comp. Neurol., V.240,* 90-102.

[116] Krukoff, TL. (1987). Coexistence of neuropeptides in sympathetic preganglionic neurons of the cat. *Peptides, V.8,* 109-112.

[117] Kummer, W; Reinecke, M; Heym, C. (1991). Neurotensin-like immunoreactivity in presumptive baroreceptor neurons innervating the guinea-pig carotid sinus. *J. Auton. Nerv. Syst., V.35,* 107-116.

[118] Kurose, M; Saeki, K. (1981). Histamine release induced by neurotensin from rat peritoneal mast cells. *Eur. J. Pharmacol., V.76,* 129-136.

[119] Lazarus, LH; Perrin, MH; Brown, MR. (1977). Mast cell binding of neurotensin. I. Iodination of neurotensin and characterization of the interaction of neurotensin with mast cell receptor sites. *J. Biol. Chem., V.252,* 7174-7179.

[120] Lazarus, LH; Perrin, MH; Brown, MR; Rivier, JE. (1977). Mast cell binding of neurotensin. II. Molecular conformation of neurotensin involved in the stereospecific binding to mast cell receptor sites. *J. Biol. Chem., V.252,* 7180-7183.

[121] Lee, YC; Allen, JM; Uttenthal, LO; Walker, MC; Shemilt, J; Gill, SS; Bloom, SR. (1984). The metabolism of intravenously infused neurotensin in man and its chromatographic characterization in human plasma. *J. Clin. Endocrinol. Metab., V.45,* 45-50.

[122] Lee, YC; Uttenthal, LO; Smith, HA; Bloom, SR. (1986). In vitro degradation of neurotensin in human plasma. *Peptides, V.7,* 383-387.

[123] Lin, B; Pilch, PF; Kandror, KV. (1997). Sortilin is a major protein component of Glut4-containing vesicles. *J. Biol. Chem., V.272,* 25145-24147.

[124] Linde, B; Rosell, S; Rokaeus, A. (1982). Intravenous administration of (Gln^4)-neurotensin induces vasoconstriction in human adipose tissue. *Ann. N.Y. Acad. Sci., V.400,* 392-393.

[125] Lucas, A; Aynsley-Green, A; Blackburn, AM; Adrian, TE; Bloom, SR. (1981). Plasma neurotensin in term and preterm neonates. *Acta Paediatr. Scand., V.70,* 201-206.

[126] Lundberg, JM; Rokaeus, A; Hokfelt, T; Rosell, S; Brown, M; Goldstein, M. (1982). Neurotensin-like immunoreactivity in the preganglionic sympathetic nerves and in the adrenal medulla of the cat. *Acta Physiol. Scand., V.114,* 153-155.

[127] Maher, E; Bachoo, M; Cernacek, P; Polosa, C. (1991). Dynamics of neurotensin stores in the stellate ganglion of the cat. *Brain Res., V.562,* 258-264.

[128] Maher, E; Bachoo, B; Polosa, C. (1994). Role of neuron soma firing in the restoration of neurotensin store in sympathetic preganglionic neuron terminals after stimulus-evoked depletion. *Brain Res., V.640,* 126-130.

[129] Maher, E; Bachoo, B; Polosa, C. (1994). In vitro and in vivo evidence of neurotensin release from preganglionic axon terminals in the stellate ganglion of the cat. *Brain Res., V.640,* 131-135.

[130] Martinez, M; Hernanz, A; Grande, C; Pallardo, LF. (1993). Plasma molecular forms of gastrin, neurotensin and somatostatin in pregnancy and gestational diabetes after an oral glucose load or a mixed meal. *Regul. Peptides, V.47,* 73-80.

[131] Mashford, ML; Nilsson, G; Rokaeus, A; Rosell, S. (1978). The effect of food ingestion on circulating neurotensin-like immunoreactivity in the human. *Acta Physiol. Scand., V.104,* 244-246.

[132] Mashford, ML; Nilsson, G; Rokaeus, A; Rosell, S. (1978). Release of neurotensin-like immunoreactivity from the gut in anaesthetized dogs. *Acta Physiol. Scand., V.104,* 375-376.

[133] Mazella, J; Chabry, J; Zsurger, N; Vincent, JP. (1989). Purification of the neurotensin receptor from mouse brain by affinity chromatography. *J. Biol. Chem., V.264,* 5559-5563.

[134] Mazella, J; Botto, JM; Guillemare, E; Coppola, T; Sarret, P; Vincent, JP. (1996). Structure, functional expression, and cerebral localization of the levocabastine-sensitive neurotensin/neuromedin N receptor from mouse brain. *J. Neurosci., V.16,* 5613-5620.

[135] McDermott, JR; Virmani, MA; Turner, JD; Kidd, AM. (1986). Peptidases involved in the catabolism of neurotensin: inhibitor studies using superfused rat hypothalamic slices. *Peptides, V.7,* 225-230.

[136] Mentlein, R; Dahms, P. (1994). Endopeptidases 24.16 and 24.15 are responsible for the degradation of somatostatin, neurotensin and other neuropeptides by cultivated rat cortical astrocytes. *J. Neurochem., V.62,* 27-36.

[137] Miller, LA; Cochrane, DE; Carraway, RE; Feldberg, RS. (1995). Blockade of mast cell histamine secretion in response to neurotensin by SR 48692, a nonpeptide antagonist of the neurotensin brain receptor. *Br. J. Pharmacol., V.114,* 1466-1470.

[138] Miller, VM; Hoffman, AM; South, FE. (1981). Cardiovascular responses to neurotensin in the Genus marmota. *Fed. Proc., V.40,* 581.

[139] Mogard, M; Bottcher, W; Kauffman, GL; Washington, J; Walsh, JH. (1986). Neurotensin-like immunoreactivity released into the portal vein by duodenal acidification in the dog. *Scand. J. Gastroenterol., V.21,* 97-103.

[140] Mogard, MH; Kobayashi, R; Lee, TD; Chen, CF; Hagiwara, M; Leung, F; Reeve, JR; Shively, JE; Walsh, JH. (1987). Neurotensin-like immunoreactivity generated by pepsin from human plasma and gastric tissue. *Regul. Peptides, V.18,* 221-232.

[141] Nakamoto, M; Tanaka, C; Taniyama, K. (1987). Release of γ-aminobutiric acid and acetylcholine by neurotensin in guinea-pig ileum. *Br. J. Pharmacol., V.90,* 545-551.

[142] Nicholls, DP; Riley, M; Elborn, JS; Stanford, CF; Shaw, C; McKillop, JM; Buchanan, KD. (1992). Regulatory peptides in the plasma of patients with chronic cardiac failure at rest and during exercise. *Eur. Heart J., V.13,* 1399-1404.

[143] Nisato, D; Guiraudou, P; Barthelemy, G; Gully, D; Le Fur, G. (1994). SR 48692, a non- peptide neurotensin receptor antagonist, blocks the cardiovascular effects elicited by neurotensin in guinea-pigs. *Life Sci., V.54,* 95-100.

[144] Norman, MU; Reeve, SB; Dive, V; Smith, AI; Lew, RA. (2003). Endopeptidases 3.4.24.15 and 24.16 in endothelial cells: potential role in vasoactive peptide metabolism. *Am. J. Physiol., V.284,* H1978-H1984.

[145] Obara, H; Kusunoki, M; Mori, K; Mikawa, K; Iwai, S. (1989). The effects of various peptides on the isolated pulmonary artery. *Peptides, V.10,* 241-243.

[146] Oishi, M; Inagaki, C; Fujiwara, M; Takaori, S; Yajima, H; Akazawa ,Y. (1981). Possible mechanisms of the triphasic effects of neurotensin on the rat blood pressure. *Jap. J. Pharmacol., V.31,* 1043-1049.

[147] Oishi, M; Ishiko, J; Inagaki, C; Takaori, S. (1983). Release of histamine and adrenaline in vivo following intravenous administration of neurotensin. *Life Sci., V.32,* 2231-2239.

[148] Onarheim, J; Helle, KB; Jorgensen, G. (1982). Neurotensin-induced increase in intestinal blood flow in the anesthetized rat. *Acta Physiol. Scand., V.114,* 505-511.

[149] Onuoha, GN; Alpar, EK; Chukwulobelu, R; Nicholls, DP. (1999). Distributions of VIP, substance P, neurokinin A and neurotensin in rat heart: an immunocytochemical study. *Neuropeptides, V.33,* 19-25.

[150] Onuoha, GN; Nicholls, DP; Alpar, EK; Ritchie, A; Shaw, C; Buchanan, K. (1999). Regulatory peptides in the heart and major vessels of man and mammals. *Neuropeptides, V.33,* 165-172.

[151] Osadchii, OE; Pokrovskii, VM; Kurzanov, AN. (1993). Modulatory action of neurotensin on the parasympathetic regulation of cardiac rhythm. *Bull. Exp. Biol. Med., V.115,* 453-455.

[152] Osadchii, OE; Pokrovskii, VM; Kompaniets, OG; Kurzanov, AN. (1996). A comparative evaluation of the cardiotropic effects of neurotensin and adrenaline in cats. *Fiziol. Zh. Im. I M Sechenova, V.82,* 104-110.

[153] Osadchii, OE; Pokrovskii, VM; Kompaniets, OG; Kurzanov, AN. (1997). The effect of neurotensin and adrenaline on sinus arrhythmia induced by vagus nerve stimulation. *Bull. Exp. Biol. Med., V.123,* 494-497.

[154] Osadchii, OE; Pokrovskii, VM; Kompaniets, OG; Kurzanov, AN. (1997). Protective effect of neurotensin in vagal arrhythmias. *Bull. Exp. Biol. Med., V.124,* 495-497.

[155] Osadchii, OE; Pokrovskii, VM. (1999). The dynamics of the chronotropic effect evoked by a single vagal stimulus in cats under the action of met-enkephalin and neurotensin. *Ross. Fiziol Zh Im I M Sechenova, V.85,* 547-53.

[156] Osadchii, OE; Kanorskii, SG; Pokrovskii, VM; Kurzanov, AN; Skibitskii, VV. (2001). The effect of regulatory peptides upon electrophysiological properties of the human heart. *Human Physiol., V.27,* 105-110.

[157] Osadchii, O; Norton, G; Deftereos, D; Badenhorst, D; Woodiwiss, A. (2005). Impact and mechanisms of action of neurotensin on cardiac contractility in the rat left ventricle. *Eur. J. Pharmacol., V.520,* 108-117.

[158] Osadchii, O; Norton, G; Deftereos, D; Badenhorst, D; Woodiwiss, A. (2006). Neurotensin-induced myocardial noradrenergic effects in spontaneously hypertensive rats. *J. Cardiovasc. Pharmacol., V.47,* 221-227.

[159] Osadchii, O; Norton, G; Deftereos, D; Muller, D; Woodiwiss, A. (2006).Impact of chronic β-adrenoceptor activation on neurotensin-induced myocardial effects in rats. *Eur. J. Pharmacol.,V.553,* 246-253.

[160] Ouru-Donat, F; Thurneyssen, O; Gonalons, N; Forgez, P; Gully, D; Le Fur, G; Soubrie P. (1995). Characterization of the effect of SR 48692 on inositol monophosphate, cyclic GMP and cyclic AMP responses linked to neurotensin receptor activation in neuronal and non-neuronal cells. *Br. J. Pharmacol., V.116,* 1899-1905.

[161] Pang, X; Alexacos, N; Letourneau, R; Seretakis, D; Gao, W; Boucher, W; Cochrane, DE; Theoharides, TC. (1998). A neurotensin receptor antagonist inhibits acute immobilization stress-induced cardiac mast cell degranulation, a corticotrophin-releasing hormone-dependent process. *J. Pharmacol. Exp. Ther., V.287,* 307-314.

[162] Park, TJ; Kim, KT. (2003). Activation of B2 bradykinin receptors by neurotensin. *Cell. Signal., V.15,* 519-527.

[163] Pedersen, JH; Fahrenkrug, J. (1986). Neurotensin-like immunoreactivities in human plasma: feeding responses and metabolism. *Peptides, V.7,* 15-20.

[164] Pelto-Huikko, M; Salminen, T; Partanen, M; Toivanen, M; Hervonen, A. (1985). Immunohistochemical localization of neurotensin in hamster adrenal medulla. *Anat. Rec., V.211,* 458-464.

[165] Pelto-Huikko, M; Salminen, T; Hervonen, A. (1987). Localization of enkephalin- and neurotensin-like immunoreactivities in cat adrenal medulla. *Histochemistry, V.88,* 31-36.

[166] Petersen, CM; Nielsen, MS; Nykjar, A; Jacobsen, L; Tommerup, N; Rasmussen, HH; Roigaard, H; Gliemann, J; Madsen, P; Moestrup, SK. (1997). Molecular identification of a novel candidate sorting receptor purified from human brain by receptor-associated protein affinity chromatography. *J. Biol. Chem., V.272,* 3599-3605.

[167] Polak, JM; Sullivan, SN; Bloom, SR; Buchan, AM; Facer, P; Brown, MR; Pearse, AG. (1977). Specific localization of neurotensin to the N cell in human intestine by radioimmunoassay and immunocytochemistry. *Nature, V.270,* 183-184.

[168] Polak JM; Bloom, SR. (1982). The central and peripheral distribution of neurotensin. *Ann. N.Y. Acad. Sci., V.400,* 75-93.

[169] Quirion, R; Rioux, F; Regoli, D. (1978). Chronotropic and inotropic effects of neurotensin on spontaneously beating auricles. *Can. J. Physiol. Pharmacol., V.56,* 671-673.

[170] Quirion, R; Regoli, D; Rioux, F; St-Pierre, S. (1979). An analysis of the negative inotropic action of somatostatin. *Br. J. Pharmacol., V.66,* 251-257.

[171] Quirion, R; Rioux, F; Regoli, D; St-Pierre, S. (1979). Neurotensin-induced vessels constriction in perfused rat hearts. *Eur. J. Pharmacol., V.55,* 221-223.

[172] Quirion, R; Rioux, F; St-Pierre, S; Regoli, D. (1979). Increased sensitivity to neurotensin in fasted rats. *Life Sci., V.25,* 1969-1973.

[173] Quirion, R; Regoli, D; Rioux, F; St-Pierre, S. (1980). The stimulatory effects of neurotensin and related peptides in rat stomach strips and guinea-pig atria. *Br. J. Pharmacol., V.68,* 83-91.

[174] Quirion, R; Regoli, D; Rioux, F; St-Pierre, S. (1980). Structure-activity studies with neurotensin: analysis of positions 9, 10 and 11. *Br. J. Pharmacol., V.69,* 689-692.

[175] Quirion, R; Rioux, F; Regoli, D; St-Pierre, S. (1980). Selective blockade of neurotensin-induced coronary vessel constriction in perfused rat hearts by a neurotensin analogue. *Eur. J. Pharmacol., V.61,* 309-312.

[176] Quirion, R; Rioux, F; Regoli, D; St-Pierre, S. (1980). Pharmacological studies of neurotensin, several fragments and analogues in the isolated perfused rat heart. *Eur. J. Pharmacol., V.66,* 257-266.

[177] Quirion, R; Rioux, F; Regoli, D; St-Pierre, S. (1980). Compound 48/80 inhibits neurotensin-induced hypotension in rats. *Life Sci., V.27,* 1889-1895.

[178] Rakovska, AD. (1993). Functional and neurochemical evidence that neurotensin-induced release of acetylcholine from Auerbach's plexus of guinea-pig ileum is presynaptically controlled via alpha 2-adrenoceptors. *Neurochem. Res., V.18,* 737-741.

[179] Reinecke, M; Almasan, K; Carraway, R; Helmstaedter, V; Forssmann, WG. (1980). Distribution patterns of neurotensin-like immunoreactive cells in the gastrointestinal tract of higher vertebrates. *Cell Tissue Res., V.205,* 383-395.

[180] Reinecke, M; Weihe, E; Carraway, RE; Forssmann, WG. (1981). Localization of neurotensin-immunorectivity in the heart by immunohistochemistry and radioimmunoassay. *Acta Anat., V.III,* 123-124.

[181] Reinecke, M; Weihe, E; Carraway, RE; Leeman, SE; Forssmann, WG. (1982). Localization of neurotensin immunoreactive nerve fibers in the guinea pig heart: evidence derived by immunohistochemistry, radioimmunoassay and chromatography. *Neuroscience, V.7,* 1785-1795.

[182] Reinecke, M; Forssmann, WG; Thiekotter, G; Triepel, J. (1983). Localization of neurotensin-immunoreactivity in the spinal cord and peripheral nervous system of the guinea-pig. *Neurosci. Lett., V.37,* 37-42.

[183] Reinecke, M; Forssmann, WG. (1984). Regulatory peptides of autonomic nerves in the guinea-pig heart. *Clin. Exp. Hypertens. A., V.6,* 1867-1871.

[184] Reinecke, M. (1985). Neurotensin. Immunohistochemical localization in central and peripheral nervous system and in endocrine cells and its functional role as neurotransmitter and endocrine hormone. *Progr. Histochem. Cytochem., V.16,* 1-173.

[185] Richelson, E; Boules, M; Fredrickson, P. (2003). Neurotensin agonists: possible drugs for treatment of psychostimulant abuse. *Life Sci., V.73,* 679-690.

[186] Rigel, DF. (1988). Effects of neuropeptides on heart rate in dogs: comparison of VIP, PHI, NPY, CGRP, and NT. *Am. J. Physiol., V.255,* H311-H317.

[187] Rigel, DF; Grupp, IL; Balasubramaniam, A; Grupp, G. (1989). Contractile effects of cardiac neuropeptides in isolated canine atrial and ventricular muscles. *Am. J. Physiol., V.257,* H1082-H1087.

[188] Rioux, F; Quirion, R; Leblanc, MA; Regoli, D; St-Pierre, S. (1980). Possible interactions between neurotensin and prostaglandins in the isolated rat portal vein. *Life Sci.,* V.27, 259-267.

[189] Rioux, F; Quirion, R; Regoli, D; Leblanc, MA; St-Pierre, S. (1980). Pharmacological characterization of neurotensin receptors in the rat isolated portal vein using analogues and fragments of neurotensin. *Eur. J. Pharmacol., V.66,* 273-279.

[190] Rioux, F; Kerouac, R; Quirion, R; St-Pierre, S. (1982). Mechanisms of the cardiovascular effects of neurotensin. *Ann. N.Y. Acad. Sci., V.400,* 56-74.

[191] Rioux, F; Kerouac, R; St-Pierre, S. (1982). Analysis of the biphasic depressor-pressor effect and tachycardia caused by neurotensin in ganglion-blocked rats. *Neuropeptides, V.3,* 113-127.

[192] Rioux, F; Kerouac, R; St-Pierre, S. (1984). Neurotensin stimulates histamine release from the isolated, spontaneously beating heart of rats. *Life Sci., V.35,* 423-431.

[193] Rioux, F; Kerouac, R; St-Pierre, S. (1985). Characterization of the histamine-releasing effect of neurotensin in the rat heart. *Peptides, V.6,* 121-125.

[194] Rioux, F; Bachelard, H; Barabe, J; St-Pierre, S. (1986). The cardiovascular response to epicardial application of neurotensin in guinea-pigs. *Peptides, V.7,* 1087-1094.

[195] Rioux, F; Lemieux, M; Kerouac, R; Bernoussi, A; Roy, G. (1989). Local application of neurotensin to abdominal organs triggers cardiovascular reflexes in guinea-pigs: possible mechanisms. *Peptides, V.10,* 647-655.

[196] Rioux, F; Lemieux, M. (1992). Hemodynamic and abdominal motor reflexes elicited by neurotensin in anaesthetized guinea-pigs. *Br. J. Pharmacol., V.106, 187-195.*

[197] Rioux, F;.Pare, D. (1983). Cardiovascular and abdominal motor responses evoked by intravenous neurotensin in guinea-pigs. *Peptides, V.14,* 227-234.

[198] Rokaeus, A; Burcher, E; Chang, D; Folkers, K; Rosell, S. (1977). Actions of neurotensin and (Gln^4)-neurotensin on isolated tissues. *Acta Pharmacol. Toxicol., V.41,* 141-147.

[199] Rokaeus, A. (1981). Studies on neurotensin as a hormone. *Acta Physiol. Scand., V.501, Suppl.,* 1-62.

[200] Rokaeus, A; Yanaihara, N; McDonald, TJ. (1982). Increased concentration of neurotensin-like immunoreactivity in rat plasma after administration bombesin and bombesin-related peptides. *Acta Physiol. Scand., V.114,* 605-610.

[201] Rokaeus, A; Fried, G; Lundberg, JM. (1984). Occurrence, storage and release of neurotensin-like immunoreactivity from the adrenal gland. *Acta Physiol. Scand., V.120,* 373-380.

[202] Rosell, S; Burcher, E; Chang, D; Folkers, K. (1976). Cardiovascular and metabolic actions of neurotensin and (Gln^4)-neurotensin. *Acta Physiol. Scand., V.98,* 484-491.

[203] Rosell, S; Rokaeus, A; Chang, D; Folkers, K. (1978). Indirect vascular actions of (Gln^4)- neurotensin in canine adipose tissue. *Acta Physiol. Scand., V.102,* 143-147.

[204] Rosell, S; Rokaeus, A. (1979). The effect of ingestion of amino acids, glucose and fat on circulating neurotensin-like immunoreactivity in man. *Acta Physiol. Scand., V.107,* 263-267.

[205] Rosell, S; Thor, K; Rokaeus, A; Nyquist, O; Lewenhaupt, A; Kager, L; Folkers, K. (1980). Plasma concentration of neurotensin-like immunoreactivity and lower esophageal sphincter pressure in man following infusion of (Gln^4)-neurotensin. *Acta Physiol. Scand.,* V.109, 369-375.

[206] Rossie, SS; Miller, RJ. (1982). Regulation of mast cell histamine release by neurotensin. *Life Sci., V.31,* 509-516.

[207] Rostene, WH; Alexander, MJ. (1997). Neurotensin and neuroendocrine regulation. *Front. Neuroendocrinol., V.18,* 115-173.

[208] Sagor, GR; Ghatei, MA; McGregor, GP; Mitchenere, P; Kirk, RM; Bloom, SR. (1981). The influence of an intact pylorus on postprandial enteroglucagon and neurotensin release after upper gastric surgery. *Br. J. Surg., V.68,* 190-193.

[209] Saito, H; Saito, S. (1980). Effects of substance P and neurotensin on somatostatin levels in rat portal plasma. *Endocrinology, V.107,* 1600-1605.

[210] Sato, M; Shiosaka, S; Tohyama, M. (1991). Neurotensin and neuromedin N elevate the cytosolic calcium concentration via transiently appearing neurotensin binding sites in cultured rat cortex cells. *Devel. Brain Res., V.58,* 97-103.

[211] Schaeffer, P; Laplace, MC; Savi, P; Pflieger, AM; Gully, D; Herbert, JM. (1995). Human umbilical vein endothelial cells express high affinity neurotensin receptors coupled to intracellular calcium release. *J. Biol. Chem., V.270,* 3409-3413.

[212] Schaeffer, P; Laplace, MC; Prabonnaud, V; Bernat, A; Gully, D; Lespy, L; Herbert, J M. (1997). Neurotensin induces the release of prostacyclin from human umbilical vein endothelial cells in vitro and increases plasma prostacyclin levels in the rat. *Eur. J. Pharmacol., V.323,* 215-221.

[213] Schaeffer, P; Laplace, M; Bernat, A; Prabonnaud, V; Gully, D; Lespy, L; Herbert, JM. (1998). SR 142948A is a potent antagonist of the cardiovascular effects of neurotensin. *J. Cardiovasc. Pharmacol., V.31,* 545-550.

[214] Schimpff, RM; Avard, C; Fenelon, G; Lhiaubet, AM; Tenneze, L; Vidailhet, M; Rostene, W. (2001). Increased plasma neurotensin concentrations in patients with Parkinson's disease. *J. Neurol. Neurosurg. Psych., V. 70,* 784-786.

[215] Selbekk, BH; Flaten, O; Hanssen, LE. (1980). The in vitro effect of neurotensin on human jejunal mast cells. *Scand. J. Gastroenterol., V.15,* 457-460.

[216] Sharma, RP; Janicak, PG; Bissette, G; Nemeroff, CB. (1997). Cerebrospinal fluid concentrations and antipsychotic treatment in schizophrenia and schizoaffective disorder. *Am. J. Psych., V.154,* 1019-1021.

[217] Shulkes, A; Chick, P; Wong, H; Walsh, JH. (1982). A radioimmunoassay for neurotensin in human plasma. *Clin. Chim. Acta., V.125,* 49-58.

[218] Shulkes, A; Fletcher, DR; Hardy, KJ. (1983). Organ and plasma metabolism of neurotensin in sheep. *Am. J. Physiol., V.245,* E457-62.

[219] Shulkes, A; Bijaphala, S; Dawborn, JK; Fletcher, DR; Hardy KJ. (1984). Metabolism of neurotensin and pancreatic polypeptide in man: role of the kidney and plasma factors. *J. Clin. Endocrinol.Metab., V.58,* 873-879.

[220] Shulkes, A; Englin, I; Read, D; Hardy, KJ. (1987). Neurotensin metabolism in the rat: contribution of the kidney. *Peptides, V.8,* 961-965.

[221] Sicard, F; Vaudry, H; Braun, B; Chartrel, N; Leprince, J; Conlon, JM; Delarue, C. (2000). Immunohistochemical localization, biochemical characterization, and biological activity of neurotensin in the frog adrenal gland. *Endocrinology, V.141,* 2450-2457.

[222] Skidgel, RA; Engelbrecht, S; Johnson, AR; Erdos, EG. (1984). Hydrolysis of substance P and neurotensin by converting enzyme and neutral endopeptidase. *Peptides, V.5,* 769-776.

[223] Snider, RM; Forray, C; Pfenning, M; Richelson, E. (1986). Neurotensin stimulates inositol phospholipid metabolism and calcium mobilization in murine neuroblastoma clone N1E-115. *J. Neurochem., V.47,* 1214-1218.

[224] Stapelfeldt, WH; Szurszewski, JH. (1989). Neurotensin facilitates release of substance P in the guinea-pig inferior mesenteric ganglion. *J. Physiol., V.411,* 325-345.

[225] Stene-Larsen, G; Helle, KB. (1979). Inotropic and chronotropic effects of neurotensin in the rat atrium and of physalaemin in the auricles of *R.esculenta. Comp. Biochem. Physiol., V.64,* 279-283.

[226] Sumners, C; Phillips, MI; Richards, EM. (1982). Central pressor action of neurotensin in conscious rats. *Hypertension, V.4,* 888-893.

[227] Sundler, F; Alumets, J; Hakanson, R; Carraway, R; Leeman, SE. (1977). Ultrastructure of the gut neurotensin cell. *Histochemistry, V.53,* 25-34.

[228] Tanaka, K; Masu, M; Nakanishi, S. (1990). Structure and functional expression of the cloned rat neurotensin receptor. *Neuron, V.4,* 847-854.

[229] Terenghi, G; Polak, JM; Varndell, IM; Lee, YC; Wharton, J; Bloom, SR. (1983). Neurotensin-like immunoreactivity in a subpopulation of noradrenaline-containing cells of the cat adrenal gland. *Endocrinology, V.112,* 226-233.

[230] Theodorsson-Norheim, E; Al-Saffar, A; Saria, A; Rosell, S. (1985). The effect of nephrectomy and ureter ligation on plasma neurotensin-like immunoreactivity levels in the conscious rat. *Acta Physiol. Scand., V.123,* 269-272.

[231] Thomas, RP; Hellmich, MR; Townsend, CM; Evers, BM. (2003). Role of gastrointestinal hormones in the proliferation of normal and neoplastic tissues. *Endocr. Rev., V.24,* 571-599.

[232] Tsuda, K; Masuyma, Y. (1993). Effects of neurotensin on norepinephrine release in blood vessels of spontaneously hypertensive rats. *Am. J. Hypertens., V.6,* 473-479.

[233] Tyler-McMahon, BM; Boules, M; Richelson, E. (2000). Neurotensin: peptide for the next millennium. *Regul. Peptides, V.93,* 125-136.

[234] Unwin, RJ; Calam, J; Peart, WS; Hanson, C; Lee, YC; Bloom, SR. (1987). Renal function during bovine neurotensin infusion in man. *Regul. Peptides, V.18,* 29-35.

[235] Vincent, JP; Mazella, J; Kitabgi, P. (1999). Neurotensin and neurotensin receptors. *Trends Pharmacol. Sci., V.20,* 302-309.

[236] Vita, N; Laurent, P; Lefort, S; Chalon, P; Dumont, X; Kaghad, M; Gully, D; Le Fur, G; Ferrara, P; Caput, D. (1993). Cloning and expression of a complementary DNA encoding a high affinity human neurotensin receptor. *FEBS Lett., V.317,* 139-142.

[237] Vita, N; Oury-Donat, F; Chalon, P; Guillemot, M; Kaghad, M; Bachy, A; Thurneyssen, O; Garcia, S; Poinot-Chazel, C; Casellas, P; Keane, P; Le Fur, G; Maffrand, JP; Soubrie, P; Caput, D; Ferrara, P. (1998). Neurotensin is an antagonist of the human neurotensin NT_2 receptor expressed in Chinese hamster ovary cells. *Eur. J. Pharmacol., V.360,* 265-272.

[238] Wagner, F; Wahl, M. (1986). Effects of neurotensin in feline pial arteries, guinea-pig ileum and portal vein of rat and guinea-pig. *Arch. Int. Pharmacodyn. Ther., V.282,* 240-251.

[239] Watson, MA; Yamada, M; Yamada, M; Cusack, B; Veverka, K; Bolden-Watson, C; Richelson, E. (1992). The rat neurotensin receptor expressed in Chinese hamster ovary cells mediates the release of inositol phosphates. *J. Neurochem., V.59,* 1967-1970.

[240] Weihe, E; Reinecke, M. (1981). Peptidergic innervation of the mammalian sinus nodes: vasoactive intestinal polypeptide, neurotensin, substance P. *Neurosci. Lett., V.26,* 283-288.

[241] Weihe, E; Reinecke, M; Forssmann, WG. (1984). Distribution of vasoactive intestinal polypeptide-like immunoreactivity in the mammalian heart. Interrelation with neurotensin- and substance P-like immunoreactive nerves. *Cell Tissue Res., V.236,* 527-540.

[242] White, RP; Robertson, JT. (1987). Pharmacodynamic evaluation of human cerebral arteries in the genesis of vasospasm. *Neurosurgery, V.21,* 523-531.

[243] Widerlov, E; Lindstrom, LH; Besev, G; Manberg, PJ; Nemeroff, CB; Breese, GR; Kizer, JS; Prange, AJ. (1982). Subnormal cerebrospinal fluid levels of neurotensin in a subgroup of schizophrenic patients: normalization after neuroleptic treatment. *Am. J. Psych., V.139,* 1122-1126.

[244] Yamada, M; Yamada, M; Watson, MA; Richelson, ER. (1993). Neurotensin stimulates cyclic AMP formation in CHO-rNTR-10 cells expressing the cloned rat neurotensin receptor. *Eur. J. Pharmacol., V.244,* 99-101.

[245] Yamada, M; Yamada, M; Lombet, A; Forgez, P; Rostene, W. (1998). Distinct functional characteristics of levocabastine sensitive rat neurotensin NT_2 receptor expressed in Chinese hamster ovary cells. *Life Sci., V.62,* PL 375-380.

[246] Yau, WM; Verdun, PR; Youther, ML. (1983). Neurotensin: a modulator of enteric cholinergic neurons in the guinea-pig small intestine. *Eur. J. Pharmacol., V.95,* 253-258.

[247] Zsurger, N; Mazella, J; Vincent, JP. (1994). Solubilization and purification of a high affinity neurotensin receptor from newborn human brain. *Brain Res., V.639,* 245-252.

In: Neuropeptide Research Trends
Editor: Bernice A. Levine, pp. 65-96
ISBN: 978-1-60021-640-4

Chapter II

Influence of *Helicobacter Pylori* Infection on Gastrin-Somatostatin Link. Are There any Differences Related to the Disease Outcome?

***Aleksandra Sokic-Milutinovic*[1], *Thomas Wex*[2], *Vera Todorovic*[3], *Milos Bjelovic*[4], *Tomica Milosavljevic*[1] *and Peter Malfertheiner*[2]**

[1] Clinic for Gastroenterology and Hepatology, Institute for Digestive Diseases, Clinical Center of Serbia, Belgrade, Serbia

[2] Clinic for Gastroenterology, Hepatology and Infectious Diseases, Otto-von-Guericke University, Magdeburg, Germany;

[3] Institute for Medical Research, Belgrade, Serbia

[4] First Surgical Clinic, Institute for Digestive Diseases, Clinical Center of Serbia, Belgrade, Serbia

Abstract

Introduction

Helicobacter pylori (Hp) causes chronic active gastritis and diffuse inflammation of gastric mucosa. While the majority of infected individuals remain asymptomatic, a subset of patients develop severe gastroduodenal pathologies, such as peptic ulcer disease, gastric cancer and low-grade gastric B cell lymphoma. We aimed to determine if previously reported *Hp* induced gastrin-somatostatin link impairment is a universal *Hp* related phenomena or if it depends on the disease outcome.

Aims and Methods

Gastrin and somatostatin (SST) protein and mRNA expression were quantified using ELISA and RT-PCR in gastric mucosa biopsies obtained during endoscopy from 70 *Hp* infected individuals and 25 *Hp* negative controls. Based on endoscopic findings and histological examination *Hp* infected patients were further divided into duodenal ulcer (DU, n=17), gastric cancer (GC, n=22) and non-ulcer dyspepsia (NUD, n=31) group.

Results

Significantly lower antral gastrin levels were detected in *Hp* infected gastric cancer patients compared to both *Hp*+DU and *Hp*+NUD patients and controls. No significant difference in antral gastrin mRNA levels was detected. Comparable and disease outcome unrelated antral SST levels were detected in all *Hp* infected individuals. However, these were significantly lower when compared to *Hp* negative controls ($p<0.05$). The same trend was observed in corpus mucosa, but with no statistical significance ($p>0.05$). Antral SST gene expression is lower in all *Hp* infected patients regardless of the disease outcome, while in corpus only NUD patients exhibited significantly lower SST mRNA levels when compared to all other groups and controls.

Conclusion

Based on the results of our study and available data in the literature we concluded that decreased gastric somatostatin synthesis and secretion is a universal *Helicobacter pylori* related phenomenon related to the presence of the infection and not to the disease outcome. On the other hand gastrin levels in *Helicobacter pylori* positive individuals depend on the disease outcome and in gastric cancer patients are probably influenced by more than one mechanism out of which some are most likely *Hp* independent.

Key words: gastrin, somatostatin, *Helicobacter pylori,* gastric cancer, duodenal ulcer.

Introduction

Gastrointestinal Diffuse Neuroendocrine System

Gastrointestinal diffuse neuroendocrine system (DNES) consists of individual endocrine cells, peptide-secreting neurons and nerve fibers localized in the gastrointestinal epithelium and adjacent connective tissue. Components of DNES synthesize and secrete regulatory peptides and amines involved in the regulation of different physiological processes. They regulate target cells' function through paracrine, juxtacrine, endocrine and neuroendocrine mechanisms, but can also exhibit autocrine and exocrine effects and play a role in neurotransmission and neuromodulation [1].

DNES regulates diverse gastrointestinal functions, such as appetite, hunger and satiety, digestion, secretion, absorption, motility and blood flow. Some of the gastrointestinal DNES secretion products exhibit trophic effects and play a role in mucosal adaptation to injury [2,3]. DNES localized in the gastric mucosa is involved in the control of gastric secretion, motility and emptying.

Gastric Endocrine Cells

Distribution and main secretory products of different gastric endocrine cells are shown in table 1 [4].

Table 1. Human gastric endocrine cells

Cell type	Antrum-pylorus (%)	Fundus-corpus (%)	Amine	Peptide
ECL	0	30-50	Histamine	Unknown
EC	10	5-15	Serotonin	Enterostatin
D	20	10-20	Unknown	Somatostatin
G	60	0	GABA	Gastrin
D_1/P/A/X	Unknown	10-25	Unknown	Unknown (endothelin; GLP-1)

Different types of endocrine cells are found in antro-pyloric and corpus mucosa, that represent two histologically and functionally distinct types of gastric mucosa. The most common antro-pyloric endocrine cell is a gastrin– producing G cell, while in the corpus the majority of endocrine cells are ECL cells.

Corpus gastric mucosa (acid-producing mucosa) contains numerous endocrine cells mainly localized in the middle and lower portion of the mucosal layer and distributed randomly between parietal and chief cells. They are "closed-type" cells thus in no direct contact with gastric lumen and its contents. Sporadic endocrine cells can also be identified in the upper third of gastric mucosa between neck glands and in *lamina propria* where they are in close contact with nerve fibers and *Schwann* cells forming the so called "neuroendocrine complex". So far, six different endocrine cell types were identified in the corpus mucosa, EC (enterochromaffine), ECL (enterochromaffine-like), D, D_1, P and X cells (table 2) [4].

Table 2. Gastric corpus mucosa endocrine cells

Cell type	Relative frequency (%)	Secretory product
ECL	30-50	Histamine, calbindin, gastrocalcin, α-HCG, enteroglucagon, PYY,TGF- α
D	10-20	Somatostatin
EC	5-15	Serotonin Enterostatin
D_1	5-10	Endothelin
P	5-10	Bombesin-like?
X	1-5	Endothelin

In antro-pyloric mucosa G, D, EC, P and D_1 cells are found. G cells are numerous (60%), followed by D (20%) and EC (10%) cells and all are "open-type" cells, whose luminal side is in direct contact with gastric lumen and its contents.

G Cell

In the stomach, G cells populate only antral mucosa. G cell is a sole cell type responding to gastrin-releasing peptide (GRP) stimulation through calcium signal and exhibiting muscarine, but no cholecystokinin receptor (CCK) [4]. Gastrin is the principal secretory product, however enkephalin, GABA and secretogranin-1 are also released from activated G cells [1].

D Cell

D cells are somatostatin producing endocrine cells. In the stomach they can be identified in both antral and corpus mucosa. In the antrum, after G and in the corpus after ECL cells they are most abundant cell population. D cell has cytoplasmic extensions originating from basolateral membrane that end on button-like formations on target cells, corpus parietal and antral G cells. Apart from different target cells, antral and corpus D cells differ regarding their intramucosal localization. Namely, antral D cell is an "open-type", while fundic D cell is "closed-type" cell. Thus, antral, but not corpus D cell is capable of responding to direct luminal stimuli [1,4,5]. It has been speculated that luminal stimuli can have indirect, sensory nerve mediated, effect on corpus D cells. According to this hypothesis, sensory nerve-endings localized on the epithelial surface react to luminal stimuli and activate D cell secretion [1,4,5].

Physiological Roles of Gastrin and Somatostatin in Gastrointestinal Tract

Main physiological effects of gastrin and somatostatin in gastrointestinal tract (GIT) are summarized in table 3.

Gastrin

Gastrin is a typical gastrointestinal hormone whose release from G cells increases gastric acid secretion from parietal cells. Its trophic effect on gastric mucosa has also been reported [6,7].

Upon G cell activation gastrin is released into circulation and its binding to gastrin/CCK-B receptor on target cells regulates gastric acid secretion. Gastrin-mediated increase in gastric acid secretion results from direct parietal cell activation and increased histamine release from ECL cells. Histamine binds to H_2 receptors and stimulates parietal cell function. Gastrin also induces somatostatin release from D cell that inhibits gastric acid secretion [8].

Nerve, hormonal and chemical mechanisms control gastrin release from antral G cells [6]. Main stimuli for gastrin secretion are amino acids released after protein hydrolysis, especially aromatic amino acids, such as phenylalanine and tyrosine. In addition, excitatory nerve fibers secrete GRP increasing intracellular Ca^{2+} in G cell thus inducing gastrin release, while cholinergic nerve fibers inhibit G cell function. Substances increasing cAMP in G cell induce gastrin gene expression and gastrin release. When intraluminal pH is below 3, gastrin secretion is inhibited trough negative feed back mechanism. H^+ released from parietal cell protonate amines in amino acids and inhibit their uptake into the G cells. Gastrin release is also inhibited by cholecystokinin binding to CCK/A receptors on D cells. Gastric distension and feeding activate, trough vagal nerve fibers, both inhibitory (D cell and somatostatin dependent) and excitatory pathways resulting in release of small amounts of gastrin from G cell [6,8].

Gastrin has a trophic effect on gastric epithelial stem cell, antral gastric mucosa, parietal and ECL cells. It activates stem cell division trough TGFα- dependent mechanism. Antral gastrin in physiological concentrations inhibits cell proliferation, probably playing a protective role against gastric cancer formation. However, prolonged gastrin hypersecretion leads to parietal cell hyperplasia and is considered important in gastric cancer development. Secretin and vagal nerve fibers inhibit trophic effects of gastrin on parietal cells [6,8].

Somatostatin (SST)

SST is released from D cells, neurons and immune cells and can act as hormone, exhibiting paracrine and endocrine effects, but also as neuropeptide, with a role of neurotransmitter and/or neuromodulator or even immunomodulator [5,8].

Somatostatin is a peptide involved in inhibition of neuroendocrine secretion throughout gastrointestinal tract (inhibits bombesin-induced gastrin release, secretion of CCK, secretin, VIP, GIP, motilin, enteroglucagon, insulin, amylin, glucagon and PP). In addition, splanchnic blood flow is decreased and gastrointestinal cell growth and proliferation inhibited trough SST mediated effects.

In the stomach, somatostatin inhibits gastric acid and pepsinogen secretion, late phase of gastric emptying and gastric motor migratory complex. As opposed to that, SST stimulates early phase of gastric emptying

SST is released from D cells as a result of luminal stimuli, neurotransmitters and neuropeptides action. Corpus D cells release SST upon CCK stimulation, while luminal pH and gastrin stimulate antral D cell secretion. Inhibition of D cell secretion occurs in response to histamine binding to H_3 receptors and stimulation of cholinergic nerve fibers.

More then one out of five so far identified subtypes of somatostatin receptors [9,10] can be identified on the same cell membrane. Majority of these receptors are complex proteins with 7 transmembrane domains and one or more inhibitory G proteins bonded to their intracellular domain. SST receptor subtype 2A is localized on myenteric and submucosal nerve plexus, nerve fibers ending on smooth muscle cells, epithelium and blood vessels, but also on ECL cells [9,10]. CCK, cholinergic agonists or molecules that directly activate protein kinase C (leads to SST receptor phosphorylation) inhibit SST binding to receptor.

Table 3. Physiological roles of gastrin and somatostatin in the gastrointestinal tract

Regulatory peptide	Main physiological effects
Gastrin	Stimulation of gastric acid secretion (direct parietal cell stimulation and indirect trough histamine release from ECL cells), SST stimulation; trophic effect on corpus gastric mucosa (increased secretion, hypertrophy and hyperplasia of ECL and parietal cells; fundic glands epithelial stem cells activation); stimulation of pancreas growth; trophic effect on endocrine pancreas after partial pancreatectomy; possible trophic effects on colon epithelium
SST	Inhibition of exocrine secretion (acid and pepsinogen in stomach; enzyme and bicarbonate in pancreas; bile-flow inhibition by mechanism independent of bile acids and ductal secretion; amylase from salivary glands); inhibition of neuroendocrine secretion in GIT (somatostatin inhibits bombesin-stimulated gastrin secretion, and also inhibits CCK, secretin, VIP, GIP, motilin, enteroglucagon secretion) and pancreas (insulin, amylin, glucagon and PP); inhibition of motility (late phase of gastric emptying; gallbladder contractions; ileum longitudinal muscle layer contractions) or stimulation (early phase of gastric emptying; bowel migratory motor complex); regulation of intestinal transport (inhibition of water and bicarbonate secretion, Ca^{++}, glucose, glycerol, fructose, xylose, lactose, amino acid and triglyceride absorption); inhibition of splanhnic blood flow; inhibition of growth and proliferation; stimulation of food intake in hungry and inhibition in fed animals

Helicobacter Pylori Infection

In 1887. *Bizzozero* and three years later *Salomon* were the first to describe morphology and motility of a spiral microorganism isolated from cat, dog and rat stomach, together with gastric mucosa changes induced by the organism in the mouse. In 1893. *Jaworowski* detected similar organism in human stomach contents, named it *Vibrio rugula* and implied its possible role in the pathogenesis of gastroduodenal diseases. Concept stating that bacteria can induce gastritis and duodenal ulcer was reintroduced in 1979. by Warren and Marshall, with their first report concerning *Campylobacter-Like Organism* published in 1983. In 1989, due to morphological and structural differences from other *Campylobacters*, new species termed *Helicobacter* was formed [11-15].

Helicobacter Pylori Characteristics

Hp is spiral, gram-negative, microaerophilic bacterium belonging to *Proteobacterium* family and *Helicobacter* species. *In vivo Hp* is spiral, of S or U shape, 2-4 μm long, 0,5 to 0,9 μm wide, with 4-7 flagella that enable movements trough gastric mucus layer. *In vitro* coccoid shape of the bacterium was reported [16-18]. Optimal pH range for *Hp* glucose metabolism is 3,5 to 8,6 [19,20]. *Helicobacter pylori* is cultivated in the presence of added CO_2, on selective medium containing blood or plasma, on 37^0C, in course of 5-7 days [15].

Epidemiology of Hp Infection

It is estimated that half of entire mankind is infected with *Helicobacter pylori*. Prevalence of the infection depends on socio-economic factors and ethnic background, but also shows distinct geographical distribution. In developing countries 80-90% of adult population is infected compared to less than 40% in developed countries [21]. Infection probably occurs during childhood, lasts for decades and in some cases throughout entire lifetime [15]. Age- related differences in *Hp* prevalence in each ethnic group are reported. Study conducted in Germany showed for example that 28% of children from immigrant families are infected and only 7% in ethnic German families. In Europe, majority of infected individuals are carrying single *Hp* strain, while in poor sanitary conditions concomitant infection with more than one strain is very common [15].

Risk factors that facilitate *Hp* infection are low socio-economic standard, inadequate hygiene, bed sharing, living in large families, inadequate drinking water, low income and mother carrying *Hp* infection [22-24].

Humans are main reservoir of *Hp* infection. The bacterium can be identified in gastric type mucosa (in the stomach and zones of ectopic gastric epithelium such as Barett esophagus, Meckel diverticulum and ectopic gastric mucosa in the rectum), but also in oral cavity (gingival region and dental plaque) and latter could be responsible for re-infection after initially successful eradication therapy. *Hp* is eliminated by stool [25,26]. Infection can also affect various animal species such as pigs, monkeys (among other *Rhesus sp*), sheep, cats and hamsters [15,27]. Previously disputed external sources of the infection undoubtedly exist and include milk, tap water, tofu, lettuce and raw chicken meet [16,22,27-29].

Infection is transmitted oro-orally and less likely trough feco-oral route [21]. It can also be iatrogenically acquired during esophagogastroduodenoscopy and transferred from mother to children [22,30].

Pathogenesis of Helicobacter Pylori Infection

Pathogenesis of *Helicobacter pylori* infection depends on bacterial strain, environmental conditions (hygienic and dietary habits, stress) and host characteristics (age, sex blood type) [14,31].

Bacterial Virulence Factors

An important role in the pathogenesis of *Helicobacter pylori* infection is played by factors necessary for the colonization and bacterial survival on gastric mucosa (urease, porins, flagella). Other factors determine bacterial strain, i.e. vacA and cagA, and in some populations are related to different outcome of *Hp* infection and clinical presentation [14,31].

1. Motility

Bacterial motility is crucial in induction and persistence of gastric mucosa colonization. *Helicobacter pylori* motility results from spiral shape that bacterium has *in vivo* and flagella [31,32]. Flagella enable penetration of bacterium through gastric mucus (aflagellar strains are unable to colonize gastric mucosa) leading to decreased efficacy of bacterial clearance by gastric peristalsis [11,31].

2. Urease Activity of the Bacterium

Urease is an enzyme involved in urea degradation to ammonia and CO_2. *Helicobacter pylori* urease is localized both intra and extracellulary. Intracellular urease, localized in bacterial cytoplasm, increases periplasmatic pH and stimulates protein synthesis in acidic environment [33,34], thus having both protective and metabolic role. Urea-derived ammonia is a substantial source of nitrogen necessary for protein synthesis. Since both bacterial metabolism and intracellular urease activity depend on pH, extracellular urease is important in decreasing microenvironment acidity enabling bacterial survival and gastric epithelium colonization. Urease-independent *Hp* induced increase of gastric pH was also described by some authors [35].

Since no transport system involved in urease exporting from bacterial cell could be detected it is presently postulated that bacterial autolysis leads to release of urease and other cytoplasmatic proteins later adsorbed on the viable bacteria membrane [36]. Apart from above mentioned protective, extracellular urease also plays a role of an adhesive factor (urease negative *Hp* mutants do not colonize gastric mucosa *in vitro*). Bacterial urease activity is essential in chronic infection development [14].

Urea-derived ammonia induces apoptosis after binding to specific proteins of class II major histocompatibility complex (MHC-II) localized on gastric epithelial cell surface, disturbing epithelial defense mechanisms and delayed ulcer healing. Ammonia and water form amonium-hydroxyde (NH_4OH) that releases cytotoxic hydroxyl residue (OH), but also leads to H^+ back-diffusion by changing pH on mucosal membrane and parietal cell surface. Therefore ammonia exhibits both direct and indirect cytotoxic effects.

Enhanced vacA mediated neutrophil and phagocyte activation, together with increased cytokine synthesis occurs in presence of ammonia, while hydrogen-peroxide (H_2O_2), as a product of activated inflammatory cells and Cl^- ion form hypochloric acid (HOCl) in the stomach. Also, monochloramine (NH_2Cl) a strong oxidative molecule, synthetised in the presence of ammonia enters and damages gastric mucosa cells. Ammonia exhibits an inhibitory effect on somatostatin release in antral mucosa [35]. Urease can also induce decrease in opsonization by degradation of C3b component of the complement bonded to bacterial cell surface [37].

3. Cytotoxin Associated Proteins - Adhesins

Helicobacter pylori is exhibiting both host and tissue tropism, colonizing only gastric-type mucosa. Tissue tropism can be partly explained by adhesins, also known as BabA, protein molecules localized on bacterial surface that act as ligands for receptors on host epithelial cells and play a role in bacterial binding to gastric mucosa. So far, several probable adhesin receptors were identified, such as: phosphatidylethanol-amine, GM3 ganglioside, Lewis B antigen, hemagglutinins (sialine acid dependant and independent) and laminin in the basal membrane [38].

4. Lipopolysaccharide

Lipopolysaccharide (LPS), a glycoprotein molecule, is an important *Helicobacter pylori* virulence factor [39,40]. LPS binds to laminin (extracellular matrix protein) weakening its interaction with integrin receptors localized on gastric epithelial cells [39,41-43]. LPS administration, as demonstrated in animal model, leads to histological findings of acute gastritis, disturbed cell cycle, cell proliferation and induction of apoptosis [39,40]. In the presence of LPS interleukin 4 and 13 (IL-4 and IL-13) up-regulate, on post-transcriptional level, synthesis of proinflammatory cytokines belonging to TNFα family and induce apoptosis. Oligosaccharide part of the LPS is involved in gastric epithelial cell recognition that enables bacterial adherence using host cell membrane receptors. LPS disturbs mucin macromolecular structure preventing its binding to the receptor localized on gastric epithelial cell. This mechanism facilitates bacterial penetration trough mucus layer, stimulates pepsinogen secretion and conversion to pepsin [39-41,44-46]. In animals, in course of LPS-induced acute gastritis, increased levels of endothelin-1, peptide involved in stress ulcer-type mucosal injury, were detected. LPS has molecular structure similar to Lewis x and y blood antigens located on gastric epithelial cells and leukocytes that could result in autoantibody formation and gastric atrophy [45,46].

5. Catalase and Superoxid Dismutase

These two enzymes neutralize effects of neutrophil originating superoxide and peroxide, preventing thus oxidative disintegration of lipid and other bacterial components [19].

6. Coccoid Forms

Coccoid forms play a role in bacterial survival in unfavorable environmental conditions, such as decreased nutritional substances, exposure to antibiotics or prolonged incubation.

7. Toxic Enzymes (Phospholipase A2 and C, Alcohol Dehydrogenase)

Phospholipase A2 and C induce degradation of the phospholipid component of mucosal barrier leading to arachidonic acid release and leukotriene and cytokine synthesis. Phospholipases trough an indirect mechanism lead to lysolecithin and platelet activating factor (PAF) synthesis, thus inducing impairment of microcirculation and ischemic tissue injury. On the other hand, alcohol dehydrogenase enters carbohydrate metabolism in the host cells leading to acetaldehyde formation that further prevents epithelial regeneration [11].

8. Proteolytic Enzymes

Helicobacter pylori possesses proteolytic enzymes that are probably involved in mucus layer glycoprotein degradation and weakening of mucus barrier. In addition, hemolytic enzymes with so far unclear role were identified in the bacterium.

9. Metabolic Enzymes

Metabolic enzymes do not have direct influence on the pathogenesis of *Helicobacter pylori* infection. [11,35].

10. Exotoxins

Almost 50% of *Helicobacter pylori* strains have the ability to synthesize vacuolization toxin (vac A) that leads to vacuolization of cytoplasm and gastric epithelial cell degeneration. VacA mechanism of action includes increased secretion of acidic hydrolases, disturbances in cytoskeleton structure with actin filaments rearrangement and weakening of intercellular junctions, together with decrease in mucus viscosity [35]. Toxin penetrates into cells, binds to and activates ATP-ase molecules. Consequently, disturbed oxidative phosphorylation leads to intracellular increase of H^+ concentration, acidosis, disturbed cell membrane permeability, passive diffusion of extracellular substances and water into the cell and cell swelling. When cell membrane breaks the cell dies. Higher prevalence of vacA positive bacterial strains was reported in patients with peptic ulcer disease, atrophic gastritis and gastric cancer [14,35,47]. Nevertheless, vacA negative strains exhibit ammonia-mediated cytotoxic effect on gastric epithelial cell [48]. Thus, presence of vacA only enhances cytotoxic effects of particular *Helicobacter pylori* strain.

Vac A gene encodes 140 kDa precursor molecule. Gene is found in all *Helicobacter pylori* strains, but only half of all strains secrete active toxin (87kDa) [49-52]. VacA toxin is activated in low pH, when oligomeric protein is cleaved into the monomers. Vac A exhibits toxic effects on epithelial cell *in vitro* and *in vivo* [40]. Different types of gastric mucosal lesions were identified in animals infected with toxin producing strains compared to animals infected with vacA negative strains [49,53]. VacA also induces mastocyte chemotactic activity with simultaneous dose-dependent increase in proinflammatory cytokine secretion from bone-marrow originating mast cells (i.e. TNF-α, IL1- β, IL-6, IL- 10 and IL-13) [54,55]. *In vitro* vacA induces disturbances of endo-lysosomal and mitochondrial function, apoptosis, anion-channels formation and changes of cytoskeleton and epithelial intercellular junctions structure [18,56-59]. Therefore, vacA plays a role in early host immune response as well as in pathogenesis of *Helicobacter pylori* induced gastritis [48-55] and peptic ulcer disease [46-54].

11. Strain Heterogeneity

Helicobacter pylori genome sequencing revealed different mechanisms responsible for strain specificity. Both variations in strain gene sequence and phenotype, influenced by the environment and host characteristics were detected. Differences in gene sequence originate from:

a. point-mutations (change in a single base pair, either purine or pyrimidine base)
b. mosaicism (simultaneous presence of variable and constitutive part of genome)
c. genes that are not part of constitutive gene sequence (i.e. cag pathogenicity island found in 60-80% of *Helicobacter pylori* strains)
d. mobile DNA and gene recombination that are not completely explained

Factors Involved in Gastric Mucosal Injury

Helicobacter pylori induced gastric mucosal injury is related to the presence of LPS, vacA, cagA, heat shock proteins (HSP) and factors responsible for chemotaxis and leukocyte activation.

VacA was described in previous section of this chapter.

CagA gene found in 60% of all *Helicobacter pylori* strains is localized in the *cag* pathogenicity island (*cag* PAI) consisting of 31 genes [49,60] and encodes synthesis of 120kDa cagA protein [36,50,60,61]. CagA protein enters host cells and induces changes in protein structure and actin cytoskeleton rearrangements [49,62-64]. Lower acid resistance of cagA-positive compared to cagA negative *Helicobacter pylori* strains places them closer to gastric epithelium (pH around 7), enabling more intense interaction with host cells resulting in extensive tissue injury. CagA+ strains induce more intense host immune response followed by higher IL-8, IL1-α and β, IL-2 and TNF-α synthesis, neutrophil activation, impaired gastric acid secretion and faster development of gastric atrophy [14,65]. Presence of CagA is related to higher risk for duodenal ulcer disease and gastric adenocarcinoma development [32,60,65,67]. CagA positive strains can cause DNA damage trough an indirect mechanism, that is an important step in carcinogenesis [66]. *Helicobacter pylori* heat shock proteins A and B represent bacterial response to stress environmental conditions and account for 3-5% of all bacterial proteins. Similar structure with host proteins implicated their role in autoimmune diseases based on the cross-reactivity mechanism. Heat shock protein termed HSP60 is involved in *Hp* induced prolonged active chronic inflammation [68] of gastric mucosa.

Tissue injury results from chemotactic and proinflammatory effect of *Helicobacter pylori* surface proteins (urease, PAF) that induce dense neutrophil and monocyte inflammatory infiltrate formation in mucosal and submucosal layer. Bacterial urease up-regulates IL-1β, IL-6 and TNFα gene expression that in return activate macrophages in host gastric mucosa [69,70]. IL-8 expression is up-regulated by local increase in cytokine concentrations and bacterial adherence on gastric epithelial cells [71].

Outcomes of Helicobacter Pylori Infection

Possible outcomes of *Helicobacter pylori* infection include in majority of individuals asymptomatic infection, NUD with histological signs of chronic gastritis, and only in some development of peptic ulcer disease or multifocal atrophic gastritis (MAG). MAG is an important step in gastric carcinogenesis and links presence of the infection to gastric adenocarcinoma. *Hp* also plays a role in pathogenesis of low-grade B-cell gastric lymphoma (MALT lymphoma).

Helicobacter Pylori Induced Gastritis

Histological signs of gastritis with diffuse inflammatory infiltrate are present in antrum and/or corpus gastric mucosa of all *Helicobacter pylori* infected individuals. Inflammatory infiltrate consists of neutrophil and mononuclear inflammatory cells. It is a chronic active inflammation followed by increased mucosal levels of proinflammatory cytokines, such as IL-8, IL-1β and TNF-α [11,22]. Possible role of IL-4 and IL-10, as anti-inflammatory cytokines, could be in lowering intensity of inflammatory response to *Helicobacter pylori*, thus enabling persistence of the bacterial infection and long-lasting chronic gastritis [66].

Helicobacter Pylori And Ulcer Formation

It has been demonstrated that *Hp* infection delays ulcer healing as a consequence of reduced mucosal microcirculation, increased cytokine release and impaired gastrin-somatostatin link [72].

Although pathogenesis of gastric and duodenal ulcer formation in course of *Hp* infection is different, first well-designed case-control study concluding that gastric ulcer patients benefit from eradication of the infection was published in 1992 [73]. When all other etiology factors involved in gastric ulcer formation are excluded 96% of idiopathic gastric ulcers are related to *Hp* infection [74,75] and eradication therapy promotes healing in course of 6 weeks in 85% of *Hp* positive gastric ulcer patients when compared to 60% healed in non-eradicated group [76]. Since only a low percent of all *Hp* infected individuals (12-13%) develop gastric ulcer, it is estimated that differences in bacterial strain, especially cagA and vacA related, are responsible for gastric ulcer formation, although role of host characteristics can not be excluded [72].

In 90- 95% of duodenal ulcer patients presence of *Hp* infection is diagnosed [76]. Duodenal ulcer develops in 15% to 25% of infected individuals [76,77]. Duodenal ulcer pathogenesis is complex, since *Hp* can only colonize zones of gastric metaplasia in the duodenal mucosa. Gastric metaplasia in the duodenum is a consequence of prolonged gastric acid hypersecretion, aiming to protect mucosa from further acid-induced injury. Therefore, two main pathophysiological mechanisms are important in duodenal ulcer formation: increased gastrin release and increased parietal cell response to gastrin. Recurrence of duodenal ulcer is registered in 4% of patients if eradication therapy is completed in comparison with 60% when no eradication therapy was prescribed [78].

Helicobacter Pylori and Gastric Cancer

Up to 1% of *Hp* infected patients develop gastric adenocarcinoma and prevalence of gastric cancer was proved to be concordant with prevalence of *Hp* infection. Two types of gastric adenocarcinoma are related to the infection, intestinal and diffuse type. Incidence of intestinal type is decreasing, and now diffuse type accounts for 30% of all gastric cancer cases. Numerous studies revealed higher incidence of *Hp* infection (50-100%) in both histological types of gastric cancer compared to general population. It is estimated that only 7% of gastric cancers develop in the stomach of *Hp* negative individuals. It is however difficult to prove causal relation between gastric cancer and *Hp* infection since both are diagnosed simultaneously. Multifocal atrophic gastritis development takes at least 20 years and is an important if not the crucial step in gastric carcinogenesis [79]. Atrophic gastritis develops in patients with high grade diffuse inflammation affecting both proximal and distal region of the stomach. This inflammatory infiltration is followed by mucosal atrophy, loss of glands, intestinal metaplasia and lower acid secretion. Individuals with multifocal atrophic gastritis can develop eiher gastric ulcer or gastric cancer. In the sequence of events during carcinogenesis dysplastic changes of metaplastic mucosa are mandatory. VacA positive strains are thought to be more carcinogenic than the others. It is probable that *Hp* mediated gastric cancer development is independent of mechanisms related to microsatelite locus instability located on chromosomes 2, 3, 17 and 18 and loss of heterozigocity of these genes [80].

Helicobacter Pylori Infection in MALT Lymphoma

Gastric MALT lymphoma is low-grade B-cell lymphoma. Results from *in vitro* studies revealed that *Hp* induces B lymphocyte proliferation. High prevalence of cagA+ *Hp* strains was demonstrated in some, but studies reporting antibodies to cagA in only 56% of MALT lymphoma patients were also published [81]. Identifying specific MALT lymphoma related *Hp* strains failed so far, leading to conclusion that host factors are of greater importance [82]. However, VacA*m2* allele is more frequently found in patients with MALT lymphoma and when combined with vacA*s1* allele it is five times more prevalent in lymphoma than gastritis patients [83].

Other Helicobacter Pylori Influenced Diseases

Helicobacter Pylori in Gastroesophageal Reflux Disease

Numerous investigations aiming to clarify possible role of *Hp* infection in the pathogenesis of gastroesophageal reflux disease (GERD) were conducted coming to conclusion that virulent *Hp* strains could have protective role against GERD. Namely, individuals with GERD were found to be more often infected with cagA and vacA negative *Hp* strains implying that in GERD cagA+ and vacA+ strains could actually have protective role [84].

Extragastric Manifestations of Helicobacter Pylori Infection

Soon after *Hp* discovery numerous studies were conducted in attempt to define extraintestinal manifestations of the infection. So far, proposed underlying mechanisms included direct bacterial effects, activation of inflammatory process, systemic effects of released cytokines and antigen mimicry between bacterial and host antigens.

Dermatological and immunological studies dealt in most cases with possible role of *Hp* infection in chronic urticaria and acne rosacea, with only one conclusive result stating that 67% of acne rosacea patients have antiboides to cagA [85].

Suggested mechanism responsible for blood vessel injury in *Hp* infected individuals involves cross-reactivity of cagA and antigens located in host blood vessels wall. Correlation of infection with cagA+ *Hp* strains and thickness of coronary arteries wall in patients with coronary disease and higher prevalence of infection with cagA+ *Hp* strains in stroke patients was reported. However, acute myocardial infarction and infection with cagA+ *Hp* seem unrelated [86-88].

Attempts were also made to relate impaired growth, refractory sideropenic anemia, monoclonal gammapathy, idiopathic thrombocytopenic purpura and rheumatoid arthritis with *Hp* infection [88].

Presence of *Hp* DNA fragments in hepatobiliary tract is not completely clarified, but probably other *Helicobacter sp* (*H.bilis* i *H.hepaticus*) are more important than *H.pylori* [86].

Methodology

Patients

We conducted an outpatient based prospective study in Institute for Digestive Diseases, Clinical Center of Serbia and included 95 individuals, 70 *Helicobacter pylori* infected patients and 25 uninfected dyspeptic controls. Patients were recruited from October 2002 till May 2004. Exclusion criteria were in concordance with the recommendations from European *Helicobacter pylori* Study Group (EHSG) [89]. In brief, age (younger than 20 and older than 80 years), pregnancy, gastric outlet obstruction, administration of antibiotics, H_2 antagonists, omeprazole and/or bismuth containing preparations within 30 days prior to endoscopy; the use of non-steroid anti-inflammatory drugs, previous gastric surgery and presence of associated diseases (hepatic, renal, cardiac, respiratory) eliminated patients from the study. Demographic factors including age, sex and smoking were analyzed (table 4).

H. pylori infection was diagnosed when rapid urease test (RUT), histology and serology were positive.

Patients were divided into 3 groups according to the endoscopic findings. In duodenal ulcer (DU) group ulcerative lesion of >5mm diameter in duodenal mucosa was identified with no signs of gastric ulcer and/or cancer. Non ulcer dyspepsia (NUD) was considered in patients with upper gastrointestinal symptoms that persisted for a month or longer and upper gastrointestinal endoscopy revealed neither peptic ulcer disease nor signs of gastric cancer. Gastric cancer (GC) group was formed after histological findings confirmed clinical suspicion of gastric adenocarcinoma. Out of 70 *Helicobacter pylori (Hp)* infected patients, in 17 DU, in 31 NUD and in 22 GC was diagnosed.

Table 4. Clinical and demographic data of patients with different outcomes of *Helicobacter pylori* infection and uninfected controls

	*Hp*neg (n=25)	*Hp*+NUD (n=31)	*Hp* +DU (n=17)	*Hp*+CA (n=22)	p
Age	44±16	53±12	45±14	65 ±11	0.000
Sex(m)	9	13	8	15	ns
Smokers	9	11	10	6	ns
Personal history of PUD	4	1	6	0	0.002
Family history of PUD	10	12	12	3	0.005
Family history of GC	0	3	1	0	ns
cagA+	n.a.	28	16	20	ns
vacA+	n.a.	11	8	8	ns

*Hp*neg- *Helicobacter pylori* negative controls; *Hp*+NUD- *H.pylori* infected non-ulcer patients, *Hp*+DU-*Helicobacter pylori* infected duodenal ulcer patients, *Hp*+CA- *H.pylori* infected gastric cancer patients, GC-gastric cancer; PUD-peptic ulcer disease.

Routine Endoscopy and H.Pylori Status

After obtaining informed consent, we performed upper endoscopy in all patients and obtained antral and corpus mucosa biopsies for routine histology, RUT, quantitative reverse transcription PCR (RT-PCR) and protein extraction. Biopsies used for histological analysis were stained using hematoxylin-eosin and modified Giemsa staining procedure and assessed separately using Sydney System classification [90] by an experienced pathologist who was blinded to the clinical data.

H.Pylori Serology and Immunoblot Assay

Blood samples were taken from all patients, sera were separated by centrifugation and stored at -20 °C until analyzed. Presence of anti-CagA and anti-VacA antibodies was determined using *Helicobacter pylori* Vira blot test kit™ (Viramed Biotech AG, Lich, Germany). Concentration of anti-*H.pylori* IgG antibodies was analyzed using Pyloriset EIA-G III™ (Orion Diagnostica, Finland) [91]. Both tests were performed according to manufacturers' instructions.

Basal Gastrin and Somatostatin Mucosal Protein Concentrations

Standard, previously described method was used for extraction of total protein lysate from antral and corpus mucosa biopsies [92]. Basal gastrin tissue levels were determined using ELISA test (Gastrin-17, RandD Systems, Minneapolis, USA). For antral and corpus SST concentration an EIA test (*Bachem*, St Helens, England, cat. no. S-1152) was used. Both tests were performed according to manufacturers' instructions.

Gastrin and Somatostatin Gastric Mucosal Gene Expression

Gastrin and somatostatin gene expression in antral and corpus mucosa was studied by quantitative reverse transcription PCR. RNA extraction, RT and quantitative RT-PCR were performed as described previously [92] by using the following primers: gastrin (5'-CAGCAGCCAGATGCACCCTTAG-3' and 5'-CTTCTTCCTCAGGCCATGGTCC-3), somatostatin (5'-CCCAGACTCCGTCAGTTTCTGC-3' and 5'-GCAGCCTGGGACAGA TCTTCAG-3') and beta-actin (5'-CATGCCATCCTGCGTCTGGACC-3' and 5'-ACATGGTGGTGCCGCCAGACAG-3')

Statistical Analysis

Results were analyzed using non parametric tests, Kruskall-Wallis test for independent samples, Mann-Whitney's, Wilcoxon's test, the chi-square or Fischer's test. P-values < 0.05 were considered significant.

Results

In patients with different outcomes of *Hp* infection and uninfected controls significant differences in antral tissue gastrin levels were detected as seen in table 5. Lower gastrin levels were found in advanced gastrin cancer patients when compared to both *Hp* infected duodenal ulcer and non-ulcer dyspepsia patients and controls ($p<0.01$). Further statistical analysis revealed no difference between non-cancer *Hp* infected patients and *Hp* negative controls.

Table 5. Antral tissue gastrin levels in patients with different outcomes of *Hp* infection compared to *Hp* negative controls

Gastrin (pg/mg)	*Hp*neg	*Hp*+NUD	*Hp*+DU	*Hp*+CA	p
Median	917.6	1540.2	1249.0	335.8	
Minimum	62.7	212.5	192.0	137.4	0.002
Maximum	5512.6	10983.0	4960.0	238725.0	

*Hp*neg- *Helicobacter pylori* negative controls; *Hp*+NUD- *H.pylori* infected non-ulcer patients, *Hp*+DU-*Helicobacter pylori* infected duodenal ulcer patients, *Hp*+CA- *H.pylori* infected gastric cancer patients.

Results obtained for antral gastrin mRNA failed to demonstrate any significant difference related to the presence or different outcomes of *Hp* infection (table 6).

Significant decrease in antral SST levels was noted in all *Hp* infected individuals compared to uninfected controls as seen in table 7 ($p<0.05$). However, no *Hp* disease outcome related differences in antral SST levels were noted. On the other hand, in corpus mucosa no significant difference was noted related to the presence of the infection or disease outcome.

Table 6. Gastrin mRNA levels in the antral mucosa of patients with different outcomes of *Helicobacter pylori* infection and healthy controls

Gastrin mRNA (a.u.)	*Hp*neg	*Hp*+NUD	*Hp*+DU	*Hp*+CA	P
Median	25.2	10.5	8.8	34.3	
Minimum	0.1	1.2	0.1	0.0	ns
Maximum	2880.0	476.3	53.4	1540.0	

*Hp*neg- *Helicobacter pylori* negative controls; *Hp*+NUD- *H.pylori* infected non-ulcer patients, *Hp*+DU-*Helicobacter pylori* infected duodenal ulcer patients, *Hp*+CA- *H.pylori* infected gastric cancer patients, ns-not significant ($p>0,05$).

Table 7. Somatostatin levels in gastric antral and corpus mucosa of patients infected with *Helicobacter pylori* infection and uninfected controls

SST (pg/ml)	*Hp*neg	*Hp*+NUD	*Hp*+DU	*Hp*+CA	P
Antrum					
Median	1.63	0.92	0.88	0.83	
Minimum	0.40	0.17	0.21	0.22	0.024
Maximum	5.67	2.55	1.73	5.04	
Corpus					
Median	1.38	0.46	0.43	0.63	
Minimum	0.20	0.05	0.17	0.22	ns
Maximum	3.39	1.78	2.50	5.04	

*Hp*neg- *Helicobacter pylori* negative controls; *Hp*+NUD- *H.pylori* infected non-ulcer patients, *Hp*+DU-*Helicobacter pylori* infected duodenal ulcer patients, *Hp*+CA- *H.pylori* infected gastric cancer patients, ns-not significant ($p>0,05$).

Significantly lower antral SST mRNA levels were noted in all *Hp* infected individuals irrespective of the disease outcome (table 8) when compared to uninfected controls. No significant differences between groups of *Hp* infected individuals were noted. On the other hand, observed inter-group differences in corpus SST mRNA levels were related to significantly lower SST mRNA levels seen in *Hp* infected non ulcer dyspepsia patients compared to all other three groups.

Table 8. Somatostatin mRNA levels in gastric antral and corpus mucosa of patients infected with *Helicobacter pylori* infection and uninfected controls

SST mRNA (a.u.)	*Hp*neg	*Hp*+NUD	*Hp*+DU	*Hp*+CA	P
Antrum					
Median	98.7	24.7	24.5	10.3	
Minimum	5.3	1.7	2.7	0.53	0.001
Maximum	1520.0	462.0	78.2	673.0	
Corpus					
Median	53.7	12.53	24.9	28.34	
Minimum	6.6	2.8	1.8	0.2	0.031
Maximum	1070.0	26.9	85.0	1540.0	

*Hp*neg- *Helicobacter pylori* negative controls; *Hp*+NUD- *H.pylori* infected non-ulcer dyspepsia patients, *Hp*+DU-*Helicobacter pylori* infected duodenal ulcer patients, *Hp*+CA- *H.pylori* infected gastric cancer patients.

Discussion

Clinical Data

The analysis of clinical data revealed, as expected, older age in gastric cancer patients compared to all other groups. Also, personal history of PUD was found to be more frequent in DU patients compared to all the others ($p<0.05$). Natural history of PUD is characterized by recurring episodes of active lesions followed by re-epithelization or scarring in patients with previously well-defined risk factors [93]. These factors include host-specific ones, like blood type O, male gender or acid output [94], infection by *H. pylori* [89,95] as well as environmental and social factors such as smoking, consumption of tea or coffee and low social class [96]. On the other hand, duodenal ulcer reduces by 40% risk for gastric cancer development [77,78].

Together with significant difference in personal, a difference in frequency of positive family history of PUD was present in our study population. The groups however did not differ with respect to any other demographic aspect and risk factor, which were shown to be associated with PUD and gastric cancer in other studies [97,98].

Gastrin in *Helicobacter Pylori* Infection

To the best of our knowledge, this is first comparative study dealing with gastrin synthesis and secretion in non-tumor antral gastric mucosa of *Hp* infected gastric cancer patients compared to patients with other outcomes of *Hp* infection and *Hp* negative controls.

We found lower gastrin concentrations in antrum of gastric cancer patients compared to all other groups of patients ($p<0.01$). However, no difference related to the presence of *Helicobacter pylori* infection, and in infected patients to the presence of duodenal ulcer was detected. Also, comparable levels of gastrin mRNA were found in antral mucosa of all patients and controls ($p>0.05$).

Antral gastrin levels in gastric cancer patients were determined by Konturek et al. [99] in a study that compared non-malignant antrum and tumor tissue gastrin levels. Significantly lower gastrin levels were found in tumor when compared to antral mucosa. Since no endoscopy was performed in controls, we were only able to conclude that their average antral gastrin levels (280±27 pg/mg) were comparable to our result in gastric cancer patients. They also showed that these levels decreased after eradication of the infection [99].

Differences in antral gastrin tissue concentrations related to the presence of *Hp* infection in non-cancer patients, were not found in study conducted by Odum et al [100] thus supporting our results. Some authors demonstrated higher antral gastrin concentrations in *Hp+* duodenal ulcer patients [101] compared to non ulcer *Hp* infected subjects, but other including our previously published results, failed to confirm this finding [7,100].

We were able to retrieve information on antral gastrin mRNA measurement related to the presence of *Helicobacter pylori* infection only in recent studies published by Konturek et al [102] .They reported a decrease in antral gastrin mRNA levels after eradication of *Helicobacter pylori* in atrophic gastritis patients [102]. Same authors investigated possible role of gastrin as a growth factor in gastric carcinogenesis and demonstrated increased gastrin synthesis and release in gastric cancer patients [103], but with no difference in antral vs. tumor gastrin gene expression [99].

Majority of authors agree that antral G cell number in course of *Helicobacter pylori* infection remains unchanged [101,104-106], but significantly lower antral G cell count in duodenal ulcer subjects was also reported [107]. Important data came from Holle et al who compared completely mapped histologic antrum from normal stomachs and patients that underwent total gastrectomy for proximal gastric cancer. They used immunohistochemistry and morphometric analysis, and counted total G cell and D cell population and revealed 20-70% lower G and 35% lower antral D cell count in gastric cancer patients.[108].

Previously it was reported that eradication of the infection leads to decrease of antral gastrin concentrations only in successfully treated non-ulcer subjects but not patients with duodenal ulcer [7] that could be explained either by impaired antral G cell function or its hyperplasia [101]. Namely, Kwan found in as many as 50% of ulcer patients signs of antral G cell hyperplasia [109]. Eradication therapy in gastric cancer patients lead to significant decrease in both antral and tumor gastrin concentration [110].

Based on the above mentioned results it is probable that posttranscriptional inhibition of gastrin synthesis occurs in gastric cancer patients. Since gastric cancer usually develops after MAG, according to Hartleb et al lower levels of both basal and stimulated serum and tissue gastrin are probably related to gastric mucosal atrophy [111] and preexisting MAG in gastric cancer patients [108]. This would imply, since gastric cancer patients exhibit higher serum gastrin levels then all other disease outcomes [110], that increased serum gastrin is of tumor-mucosa origin.

Even before discovery of *Helicobacter pylori*, role of impaired gastric acid secretion mechanisms and increased parietal cell mass [8] in pathogenesis of duodenal ulcer was recognized [8]. Namely, Kaneko et al reported almost two-fold higher parietal cell number in duodenal ulcer subjects when compared to healthy controls [8]. *Hp* discovery further clarified pathogenesis of duodenal ulcer disease. *Hp* infection induces hypergastrinemia and consequently increases gastric acid secretion. Namely, gastrin has direct stimulatory effect, but also potentates the effects of histamine and acetylcholine on parietal cell. Thus increased gastrin release leads to increased acid output that is one of the principal steps in ulcerogenesis. In humans gastrin release is regulated by GRP, epinephrine, luminal gastric contents (proteins, calcium and hydrogen ion concentration) and somatostatin. Among these, SST and low pH are principal inhibitors of gastrin secretion. Numerous studies demonstrated that *Hp* infection can be associated with duodenal ulcer formation with increased gastrin and lower somatostatin synthesis and secretion.

After it was demonstrated that *Hp* increased both basal and CGRP/ bombesin stimulated gastrin secretion [13], three major groups of possible underlying mechanisms were proposed, stating that hypergastrinemia results from bacterial products activity, presence of proinflammatory cytokines and impaired function of nerve fibers involved in the regulation of gastric acid secretion.

Bacterial products increase antral gastrin synthesis and secretion both trough direct or indirect mechanisms. According to one of proposed hypotheses ammonia released as a result of bacterial urease activity in contact with CO_2 present in the stomach, forms monochloramine that upon entering G cell and its secretory granules induces increase in intracellular pH, G cell degranulation and gastrin release into the adjacent mucosa. However, there are also opinions that the effect of ammonia is indirect and results from increased pH on gastric mucosal surface [112,113], but some authors consider this mechanism as irrelevant in the pathogenesis of hypergastrinemia [114]. Some attribute principal role in increased gastrin release to *Helicobacter pylori*'s LPS binding to SST receptor on antral G cell thus preventing inhibitory effects of SST. Also, LPS is known to enhance histamine release from ECL cell [115].

Proinflammatory cytokines, IL1-β, IL-8 and TNF-α, and according to some authors interferon-γ and PAF *in vitro* can stimulate gastrin release from G cells [13,112,116]. In the assessment of proinflammatory cytokine role in inducing hypergastrinemia an important study was published by Anike Hamlet et al. It is well-known that non-steroid antiinflammatory drugs and aspirin affect gastric mucosa by mechanisms independent of those involved in *Hp* induced tissue injury. Since aspirin-induced mucosal injury does not lead to a significant increase in gastrin concentrations Hamlet et al analyzed mucosal levels of different proinflammatory cytokines and concluded that both aspirin use and *Helicobacter pylori* infection increase mucosal IL1-β, IL-8 and IL-6. Nevertheless, increased TNF-α and IF- γ levels were solely *Hp* related [117] and later their direct stimulatory effect on gastrin release from G cells [13,118] was reported. It is also possible that other proinflammatory mediators involved in chronic inflammation (IL-2, IL-4, IL-10, IL-12), but not observed in course of acute aspirin-induced inflammation, could promote gastrin release, that was indirectly supported by the findings of persisting hypergastrinemia after eradication therapy that correlated with presence of neutrophil inflammatory infiltrate and mononuclear inflammatory cells. Only when either is no longer detectable gastrin levels return to normal [111,112,117].

Hypergastrinemia is probably not a single underlying factor responsible for impaired gastric acid secretion in *Hp* infected individuals [119]. Namely, in course of *Hp* infection two- to four-fold increase in basal acid output (BAO-basal acid output), increased maximal gastrin- or GRP-stimulated output (MAO-maximal acid output), and an increase in BAO/MAO ratio occurs. Results of some studies [79] revealed that gastrinemia and BAO are normalized 4 weeks after completing eradication therapy and remain unchanged during 1 year of follow up, while enhanced response to exogenous gastrin stimulation persists [8].

Possible explanation of prolonged response to stimuli inducing gastric acid secretion would be the influence of proinflammatory mediators on afferent nerves in the gastric mucosa. It is probably a result of direct neurotoxic effect of bacterial products or decreased nerve fiber density in gastric mucosa. Suggested mechanism would involve so far unidentified *Hp* derived toxin with a stimulatory effect on inhibitory CGRP and substance P releasing nerve fibers in gastric mucosa. According to this hypothesis, stimulation of nerve fibers could lead to a transient decrease in gastric acid secretion necessary for the acute phase of *Hp* infection during which the bacterium colonizes gastric epithelium. Prolonged stimulation leads to desensitation or depletion of these fibers resulting in increased gastric acid secretion. Our previous results support this hypothesis since in *Hp* infected individuals with gastritis we identified lower density of CGRP immunoreactive nerve fibres when compared to controls [120].

Somatostatin in *Helicobacter Pylori* Infection

After it was reported that *Helicobacter pylori* infection altered gastric acid secretion different studies aimed at its possible influence on gastric neuroendocrine cells, among others somatostatin- producing antral and corpus D cells.

We demonstrated that *Helicobacter pylori* infection impairs gastric D cell function. Namely, in antrum lower SST concentrations, together with a decrease in SST mRNA levels were detected in all *Hp* infected individuals, irrespective of the disease outcome, presence of duodenal ulcer or gastric cancer . In corpus, same trend of lower SST protein level was observed, while in NUD patients compared to other *Hp*+ patients and *Hp* negative controls significantly lower SST mRNA was observed.

Antral D cell exhibits regulatory functions in the gastric mucosa mainly trough paracrine mechanisms, thus determination of gastric tissue SST levels can reveal changes related to presence of *Hp* infection. It was previously reported that antral SST levels in the *Hp* infected individuals were reversibly decreased and normalized after eradication of the infection [8,100,121,122], both in adult and pediatric patients [101]. Corpus tissue SST concentrations are according to Zavros et al. [123] and our results not changed in course of *Hp* infection.

Possible explanation is that the majority of patients with NUD and DU has higher grade or even limited inflammation of antral mucosa.

Two independent studies revealed lower antral SST mRNA levels in all *Hp* infected individuals [123,124]. Namely, Zavros et al. [123] demonstrated six-fold decrease of antral SST mRNA in *Hp* infected individuals compared to *Hp* negative controls, thus confirming our results.

Lower number of SST immunoreactive cells was detected by many authors in *Hp* infected individuals [104,105,107,124-126]. After successful eradication of the infection majority [101,106,124,126], but not all [111] authors demonstrated an increase in antral D cell number. Previously described results of Holle et al. support this finding in gastric cancer patients where total antral D cell population count revealed 35% lower number in gastric cancer patients [108].

Six weeks after eradication therapy number of D cells in successfully treated patients does not reach that found in the uninfected individuals implying that complete recovery of D cells occurs only after longer period of at least 6 months [126].

That presence of the *Hp* infection is the main cause of reduction in antral D cell number was previously confirmed by the finding that only successful eradication of the infection lead to observed increase in antral SST levels and D cell number, while in individuals in whom eradication therapy failed no such change was detected [101,126]. Previous study that we conducted using electron microscopy revealed suppressed gastric D cell secretory activity. Namely, D cells of *Hp* infected individuals were smaller with lower number and diameter of secretory granules when compared to the uninfected individuals. Same study demonstrated that *Hp* infection inhibits D cell, since secretory apparatus in infected individuals (endoplasmatic reticulum and Golgi apparatus) had similar morphological characteristics in all patients unrelated to *Hp* status. Six months after eradication of the infection D cells in successfully treated individuals resembled those in the uninfected group [126].

Currently few different hypotheses were suggested aiming to explain mechanism of this *Hp* induced change in SST expression. Konturek et al demonstrated that in *Hp* infected duodenal ulcer individuals cholecystokinin mediated release of SST is disturbed [11,72] and that infusion of GRP in *Hp* infected patients stimulates SST mRNA synthesis but to a lower extent than in uninfected individuals. This implies that *Helicobacter pylori* suppresses antral D cell function at least partly by a mechanism independent of antral G cell function and gastrin [127]. Mechanism is unclear, but it might include TNF-α effects on D cells, since *in vitro* studies demonstrated that 24 hour incubation of D cell with this cytokine decreases up to 40% cholecystokinin-stimulated release of SST.

It is not easy to clarify the underlying mechanism of decreased SST secretion in course of *Helicobacter pylori* infection since SST is released from many different organs and is also very often acting locally. Possible mechanism could include bacterial urease activity, since urease is involved in the alkalization of gastric contents [125]. In addition, urea-derived ammonia as a result of urease activity could be influencing SST secretion. Study by Kaneko et al revealed inverse correlation between levels of ammonia in gastric juice and antral SST concentrations in *Hp* infected individuals [8]. Same group of authors supported this finding in an elegant study on animal model proving that only oral administration of ammonia in concentrations similar to those found in *Hp* infected individuals leads to decrease in SST release and lower antral D cell number [8]. That this a very low range dose dependent mechanism specific for ammonia was suggested after an *in vitro* study since different concentration or replacing ammonia by sodium hydroxide did not lead to changes in antral SST release [8]. There is also evidence that apart from direct mechanism, ammonia can exhibit paracrine effect on type 2 SST receptor inhibiting thus SST release [8]. Bacterial strain is also of considerable importance since cagA+ and vacA *s1m1* positive strains induce more intense SST suppression [101,129].

As a possible mechanism of SST suppression effect of histamine catabolic products derived form *Hp*, such as N^{α}-methyl histamine [130] was suggested, but studies on isolated D cells failed to confirm this hypothesis [131].

Plausible hypotheses confirmed on animal models implicated proinflammatory cytokines, such as TNF-α in lowering intracellular contents and percentage of stored SST released after D cell activation [132]. Beales et al demonstrated complex role of cytokines showing that TNF-α increases SST release during first two hours of incubation in a dose dependent manner, and that IL-8 potentate this effect. On the other hand, pre-exposure of D cell to medium that contains TNF-α inhibited all the above mentioned effects [133]. Latest results form Konturek et al showed reduced gastrin inhibition as a result of lower SST synthesis, either because of the direct or indirect (cytokine mediated) effect of *Hp* on gastric D cell function [72]. After successful eradication of the infection increase in D cell density is followed by a decrease in antral gastrin concentrations [125].

It was also demonstrated that SST *in vtiro* in concentration close to those found in gastric juice has an inhibitory effect on *Hp* proliferation using a cGMP dependent mechanism [134]. Cytokines, especially TNF-α lower intracellular SST concentration [132], while chronic exposure of D cells in culture to nitric oxide (NO) or to a substrate of nitric oxide synthase (NOS), L-arginine, lowers SST production, implying that NO released during chronic inflammation is limiting SST release [135].

Our unpublished data suggest that inflammatory infiltrate density, together with and increased gastrin, IL1-β, IL-8 and TNF-α levels, are related to a decrease in antral SST concentration. Other authors also found that both the activity of inflammation and mucosal IL-8 concentrations in duodenal ulcer patients are inversely related to gastric mucosal SST levels [136]. Peluso [137] demonstrated that SST in physiological concentrations is able to reduce macrophage and monocyte response to bacterial LPS by influencing IL1-β, IL-6, IL-8 and TNF-α secretion, while acute *Hp* infection leads to an almost immediate decrease in SST levels that in a week will result in increased IL1-β and IL-12 concentrations [72].

Conclusion

Based on the results of our study and available data in the literature we concluded that decreased gastric somatostatin synthesis and secretion is a universal *Helicobacter pylori* related phenomenon related to the presence of the infection and not to the disease outcome. On the other hand, gastrin levels in *Helicobacter pylori* positive individuals depend on the disease outcome and in gastric cancer patients are probably influenced by more then one mechanism out of which some are most likely *Hp* independent.

References

[1] Dayal Y. Neuroendocrine cells of the gastrointestinal tract: introduction and historical perspective. In: Dayal Y, editor. *Endocrine Pathology of the Gut and Pancreas*. Boca-Raton: CRC Press; 1991. p 1-31.

[2] De Vault K, Rattan S. Physiological role of neuropeptides in the gastrointestinal smooth muscle sphincters: neuropeptide and VIP-nitric oxide interactions. In: Walsh JH; Dockrey GJ, editors. Gut Peptides: *Biochemistry and Physiology*. New York: Raven Press Ltd: 1994. p 715-46.

[3] Holst JJ, Schmidt P. Gut hormones and intestinal function. *Baillieres Clin Endocrinol Metab,* 1994 Jan; 8(1):137-64.

[4] Sachs G, Zeng N, Prinz C. Physiology of isolated gastric endocrine cells. *Annu. Rev. Physiol,* 1997;59:243-56.

[5] Chiba T, Yamada T. Gut somatostatin. In: Walsh JH, Dockrey GJ, editors. Gut Peptides: *Biochemistry and Physiology*. New York: Raven Press Ltd: 1994. p 123- 45.

[6] Dockray GJ. Gastrin and gastric epithelial physiology. *J. Physiol,* 1999 Jul 15; 518 (Pt 2):315-24.

[7] Sokic Milutinovic A, Todorovic V, Milosavljevic T, Micev M, Drndarevic N, Mitrovic O. Gastrin and antral G cells in course of *Helicobacter pylori* eradication: 6 months follow up study. *World J. Gastroenterol,* 2005 Jul 21;11(27):4140-7.

[8] Kaneko H, Konagaya T, Kusugami K: *Helicobacter pylori* and gut hormones. *J. Gastroenterol,* 2002;37(2):77-86.

[9] Prinz C, Sachs G, Walsh JH, Coy DH, Wu SV. The somatostatin receptor subtype on rat enterochromaffin-like cells. *Gastroenterology,* 1994 Oct;107(4):1067-74.

[10] Sternini C, Wong H, Wu SV, De Giorgio R, Yang M, Reeve JJr, et al. Somatostatin 2A receptor is expressed by enteric neurons and by interstitial cells of Cajal and enterochromaffine-like cells of the gastrointestinal tract. *J. Comp. Neurol,* 1997 Sep 29;386(3):396-408.

[11] Konturek, PC. Physiological, immunohystochemical and molecular aspects of gastric adaptation to stress, aspirin and to H. *pylori* derived gastrotoxins. *J. Physiol. Pharmacol,* 1997 Mar; 48(1): 3-42.

[12] Feldman M, Cryer B, Lee E. Effects of *Helicobacter pylori* gastritis on gastric secretion in healthy human beings. *Am. J. Physiol.* 1998 Jun; 274(6 Pt 1):G1011-17.

[13] Calam J, Gibbons A, Healy ZV, Bliss P, Arebi N. How does *Helicobacter pylori* cause mucosal damage? Its effect on acid and gastrin physiology. *Gastroenterology,* 1997 Dec;113(6 Suppl): S43-9

[14] Peek RM, Blaser MJ. Pathophysiology of *Helicobacter pylori*- induced gastritis and peptic ulcer disease. *Am. J. Med,* 1997 Feb; 102(2): 200-7.

[15] Bobrzynski A. Hormonal, secretory and morphological alterations in gastric mucosa in the course of *Helicobacter pylori* eradication in patients with duodenal ulcer and non-ulcer dyspepsia. *J. Physiol. Pharmacol,* 1997 Jun; 48 Suppl 3: 1-56.

[16] Fan XG, Chua A, Li TG, Zeng QS. Survival of *Helicobacter pylori* in milk and tap water. *J. Gastroenterol. Hepatol.* 1998 Nov;13(11):1096-8.

[17] Figura N, Cetta F, Angelico M, Montalto G, Cetta D, Pacenti L, et al. Most *Helicobacter pylori*-infected patients have specific antibodies, and some also have *H. pylori* antigens and genomic material in bile. Is it a risk factor for gallstone formation. *Dig. Dis. Sci,* 1998 Apr;43 (4):854-62.

[18] Figura N, Vindigni C, Covaci A, Presenti L, Burroni D, Vernillo R, et al. Cag A positive and negative *Helicobacter pylori* strains are simultaneously present in the stomach of most patients with non- ulcer dyspepsia: relevance to histological damage. *Gut,* 1998 Jun; 42(6): 772-8.

[19] Kelly DJ. The physiology and metabolism of the human gastric pathogen *Helicobacter pylori*. *Adv. Microbiol. Physiol,* 1998;40:137- 89.

[20] Rektorschek M, Weeks D, Sachs G, Melchers K. Influence of pH on metabolism and urease activity of *Helicobacter pylori*. *Gastroenterology,* 1998 Sep; 115(3): 628-41.

[21] Perez-Perez GI, Rotenbacher D, Brenner H. Epidemiology of *Helicobacter pylori* infection. *Helicobacter,* 2004;9 Suppl 1 :1-6.

[22] Malaty HM, Nyren O. Epidemiology of *Helicobacter pylori* infection. *Helicobacter,* 2003;8 Suppl 1:8-12.

[23] Ertem D, Harmanci H, Pehlivanoglu E. *Helicobacter pylori* infection in Turkish preschool and school children. Role of socio-economic factors and breast feeding. *Turk J. Pediatr,* 2003 Apr-Jun;45(2):114-22.

[24] Maherzi A, Bovaziz-Abed A, Fendri C, Oubich F, Konbaa C, Fanchere JL. *Helicobacter pylori* infection: prospective study for asymptomatic Tunisian children. *Arch. Pediatr,* 2003 Mar;10(3):204-7.

[25] Genta RM, Gurer IE, Graham DY, Krishnan B, Segura AM, Gutierrez O, et al. Adherence of *Helicobacter pylori* to areas of incomplete intestinal metaplasia in gastric mucosa. *Gastroenterology,* 1996 Nov;111(5):1206-11.

[26] Makristathis A, Hirsch AM, Lehours P, Megraud F. Diagnosis of *Helicobacter pylori* infection. *Helicobacter,* 2004;9 Suppl 1:7-14.

[27] Mitchell H, Megraud F. Epidemiology and diagnosis of *Helicobacter pylori* infection. *Helicobacter,* 2002;7 Suppl 1:8-16.

[28] Peterson WL, Graham DY. *Helicobacter pylori*. In: Feldman M, Scharschmidt BF, Sleisinger MH, editors. Sleisinger and Fortrands Gastrointestinal and Liver Diease, 6th edition. W.B. Philadelphia: Saundres Company:1998. p. 604-20.

[29] Karita M, Teramukai S, Matsumoto S. Risk of *Helicobacter pylori* transmission from drinking well water is higher than that from infected intrafamilial members in Japan. *Dig. Dis. Sci,* 2003 Jun; 48(6):1062-7.

[30] Triantafillidis JK, Gikas A, Hyphantis T Cheracakis P, Rokkas T, Konstantellou E. *Helicobacter pylori* infection in hospital workers over a 5-year period Correlation with demographic and clinical parameters. *J. Gastroenterol,* 2002;37(12):1005-13.

[31] Josenhans C, Suerbaum S. Flagella and motility of *Helicobacter pylori*. In: Moran AP, O Morain CA, editors. Pathogenesis and Host Response in *Helicobacter pylori* Infections. Bad Hamburg: Normed- Verlag; 1997. p 16-25.

[32] Blaser MJ. Heterogeneity of *Helicobacter pylori*. *Eur. J. Gastroenterol. Hepatol,* 1997 Apr; 9 Suppl 1: S3-6.

[33] Mobley HLT, Fulkerson JJ, Hendricks JK, Garner RM. Expression of catalytically active urease by *Helicobacter pylori*. In: Moran AP, O Morain CA, editors. Pathogenesis and Host Response in *Helicobacter pylori* Infections. Bad Hamburg: Normed- Verlag; 1997. p 58-66.

[34] Scott DR, Weeks D, Hong C, Postius S, Melchers K, Sachs G. The role of internal urease in acid resistance of *Helicobacter pylori*. *Gastroenterology,* 1998 Jan; 114(1): 58- 70.

[35] Figura N, Vindigni C, Presenti L, Carducci A. New acquisitions in *Helicobacter pylori* characteristics. *Ital. J. Gastroenterol. Hepatol,* 1998 Oct; 30 Suppl 3: S254- 8.

[36] Kusters JG, van Vliet AH, Kuipers EJ. Pathogenesis of *Helicobacter pylori* infection. *Clin. Microbiol. Rev,* 2006 Jul; 19(3): 449-90.

[37] Rokita E, Makristathis A, Presteri E, Rotter ML, Hirschl AM. *Helicobacter pylori* urease significantly reduces opsonization by human complement. *J. Inf. Dis,* 1998 Nov; 178(5):1521-5.

[38] Ilver D, Arnqvist A, Ogren J, Boren T. Adherence of *Helicobacter pylori*: mechanisms and relevance to pathogenesis. In: Moran AP, O Morain CA, editors. Pathogenesis and Host Response in *Helicobacter pylori* Infections. Bad Hamburg: Normed- Verlag; 1997. p 16-25.

[39] Slomiany BL, Piotrowski J, Slomiany A. Effect of sucralfate on gastric mucosal inflammatory responses induced by *Helicobacter pylori* lipopolysaccharide. *Scand. J. Gastroenterol,* 1998 Sep; 33(9):916-22.

[40] Slomiany BL, Piotrowski J, Slomiany A. Induction of caspase-3 and nitric oxide synthase-2 during gastric mucosal inflammatory reaction to *Helicobacter pylori* lipopolisaccharide. *Bioch. Mol. Biol. Int,* 1998 Dec; 46(5):1063-70.

[41] Luo YH, Yan J, Mao YF. *Helicobacter pylori* lipopolysaccharide: biological activities in vivo and in vitro, pathological correlation to human chronic gastritis and peptic ulcer. *World J. Gastroenterol,* 2004 Jul 15; 10(14):2055-9.

[42] Piotrowski J, Piotrowsky E, Skrodzka D, Slomiany A, Slomiany BL. Induction of acute gastritis and epithelial apoptosis by *Helicobacter pylori* lipopolysaccharide. *Scand. J. Gastroenterol,* 1997 Mar; 32(3):203-11.

[43] Bertram-Drogatz PA, Sobek-Klocke I, Moller C, Wingbermuhle D, Beil W, Sewing KF, et al. Growth characteristics and influence of antibiotics on rough/smooth phenotypic variants of *Helicobacter pylori*. *Eur. J. Clin. Microbiol. Infect. Dis,* 1999 Jul; 18(7):490-5.

[44] Aspinall GO, Moran AP. *Helicobacter pylori* lipopolysaccharide structure and mimicry of Lewis blood group antigens. In: Moran AP, O Morain CA, editors. Pathogenesis and Host Response in *Helicobacter pylori* Infections. Bad Hamburg: Normed- Verlag; 1997. p 34-43.

[45] Slomiany BL, Piotrowski J, Slomiany A. Gastric mucosal inflammatory responses to *Helicobacter pylori* lipopolysaccharide: down regulation of nitric oxide synthase-2 and caspase-3 by sulgulcotide. *Bioch. Bioph. Res. Com,* 1999 Jul 22; 261(1):15-20.

[46] Slomiany BL, Piotrowski J, Slomiany A. Involvement of endothelin-1 in up- regulation of gastric mucosal responses to *Helicobacter pylori* lipopolisaccharide. *Bioch. Bioph. Res. Com,* 1999 Apr 29; 258(1):17-20.

[47] Radosz-Komonievska H, Beh T, Jozwiak J, Martirosian G. Pathogenicity of *Helicobacter pylori* infection. *Lin. Microbiol. Infect,* 2005 Aug; 11(8):602-10.
[48] Sommi P, Ricci V, Fiocca R, Romano M, Ivey KJ, Cova E, et al. Significance of ammonia in the genesis of gastric epithelial lesions induced by *Helicobacter pylori*: an in vitro study with different bacterial strains and urea concentrations. *Digestion,* 1996; 57(5): 299-304.
[49] Megraud F. Impact of *Helicobacter pylori* virulence on the outcome of gastroduodenal diseases: lessons from the microbiologist. *Dig. Dis,* 2001;19(2):99-103.
[50] Marais A, Mendez GL, Hazel S, Megraud F. Metabolism and genetics of *Helicobacter pylori*: the Genomic Era. *Microb. Mol. Biol. Rev,* 1999 Sep; 63(3):642-74.
[51] Cover TL, Tummuru MK, Cao P, Thompson SA, Blaser MJ. Divergence of genetic sequences for the vacuolating among *Helicobacter pylori* strains. *J. Biol. Chem,* 1994 Apr 8;269(14):10566-73.
[52] Rudi J, Kolb C, Maiwald M, Kuck D, Sieg A, Galle P, et al. Diversity of *Helicobacter pylori* vacA and cagA protein expression, cytotoxin production and associated diseases. *J. Clin. Microbiol,* 1998 Apr; 36(4):944-8.
[53] Ghiara P, Marchetti M, Blaser MJ, Tummuru MK, Cover TL, Segal ED, et al. Role of *Helicobacter pylori* virulence factors vacuolating cytotoxin, cagA, and urease in a mouse model of disease. *Infect. Immun,* 1995 Oct;63(10):4154-60.
[54] Cover TL, Cao P, Lind CD. Correlation between vacuolisating cytotoxin production by *Helicobacter pylori* isolates in vitro and in vivo. *Infect. Immun,* 1993 Dec;61(12):5008-12.
[55] Supajatura V, Ushio H, Wada A, Yahiro K, Okumura K, Ogawa H, et al. Cutting edge: vacA a vacuolating cytotoxin of *Helicobacter pylori* directly activates mast cell for migration and production of proinflammatory cytokines. *J. Immunol,* 2002 Mar 15; 168(6): 2603-7.
[56] Shirasaka D, Aoyama N, Sakashita M, Kuroda K, Maekawa S, Marwa C, et al. Relationship between gastric ulcer and *Helicobacter pylori* vacA detected in gastric juice using bead ELISA method. *Helicobacter,* 2002 Oct; 7(5): 281-6.
[57] Pai R, Sasaki E, Tarnawski AS. *Helicobacter pylori* vacuolating cytotoxin alters cytoskeleton associated proteins and interferes with re epithelization of wounded gastric epithelial monolayers. *Cell Biol. Int,* 2000; 24(5):291-301.
[58] Kimura M, Goto S, Wada A, Yahiro K, Nidome T, Hataheyama T. Vacuolating cytotoxin purified from *Helicobacter pylori* causes mytochondrial damage in human gastric cells. *Microb. Pathog,* 1999 Jan; 26(1):45-52.
[59] Papini E, Satin B, Norais N, de Bernard M, Telford JL, Rappuoli R. Selective increase of the permeability of polarized epithelial cell monolayers by *Helicobacter pylori* vacuolating cytotoxin. *J. Clin. Invest,* 1998 Aug 15; 102(4):813-20.
[60] Audibert C, Burucoa C, Janvier B, Fauchere JL. Implication of the structure of the *Helicobacter pylori* cag pathogenicity island in induction of interleukin 8 secretion. *Infect. Immun,* 2001 Mar; 69(3):1625-9.
[61] Hsu P, Hwang I, Cittely D, Lai K, El-Zimaity H, Gutierrez O, et al. Clinical presentation in relation to diversity within the *Helicobacter pylori* cag pathogenicity island. *Am. J. Gastroenterology,* 2002 Sep; 97(9): 2231-8.

[62] Asahi M, Azuma T, Ito S, Ito Y, Suto H, Nagai Y, et al. *Helicobacter pylori* cagA protein can be tyrosine phosphorilated in gastric epithelial cells. *J. Exp. Med,* 2000 Feb; 191(4):593-602.

[63] Odenbreit S, Puls J, Sedlamier B, Gerland E, Fischer W, Haas R. Translocation of *Helicobacter pylori* cagA into gastric epithelial cells by type IV secretion. *Science,* 2000 Feb 25; 287(5457):1497-1500.

[64] Nedrud J, Blanchard S, Czinn S. *Helicobacter pylori* inflammation and immunity. *Helicobacter,* 2002; 7 Suppl 1: 24-9.

[65] Vilaichone RK, Mahachai V, Tumwasorn S, Wu J, Graham DY, Yamaoka Y. Gastric mucosal cytokine levels in relation to host interleukin 1 polymorphisms and *Helicobacter pylori* cag genotype. *Scand. J. Gastroenterol,* 2005 May; 40(5): 530-9.

[66] Papa A, Danese S, Sgambato A, Ardito R, Zannoni G, Rinelli A, et al. Role of *Helicobacter pylori* CagA+ infection in determining oxidative DNA damage in gastric mucosa. *Scand. J. Gastroenterol,* 2002 Apr;37(4):409-13.

[67] Warburton V, Everett S, Mapstone N, Axon A, Hawkey C, Dixon M. Clinical and histological association of cagA and vac A genotypes in *Helicobacter pylori* gastritis. *J. Clin. Pathol,* 1998 Jan;51(1):55-60.

[68] Vorobjova T, Ananieva O, Maaroos H, Sipponen P, Villako K, Utt M, et al. Seropositivity to *Helicobacter pylori* heat shock protein 60 is strongly associated with intensity of chronic inflammation, particularly in antrum mucosa: an extension of an 18-year follow-up study of chronic gastritis in Saaremaa, Estonia. *FEMS Immunol. Med. Microbiol,* 2001 Mar;30(2):143-9.

[69] Furukawa K, Takahashi T, Arai F, Matsushima K, Asakura H. Enhanced mucosal expression of interleukin-6 mRNA but not of interleukin-8 mRNA at the margin of gastric ulcer in *Helicobacter pylori*-positive gastritis. *J. Gastroenterol,* 1998 Oct; 33(5):625-33.

[70] Harris PR, Ernst PB, Kawabata S, Kiyono H, Graham MF, Smith PD. Recombinant *Helicobacter pylori* urease activates primary mucosal macrophages. *J. Inf. Dis,* 1998 Nov; 178(5):1516-20.

[71] Shimada T, Terano A. Chemokine expression in *Helicobacter pylori*-infected gastric mucosa. *J. Gastroenterol,* 1998 Oct; 33(5):613-7.

[72] Konturek PC, Brzozowski T, Konturek SJ, Stachura J, Karczewska E, Pajdo R, et al. Mouse model of *Helicobacter pylori* infection: studies of gastric function and ulcer healing. *Aliment Pharmacol. Ther,* 1999 Mar; 13(3):333-46.

[73] Graham DY, Lew G, Malaty HM, Evans DG, Evans DJ, Klein PD, et al. Factors influencing the eradication of *Helicobacter pylori* with triple therapy. *Gastroenterology,* 1992 Feb;102(2):493-96.

[74] Labenz J, Borsch G. Evidence for the essential role of the *Helicobacter pylori* in gastric ulcer disease. *Gut,* 1994 Jan;35(1):19-22.

[75] Tsuji H, Kohli Y, Fukumitsu S, Morita K, Kaneko H, Ohkawara T, et al. *Helicobacter pylori*-negative gastritis and duodenal ulcers. *J. Gastroenterol,* 1999 Aug;34(4):455-60.

[76] Hamlet A, Thoreson A-C, Nilsson O, Svennerholm A-M, Olbe L. Duodenal *Helicobacter pylori* infection differs in cagA genotype between asymptomatic subjects and patients with duodenal ulcers. *Gastroenterology,* 1999 Feb; 116(2): 259-68.

[77] Prinz C, Schwendy S, Voland P. H pylori and gastric cancer: shifting the global burden. *World J. Gastroenterol,* 2006 Sep 14;12(34):5458-64

[78] Crew KD, Neugut AI. Epidemiology of gastric cancer. *World J. Gastroenterol,* 2006 Jan 21;12(3):354-62

[79] Savarino V, Mela GS, Zentilin P, Lapertosa G, Bisso G, Mele MR, et al. Histological and functional recovery in patients with multifocal atrophic gastritis after eradication of *Helicobacter pylori* infection. *Ital. J. Gastroenterol. Hepatol,* 1999 May;13(5):4-8.

[80] Arai H, Yoshii S, Maruyama Y, Furuta T, Kaneko E, Hanai H, et al. Microsatelite instability and loss of heterozygosity in Japanese gastric carcinoma with consideration of *Helicobacter pylori* infection. In: Kim JP, Min JS, Mock YJ, editors. Proceedings of the 3rd International Gastric Cancer Congress, Soeul, Korea, 27-30 April, 1999. Bologna; Monduzzi Editore Sp.A: 1999. p 203-5.

[81] Delchier JC, Lamarque D, Levy M, Tkoub EM, Copie-Bergman C, Deforges L, et al. *Helicobacter pylori* and gastric lymphoma: high seroprevalence of CagA in diffuse large B-cell lymphoma but not in low-grade lymphoma of mucosa-associated lymphoid tissue type. *Am. J. Gastroenterol,* 2001Aug;96(8):2324-8.

[82] Lamarque D, Gilbert T, Roudot-Thoraval F, Deforges L, Chaumette MT, Delchier JC. Seroprevalence of eight *Helicobacter pylori* antigens among 182 patients with peptic ulcer, MALT gastric lymphoma or non-ulcer dyspepsia. Higher rate of seroreactivity against CagA and 35-kDa antigens in patients with peptic ulcer originating from Europe and Africa. *Eur. J. Gastroenterol. Hepatol,* 1999 Jul; 11(7): 721-6.

[83] Koehler CI, Mues MB, Dienes *HP*, Kriegsmann J, Schirmacher P, Odenthal M. *Helicobacter pylori* genotyping in gastric adenocarcinoma and MALT lymphoma by multiplex PCR analyses of paraffin wax embedded tissues. *Mol. Pathol,* 2003 Feb;56(1):36-42.

[84] Arents NL, van Zwet AA, Thijs JC, Kooistra-Smid AM, van Slochteren KR, Degener JE, et al. The importance of vacA, cagA, and iceA genotypes of *Helicobacter pylori* infection in peptic ulcer disease and gastroesophageal reflux disease. *Am. J. Gastroenterol,* 2001 Sep;96(9):2603-8.

[85] Szlachcic A, Sliwowski Z, Karczewska E, Bielanski W, Pytko-Polonczyk J, Konturek SJ. *Helicobacter pylori* and its eradication in rosacea. *J. Physiol. Pharmacol,* 1999 Dec; 50(5):777-86.

[86] Gasbarrini A, Carloni E, Gasbarrini G, Menard A. *Helicobacter pylori* and extragastric diseases-Other *Helicobacters*. *Helicobacter,* 2003;8 suppl 1:S68-S76.

[87] Pellicano R, Parravicini P, Bigi R, Gandolfo N, Aruta E, Gai V, et al. Infection by *Helicobacter pylori* and acute myocardial infarction. Do cytotoxic strains make a difference? *New Microbiol,* 2002 Jul; 25(3):315-21.

[88] Perez-Perez GI, Peek RM, Legath AJ, Heine PR, Graff LB. The role of CagA status in gastric and extragastric complications of *Helicobacter pylori*. *Physiol. Pharmacol,* 1999 Dec;50(5):833-45.

[89] Malfertheiner P, Megraud F, O' Morain CO, Bell D, Bianchi-Porro G, Deltenre M, et al. Current European concepts in the management of *Helicobacter pylori* infection- the Maastricht consensus report. *Eur. J. Gastroenterol. Hepatol,* 1997 Jan; 9(1):91-2.

[90] Dixon MF, Genta RM, Yardley JH, Correa P. Classification and grading of gastritis. The updated Sydney System. International workshop on the histopathology of gastritis, Houston 1994. *Am. J. Surg. Pathol,* 1996 Oct; 20(10):1161-81.

[91] Weijnen CF, Hendriks HA, Hoes AW, Verweij WM, Werheij TJM, de Wit NJ. New immunoassay for the detection of *Helicobacter pylori* infection compared with urease test, 13C test and histology: validation in primary care setting. *J. Microbiol. Meth,* 2001 Sep;46(3):235-40.

[92] Wex T, Treiber G, Lendeckel U, Malfertheiner P. A two step method for the extraction of high-quality RNA from endoscopic biopsies. *Clin. Chem. Lab. Med,* 2003 Aug; 41(8): 1033-7.

[93] Carattoli A, Pezzella C, Pitroiust A, Galante A, Pezzotti P, Luzzi I. Cytotoxin-associated genA and vacuolating cytotoxinA in human isolates of *Helicobacter pylori* and their association with the clinical status of ulcer disease. *Eur. J. Gastroenterol. Hepatol,* 2000 Nov;12(11):1207-1213

[94] Cover TL, Glupczynski Y, Lage AP, Burette A, Tummuru MKR, Perez-Perez GI, et al. Serologic detection of infection with cagA+ *Helicobacter pylori* strains. *J. Clin. Microbiol,* 1995 Jun;33(6):1496-500.

[95] Hunt RH, Lam SK. *Helicobacter pylori*: from art to science. *J. Gastroenterol. Hepatol,* 1998 Jan;13(1):21-8.

[96] Suadicani P, Hein OH, Gyntelberg F. Genetic and life-style determinants of peptic ulcer. *Scand. J. Gastroenterol. Hepatol,* 1999 Jan;34(1):12-7.

[97] Stone MA, Barnett DB, Mayberry JF. Lack of correlation between self-reported symptoms of dyspepsia and infection with *Helicobacter pylori* in a general population sample. *Eur. J. Gastroenterol. Hepatol,* 1998 Apr;10(4):301-4.

[98] Bielanski W, Plonka M, Dobrzanska M, Kaminska A, Sito E, Ziemniak W, et al. Is *Helicobacter pylori* prevalence in asymptomatic Polish population different from symptomatic patients? *Gut,* 1999 Oct; 45(Suppl 5):42.

[99] Konturek PC, Hartwich A, Zuchowicz M, Labza H, Pierzchalski P, Karczewska E, et al. *Helicobacter pylori*, gastrin and cyclooxigenases in gastric cancer. *J. Physiol. Pharmacol,* 2000 Dec;51(4 Pt 1):737-49.

[100] Odum L, Petersen HD, Andersen IB, Hansen BF, Rehfeld JF: Gastrin and somatostatin in *Helicobacter pylori* infected antral mucosa. *Gut,* 1994 May; 35(5):615-8.

[101] Queiroz DM, Mendes EN, Rocha GA, Moura SB, Resende LM, Barbosa AJ: Effect of *Helicobacter pylori* eradication on antral gastrin– and somatostatin–immunoreactive cell density and gastrin and somatostatin concentrations. *Scand. J. Gastroenterol,* 1993 Oct; 28(10):858-64.

[102] Konturek PC, Rembiasz K, Konturek SJ, Stachura J, Bielanski W, Galuscha K, et al. Gene expression of ornithine decaboxylase, cyclooxigenase 2 and gastrin in atrophic gastric mucosa infected with *Helicobacter pylori* before and after eradication therapy. *Dig. Dis. Sci,* 2003 Oct;48(10):36-46.

[103] Konturek PC, Konturek SJ, Sulekova Z, Meixner H, Bielanski W, Starzynska T, et al. Expression of hepetocyte growth factor, transforming growth factor alpha, apoptosis related proteins Bax and Bcl-2 and gastrin in human gastric cancer. *Aliment Pharmacol. Ther,* 2001 Jul;15(7):989-99.

[104] Kamada T, Haruma K, Kawaguchi H, Yoshihara M, Sumii K, Kajiyama G: The association between antral G and D cells and mucosal inflammation, atrophy, and *Helicobacter pylori* infection in subjects with normal mucosa, chronic gastritis and duodenal ulcer. *Am. J. Gastroenterol,* 1998 May; 93(5):748-52.

[105] Park SM, Lee HR, Kim JG, Park JW, Jung G, Han SH, et al: Effect of *Helicobacter pylori* infection on antral gastrin and somatostatin cells and on serum gastrin concentrations. *Korean J. Int. Med,* 1999 Jan; 14(1):15-20.

[106] Tzaneva M: Light and microscopic immunohistochemical investigation on G and D cells in antral mucosa in *Helicobacter pylori* related gastritis. *Exp. Toxicol. Pathol,* 2001 Feb; 52(6):523-28.

[107] Chamouard P, Walter P, Wittersheim C, Demuynck P, Meunier O, Baumann R. Antral and fundic D-cell numbers in *Helicobacter pylori* infection. *Eur. J. Gastroenterol. Hepatol,* 1997 Apr;9(4):361-5.

[108] Holle GE, Stiegelmeier M, Riedel J, Ringel R, Siewert R, Eisenmenger W, et al. Histologic, immunohistichemical and morphometric study of the mucosa of the total gastric antrum in patients with carcinoma of the stomach. *Surg. Gynecol. Obstet,* 1993 Jan;176(1):65-72.

[109] Kwan CP, Tytgat GN: Antral G cell hyperplasia: a vanishing disease? *Eur. J. Gastroenterol. Hepatol,* 1995 Nov; 7(11):1099-103.

[110] Konturek PC, Konturek SJ, Bielanski W, Karczewska E, Pierczhalski P, Duda A, et al. Role of gastrin in gastric carcinogenesis in *Helicobacter pylori* infected humans. *J. Physiol. Pharmacol,* 1999 Dec;50(5):857-73.

[111] Hartleb M, Wandzel P, Waluga M, Matyszczyk B, Boldys H, Romanczyk T. Non endoscopic diagnosis of multifocal atrophic gastritis; efficacy of serum gastrin-17, pepsinogens and *Helicobacter pylori* antiobodies. *Acta Gastroenterol. Belg,* 2004 Oct-Dec;67(4):320-6

[112] Graham DY, Lew GM, Lechago J: Antral G-cell and D-cell numbers in *Helicobacter pylori* infection: effect of *H. pylori* eradication. *Gastroenterology.* 1993 Jun; 104(6):1655-60.

[113] Chittajalu RS, Neithercut WD, Macdonald AM, McColl KEL. Effect of increasing *Helicobacter pylori* ammonia production by urea infusion on plasma gastrin concentrations. *Gut,* 1991 Jan;32(1):21-4.

[114] Sidebotham RL, Barnon JH. Hypothesis. *Helicobacter pylori*, urease, mucus and gastric ulcer. *Lancet,* 1990 Jan 27;335(8683):193-5.

[115] Padol IT, Moran AP,Hunt RH. Effect of Purified Lipopolysaccharides from Strains of *Helicobacter pylori* and *Helicobacter felis* on Acid Secretion in Mouse Gastric Glands In Vitro . *Infect. Immun,* 2001 Jun;69(6): 3891-6.

[116] Kramling HJ, Enders G, Teichman RK, Demmel T, Merkle R, Brendel W. Antigen induced gastrin release: an immunologic mechanism of gastric antral mucosa. *Adv. Exp. Biol,* 1987;216A:417-9

[117] Hamlet A, Lindholm C, Nilsson O, Olbe L. Aspirin induced gastritis, like *Helicobacter pylori* gastritis disinhibits acid secretion in humans: relation to cytokine expression. *Scand. J. Gastroenterol,* 1998 Apr;33(4):346-56.

[118] Weigert N, Schaffer K, Shusdziarra V, Classen M, Shepp W. Gastrin secretion from primary cultures of rabbit antral G cells: stimulation by inflammatory cytokines. *Gastroenterology.* 1996 Jan; 110(1):1744-52.

[119] Sawada,M, Dickinson CJ. The G cell. *Annu. Rev. Physiol.* 1997;69:273-98

[120] Sokic A, Todorovic V, Micev M, Milosavljevic T, Bulajic M, Spuran M, et al. *Helicobacter pylori* infection and calcitonin gene-related peptide (CGRP) and substance P (SP) in the human gastric mucosa. *Helicobacter pylori*: Basic mechanisms to clinical cure 2000. Southampton (Bermuda), 2000, p. 30

[121] Haruma K, Sumii K, Okamoto S, Yoshihara M, Kajiyama G, Wagner S: *Helicobacter pylori* infection associated with low antral somatostatin content in young adults: implications for the pathogenesis of hypergastrinemia. *Scand. J. Gastroenterol,* 1995 Jun; 30(6):550-3.

[122] Götz JM, Veenendaal RA, Biemond I, Muller ES, Veselic M, Lamers CB. Serum gastrin and mucosal somatostatin in *Helicobacter pylori* -associated gastritis. *Scand. J. Gastroenterol,* 1995 Nov; 30(11):1064-8.

[123] Zavros Y, Paterson A, Lambert J, Shulkes A. expression of progastrin- derived peptides and somatostatin in fundus and antrum of nonulcer dyspepsia subjects with and without *Helicobacter pylori* infection. *Dig. Dis. Sci,* 2000 Oct; 45(10):2058-64.

[124] Moss SF, Legon S, Bishop AE, Polak JM, Calam J. Effects of *Helicobacter pylori* on gastric somatostatin in duodenal ulcer disease. *Lancet.* 1992 Oct 17; 340(8825):930-2.

[125] Tham TC, Chen L, Dennison N, Johnston CF, Collins JS, Ardil JES, et al. Effect of *Helicobacter pylori* eradication on antral somatostatin cell density in humans. *Eur. J. Gastroenterol. Hepatol,* 1998 Apr;10(4):289-91.

[126] Sokic Milutinovic A, Todorovic V, Milosavljevic T, Micev M, Spuran M, Drndarevic N. Somatostatin and D cells in patients with gastritis in the course of *Helicobacter pylori* eradication: a six-month follow-up study. *Eur. J. Gastroenterol. Hepatol,* 2003 Jul;15(7):755-66.

[127] Gibbons AH, Legon S, Walker MM, Ghatei M, Calam J. The effect of gastrin-releasing peptide on gastrin and somatostatin messenger RNA in humans infected with *Helicobacter pylori* . *Gastroenterology,* 1997 Jun; 112(6): 1940-7.

[128] Iyo T, Kaneko H, Konagaya T, Mori S, Katera H, Uruma M. Effect of intragastric ammonia on gastrin-, somatostatin- and somatostatin receptor subtype 2 positive cells in rat antral mucosa. *Life Sci,* 1999;64(26):2497-504.

[129] Kim JH, Park HJ, Cho JS, Lee KS, Lee SI, Park IS, et al: Relationship of cagA to serum gastrin concentrations and antral G, D cell densities in *Helicobacter pylori* infection. *Yonsei Med. J,* 1999 Aug;40(4): 301-6.

[130] Courillon-Mallet A, Launay JM, Roucayrol AM. *Helicobacter pylori* infection: physiopathologic implications of N-alpha methyl histamine. *Gastroenterology,* 1995 Apr;108(4):959-66.

[131] Belaes ILP, Calam J, Post L, Srinivasan S, Yamada T, Del Valle J. Effect of transforming growth factor alpha and interleukin 8 on somatostatin release from canine fundic D cell. Gastroenterology. 1997 Jan;112(1):136-43.

[132] Lamarque D, Tran Van Nhieu J, Breban M. What are the gastric modifications induced by acute and chronic *Helicobacter pylori* infection? *Gastroenterol. Clin. Biol.* 2003 Mar;27(3 Pt 2):391-400

[133] Belaes ILP, Calam J, Post L, Srinivasan S, Yamada T, Del Valle J. Effect of transforming growth factor alpha and interleukin 8 on somatostatin release from canine fundic D cell. *Gastroenterology.* 1997 Jan;112(1):136-43

[134] Yamashita K, Kaneko H, Yamamoto S, Konagaya T, Kusugami G, Mitsuma T. Inhibitory effect of somatostatin on *Helicobacter pylori* proliferation in vitro. *Gastroenterology,* 1998 Nov;115(5):1123-30.

[135] Arebi N, Healy ZV, Bliss PW, Ghatei M, Van Noorden, Playford RJ. Nitric oxide regulates the release of somatostatin from culture gastric rabbit primary D cells. *Gastroenterology,* 2002 Aug;123(2):566-76.

[136] Yamamoto S, Kaneko H,, Konagaya T, Mori S, Kotera H, Hayakawa T, et al. Interactions among gastric somatostatin, interleukin8 and mucosal inflammation in *Helicobacter pylori* positive peptic ulcer patients. *Helicobacter,* 2001 Jun; 6(2): 136-45.

[137] Peluso G, Petillo O, Melone MAB, Mazzarella G, Ranieri M, Tajan GF. Modulation of cytokine production in activated human monocytes by somatostatin. *Neuropeptides,* 1996 Oct;30(5):443-51.

In: Neuropeptide Research Trends
Editor: Bernice A. Levine, pp. 97-119
ISBN: 978-1-60021-640-4

Chapter III

Neuropeptides and Cancer: Focus on Substance P/Neurokinin-1 Receptor System Research

***M. Muñoz**[*1], **M. Rosso**[1] and **R. Coveñas**[2]*
[1] Pediatric Intensive Care Unit, "Virgen del Rocío"
University Children's Hospital, Sevilla, Spain
[2] Laboratory of Neuroanatomy of the Peptidergic Systems,
Institute of Neurosciences of Castilla y León (INCYL), Salamanca, Spain

Abstract

Neuropeptides, which show a widespread distribution in both the central and peripheral nervous systems, are neuroactive substances that acting as neurotransmitters, neuromodulators (paracrine and autocrine actions) and neurohormones are involved in numerous physiological actions. Thus, for example, neuropeptides have been implicated in inflammatory and immune processes, as well as in angiogenesis, mitogenesis, cancer and metastasis.

In this sense, it is known that after binding to the neurokinin (NK)-1 receptor, substance P, a neuropeptide that belongs to the tachykinin family of peptides and that shows a widespread distribution in the nervous system, promotes proliferation of cancer cells (neuroblastoma, glioma, retinoblastoma, melanoma...) favoring tumour growth. It is also known that in some tumours, tachykinins facilitate tumour metastasis. Moreover, cancer cells express more NK-1 receptors than normal cells and NK-1 receptor antagonists exert an antitumoural action on numerous cancer cell lines, being the cell death observed produced by apoptosis. These data suggest that the NK-1 receptor is a new and promising target in the treatment of cancer and that the development of antagonist molecules of the NK-1 receptor represents an important opportunity for exploiting these molecules as novel therapeutic agents.

* Dr. Miguel Muñoz; Hospital Infantil Virgen del Rocío; Unidad de Cuidados Intensivos Pediátricos; Av. Manuel Siurot s/n; 41013 – Sevilla; Spain; Phone number: 34-955012965; E-mail: mmunoz@cica.es; Fax number: 34-955012921

According to this, in this review the following aspects will be discussed: 1) Substance P/NK-1 receptor system: a) presence of substance P in human tumours; b) presence of NK-1 receptors in human tumours; 2) Cancer progression and mitogenic action of substance P: a) mitogenic action of substance P; b) angiogenesis; c) metastasis; d) substance P, inflammation, pain and cancer; e) substance P, emotional behaviour and cancer; and 3) Future perspectives: the NK-1 receptor as a new target in the treatment of cancer: a) NK-1 receptor antagonists; b) antitumoural action of NK-1 receptor antagonists.

Key words: cancer; neuropeptides; peptides; Substance P; NK-1 receptor; NK-1 receptor antagonists; angiogenesis; metastasis.

Introduction

In recent years, the expression and secretion of peptides by tumours has gained increasing interest since peptides have been shown to influence tumour proliferation (see Muñoz et al., 2006b, 2007). Thus, it is known that cancer cells express receptors for peptides (e.g., somatostatin, galanin, substance P, cholecystokinin...) and there are numerous data suggesting that the substance P/neurokinin-1 receptor system could play an important role in the development of cancer (see Hening et al., 1995; Palma et al., 1999: Singh et al., 2000; Friess et al., 2003; Reubi, 2003; Berger et al., 2005; Muñoz et al., 2005a, 2006a, 2007; Schulz et al., 2006).

In this sense, it is known that substance P exerts a mitogenic action on several human cancer cell lines (neuroblastoma, glioma, melanoma, retinoblastoma, pancreas carcinoma) (see Muñoz et al., 2004b, 2005a, b, 2006a, 2007), acting through the neurokinin-1 (NK-1) receptor; and that NK-1 receptor antagonists (e.g., L-733,060...) exert an antitumoural action against human cancer cell lines (see Muñoz et al., 2004a, 2005a). These data indicate that substance P may be a universal mitogen in NK-1 receptor-expressing tumour cell types and that the NK-1 receptor may be a new and promising target in the treatment of human cancer and metastasis, since it seems that NK-1 receptor antagonists act as a broad-spectrum antitumoural agent (see Muñoz et al., 2006b, 2007).

According to the above mentioned, our focus in this work is to review the involvement of the substance P/NK-1 receptor system in cancer.

Substance P/NK-1 Receptor System

Substance P (SP) is an undecapeptide that belongs to the tachykinin family of peptides; it is derived from the preprotachykinin A gene; it acts as a neurotransmitter or a neuromodulator in the nervous system; and it is hydrolysed by endopeptidase-24.11 (Turner, 1987). Numerous radioimmunological and immunohistochemical studies have demonstrated a widespread distribution of this tachykinin in the mammalian central and peripheral nervous systems (Cuello and Kanazawa, 1978; Hökfelt et al., 1978; Mai et al., 1986; Burgos et al., 1988).

The biological action of SP being preferentially mediated by the NK-1 receptor coupled with G-protein (MacDonald and Boyd, 1989), since SP is the natural ligand with the highest affinity for the NK-1 receptor. In this sense, the NK-1 receptor is defined as the mediator of the biological activities encoded by the C-terminal sequence of tachykinins, for which SP is a more potent agonist than neurokinin A or neurokinin B (see Quartara and Maggi, 1997). Numerous studies have shown a widespread distribution of the NK-1 receptor in the mammalian central nervous system and in peripheral tissues (Wolf et al., 1985; Dam and Quirion, 1986; Danks et al., 1986; Saffroy et al, 1988; Maeno et al., 1993; Nakaya et al., 1994; Whitty et al., 1995).

After binding to the NK-1 receptor, SP has been involved in the following biological functions: dilatation of the arterial system, sensory perception, regulation of the cardiovascular system, movement control, salivation and micturition, neuronal survival and degeneration, regulation of respiratory mechanisms, gastric motility, and in inflammation, pain and depression (von Euler and Gaddum, 1931; Unger et al., 1981; Pernow, 1983; Hershey et al., 1991; Rupniak et al., 1996; Kramer et al., 1998; Quartara and Maggi, 1998; Samsam et al., 2000; Harrison and Geppetti, 2001; Hökfelt et al., 2001; Bang et al., 2003; Varty et al., 2003).

Presence of Substance P in Human Tumours

SP has been found in human neuroblastoma cells of the parotid gland and it is expressed in almost 90% of metastatic neuroblastoma cells in the bone marrow, in 68% of primary invasive malignant melanomas, in 40% of metastatic melanomas, in 60% of melanomas in situ, in 58% of atypical nevi and in 40% of spindle and epithelioid nevi, but SP was not detected in acquired benign melanocytic nevi (Shrestha et al., 1994; Khare et al., 1998; Nowicki and Miskowiak, 2002). Thus, for example, in the atypical nevi and in the spindle and epithelioid nevi, most of the tumour cells expresing SP were located at the dermal-epidermal junction overlying areas of papillary dermal fibrosis. Moreover, SP has been found in human malignant glioma and retinoblastoma (Tarkkanen et al., 1983; Palma et al., 2000). In human glioma U373 MG xenograft, SP was observed in the peritumoural and in the tumour mass, especially in the periphery (Palma et al., 1999, 2000) and it is also known that the expression of preprotachykinin A was increased in breast cancer in comparison with normal mammary epithelial cells (Singh et al., 2000). In addition, it is also known that the level of SP was increased 7-fold in breast cancer cells in comparison with normal mammary epithelial cells and immunohistochemical studies have shown positive staining for SP in malignant tissues and negative staining in benign tissues.

Presence of NK-1 Receptors in Human Tumours

NK-1 receptors have been localized and characterized in primary human neoplastic cells of 9/12 astrocytomas, 10/10 glioblastomas, 10/12 medullary thyroid carcinomas, 8/16 breast carcinomas and 4/5 ganglioneuroblastomas. By contrast, NK-1 receptors were not or only rarely identified on non-small-cell carcinomas of the lung (1/16), neuroblastomas (0/8), colon (1/21) or pancreas (1/9) adenocarcinomas, or on malignant lymphomas (3/18) (see Hennig et al., 1995). However, in the great majority of the investigated tumours, NK-1 receptors were found on intra- and peritumoural blood vessels. The presence of NK-1 receptors, in such blood vessels and in tumour cells, could have clinical implications (see Hennig et al., 1995). Moreover, it has been reported the expression of the NK-1 receptor in all the breast cancer cell lines and breast biopsies studied. Thus, in this cell line the number of NK-1 receptors was increased significantly when compared with normal mammary epithelial cells. It is also known that the expression of the NK-1 receptor was increased in malignant breast biopsies in comparison with that found in benign tissue. Furthemore, using an in situ hybridization technique, it has been reported that mRNA NK-1 receptor expression was increased in all malignant tissues (e.g., breast biopsies...) studied, being such expression not detected in benign tissues (Singh et al., 2000). Recently, in human glioma, neuroblastoma and pancreas carcinoma cell lines the expression of NK-1 receptors has been reported (Muñoz et al., 2005a, 2006a, 2007). In other study the distribution of NK-1 receptors was investigated in seventy-two formalin-fixed, paraffin-embedded human tumours showing, for example, that NK-1 receptors were frequently expressed in glioblastomas, and in breast and pancreatic carcinomas. Immunoreactive NK-1 receptors were clearly confined to the plasma membrane and uniformly present on nearly all tumour cells studied. In this sense, NK-1 receptor visualization with immunohistochemical method will facilitate identification of tumours with a sufficient receptor overexpression for diagnostic or therapeutic intervention using NK-1 receptor antagonists (Schultz et al., 2006).

It is also known that the NK-1 receptor expression is increased 25-36-fold in human pancreatic cancer cell lines in comparison with normal controls, and that tumour samples of patients with advanced tumour stages exhibit significantly higher NK-1 receptor levels (see Friess et al., 2003). Thus, the number of NK-1 receptors expressed in the human normal cells is lower (e.g., human blood T-lymphocytes express 7,000-10,000 NK-1 receptors/cell) (Payan et al. 1986) than that expressed in human tumour cells (e.g., astrocytoma cells express 40,000 NK-1 receptors/cell) (Fowler et al., 1994). It is also known that high levels of NK-1 receptors are expressed in human glioma cell lines U373 MG (Palma et al., 1999) and UC11 MG (Johnson et al., 1991).

At the present, different isoforms of the NK-1 receptor have been detected in both rat (isoform sizes ranging from 46 kDa to 54 kDa) (Nakata et al., 1988; van Ginkel et al., 1996) and human tissues (33 kDa, 38 kDa and 58 kDa) (Organist et al., 1988; McGillis et al, 1990; van Ginkel et al., 1996; Caberlotto et al., 2003). Recently, in several human tumour cell lines the presence of four isoforms (75, 54-58, 46, 33-38 kDa) of the NK-1 receptor has been described. Thus, in IMR-32 and KELLY neuroblastoma cell lines a band of 33 kDa and a major 54 kDa band are observed, and in IMR-32 cells an additional isoform of about 75 kDa is detected (Muñoz et al., 2007), whereas in neuroblastoma SKN-BE(2) cell line a major 54 kDa band is observed (Muñoz et al., 2005a). Moreover, in the glioma GAMG cell line two bands of 33-38 and 54 kDa is detected (Muñoz et al., 2005a), and in pancreas carcinoma CAPAN-1 and PA-TU 8902 cell lines bands of 75, 58, 46, 34 kDa were reported in both cell lines (Friess et al., 2003; Muñoz et al., 2006a). In the future, the functional roles of the different isoforms of the NK-1 receptor observed in human cancer cells must be investigated.

The stimulation of NK-1 receptors can generate several second messengers which, in turn, can trigger a wide range of effector mechanisms that are responsible for regulating cellular excitability and function. In this sense, it is known that the following second-messenger systems are activated by NK-1 receptor agonists: a) arachidonic acid mobilization via phospholipase A2; b) stimulation, via phospholipase C, of phosphatidyl inositol turnover, leading to calcium mobilization from both intra- and extracellular sources; and c) cAMP accumulation via stimulation of adenylate cyclase (Mitsuhashi et al., 1992; Nakajima et al., 1992; Takeda et al., 1992; Seabrook and Fong, 1993; García et al., 1994; Mochizuki-Oda et al., 1994). Moreover, it is known that the activation of NK-1 receptors by SP or by its agonists produces in human astrocytoma U373 MG and UC11 MG cell lines the following actions: mobilizes intracellular calcium, induces the activation of phospholipase D, the release of interleukins (IL-8 and IL-6), taurine and glutamate, the formation of inositol phosphates, it stimulates glycogen breakdown and it influences both glutamate and K^+ transport (Johnson et al., 1991; Lee et al., 1992; Johnson and Johnson, 1993; Bordey et al., 1994; Medrano et al., 1994; Oury-Donat et al., 1994; Palma and Manzini, 1998; Torrens et al., 1998).

Cancer Progression and Mitogenic Action of Substance P

Mitogenic Action of Substance P

It has been reported that, by activating NK-1 receptors, SP stimulates mitogenesis in several tumour (neuroblastoma, melanoma, retinoblastoma, pancreas carcinoma, glioma...) and normal cell types, probably via autocrine/paracrine mechanisms (Payan et al., 1983; Nilsson et al., 1985; Lotz et al., 1987; Ziche et al., 1990a; Luo et al., 1996; Sharif et al., 1996; Palma et al., 1999; Muñoz et al., 2004b, 2005a, b, 2006a, 2007).

It is also known the presence of NK-1 receptors in U373 MG, U138 MG, UC11 MG and GAMG human glioma cell lines; that SP is expressed in almost 90% of metastatic human neuroblastoma cells (Nowicki and Miskowiak, 2002) and in primary retinoblastoma (Tarkkanen et al., 1983); and that astrocytoma/glioma primary tumours express more NK-1 receptors than that observed in astrocytoma/glioma cell lines in culture (Luo et al., 1996).

It has been reported that tachykinins (e.g., SP), via NK-1 receptors, increase DNA synthesis and cellular proliferation (e. g., astrocytoma cell line U-373 MG) (Eistetter et al., 1992; Luo et al., 1996; Sharif et al., 1996; Palma at al., 1999). This observation is in agreement with the findings reported by other works, since it has been observed that nanomolar concentrations of SP stimulates the mitogenesis of human GAMG glioma, CAPAN-1 and PA-TU 8902 pancreas carcinoma and SKN-BE(2), KELLY and IMR-32 neuroblastoma cells (these cell lines express NK-1 receptors) (Muñoz et al., 2005a, 2006a, 2007). It is known that SP act specifically through the NK-1 receptor in the mentioned cancer cell lines because L-733,060 or L-732,138 (NK-1 receptor antagonists) growth inhibition of human GAMG glioma, MEL HO, COLO 679, COLO 858 melanoma, WERI-Rb-1 and Y-79 retinoblastoma, CAPAN-1 and PA-TU 8902 pancreas carcinoma, and SKN-BE(2), KELLY and IMR-32 neuroblastoma cell lines, is partially reversed by the administration of a nanomolar dose of exogenous SP (Muñoz et al., 2004b, 2005a, b, 2006a, 2007).

In sum, these data indicate that activation of NK-1 receptors by SP induces mitogenesis in several human cancer cell lines (Luo et al., 1996; Palma et al., 1999, 2000; Muñoz et al., 2004b, 2005a, b, 2006a, 2007), an effect in which the activation of the mitogen-activated protein kinase signalling pathway could occur (see Luo et al., 1996).

Angiogenesis

Neovascularization or neoangiogenesis is a sequential process, with early endothelial proliferation followed by new vessel formation and increased blood flow, with maturation of endogenous neurovascular regulatory systems occurring late in this process in inflamed tissues. The growth of new vessels from a pre-existing vasculature, is a common feature of chronic inflammation (early neoangiogenesis is a key step in the transition from acute to persistent inflammation) and wound healing. Moreover, neoangiogenesis, a hallmark of tumour development, has been associated with increased tissue innervations and expression of NK-1 receptors (Singh et al., 2000).

It has been reported the effect of SP, a main mediator of neurogenic inflammation, on the growth of capillary vessels in vivo, and on the proliferation of cultured endothelial cells in vitro, as well as it has been demonstrated that NK-1 receptor agonists also induced neovascularization. It is also known that the proliferation of endothelial cells increased in a concentration-dependent manner (maximal activity of the NK-1 receptor agonists between 1 nM and 10 nM), whereas the action of selective NK-2 and NK-3 receptor agonists has not significant effect on the proliferation of endothelial cells. However, NK-1 receptor antagonists blocked the proliferative action of SP. These findings indicate that this peptide can directly stimulate the process of neovascularization, probably through induction of endothelial cell proliferation (Ziche et al., 1990b).

Moreover, in a rat inflammatory model, angiogenesis can be enhanced by the administration of SP. In this sense, SP-containing sensory nerves and specific, high affinity SP binding sites with characteristics of the NK-1 receptor were localized in microvessels. These data indicate that SP-enhanced angiogenesis results from a direct action on microvascular NK-1 receptors.

The nervous system has long been known to play a role in inflammation, in part through the release of SP from peripheral nerve terminals (neurogenic inflammation). As indicated previously, it is known that the application of SP can stimulate vessel growth in a variety of angiogenesis assays; that SP enhances endothelial cell proliferation; and that in vivo angiogenesis is mimicked by selective NK-1 receptor agonists (Ziche et al., 1990b). Thus, the SP/NK-1 receptor system controls the neoangiogenesis, but in addition (see below) this system could also regulate the growth of the tumoural mass, since NK-1 receptors are overexpressed in tumoural cells and in the tumoural and peritumoural tissues (Hennig et al., 1995; Friess et al., 2003).

Metastasis

The value of surgical removal of tumours was recognised hundreds of years ago. Because of these successful operative procedures, complications in cancer treatment have shifted; more than 90% of deaths from cancer are not from the primary tumour but from development of metastasis (Sporn, 1996). Therefore, a major aim in the treatment of cancer must be to inhibit the spreading of tumour cells, thus inhibiting development of metastasis. A model for the development of metastasis is based on the finding that G protein-coupled receptors, also known as serpentine receptors or seven transmembrane receptors, regulate the migratory activity of tumour cells similar to the recruitment and homing of leucocytes. Furthermore, ligands (e.g., neurotransmitters...) to these receptors can induce directed (chemotactic) migration (Entschladen et al., 2002). This means that the active migration of tumour cells, a crucial requirement for metastasis development and cancer progression, could be regulated by signal substances including neurotransmitters. Thus, these substances induce a metastatogenic tumour cell type by directly regulating gene expression and by increasing migratory activity, which can be prevented by established neurotransmitter antagonists. In this sense, it has been described that tumour cell (MDA-MB-468 breast and PC-3 prostate carcinoma cells) migration is induced by noradrenalin, dopamine and SP and that this process can be inhibited by using, respectively, specific and clinically established adrenoceptor, D_2 receptor, or NK-1 receptor antagonists. It is also known that all the investigated neurotransmitters significantly activated the cyclic adenosine-monophosphate response element binding protein (Lang et al., 2004). In sum, tumour-cell migration is a prerequisite for invasion and metastasis, and it is dependent of the signal substances of the immune and neuroendocrine systems. Thus, the growth of the tumoural mass, the peritumoural infiltration and the metastasis could be regulated by the SP/NK-1 receptor system that is overexpressed in tumour cells and in tumoural and peritumoural tissue (inflammatory cells, fibroblasts, blood vessels, nerves, ganglia...) (Hennig et al., 1995; Friess et al., 2003). This functional interplay is a further complexity of metastatic development and progression; however it opens new pathways for specific chemopreventive inhibition of tumour-cell invasion and metastasis.

Substance P, Inflammation, Pain and Cancer

SP released from peripheral terminals of primary sensory neurons has been implicated in neurogenic inflammation (terminals containing SP are located close to blood vessels; the release of SP in peripheral tissues reproduces many of the physiological changes seen in acute inflammation (plasma extravasation...)), vasodilatation (SP is a potent vasodilator in several peripheral tissues), vascular permeability (SP increases permeability), mast-cell degranulation, and modulation of immune cell function (Holzer, 1988; Lembeck, 1988; Maggi and Meli, 1988). Several lines of evidence indicate that SP plays a role in immune responses in both peripheral tissues and in the central nervous system, as well as in pain (see Mantyh, 1991; Harrison and Geppetti, 2001) and that a dramatic up-regulation in the expression of NK-1 receptors has been reported during painful chronic conditions (Chahl and

Ladda, 1976; Lembeck and Holzer, 1979; Furness et al., 1982; Lundberg et al., 1984). Thus, the NK-1 receptor antagonist L-733,060 has been used as antiinflammatory (Bang et al., 2003) or analgesic (Rupniak et al., 1996) agent.

The SP/NK-1 receptor system can modulate the immune function from sensory nerve/neurogenic inflammation. It is known that activation of the SP/NK-1 receptor system produces alterations in the humoural (Pascual et al., 1991) and cellular, normal (Lieb et al., 1997; Feistritzer et al., 2003), and tumoural (Luo et al., 1996; see Muñoz et al., 2006b) immune response. In this sense, it has been reported that tachykinins are capable of modifying the response of a variety of inflammatory cells, including mast cells, granulocytes, lymphocytes and monocytes and macrophages, and that oedema formation is induced by lower concentrations of SP and is blocked by NK-1 receptor antagonists (Walsh et al., 1995). Moreover, SP regulation of immune cell function could originate not only from neuronal sources (sensory nerves, neurogenic inflammation), but also from non-neuronal elements, such as eosinophils and macrophages, since in these cells the expression of both SP and NK-1 receptors is up-regulated during inflammation (Weinstock et al., 1988; Cook et al., 1994).

An increase in the SP level has been described during intestinal inflammation, and a significant correlation between the degree of inflammation and the clinical status of the disease has been indicated (Keranen et al., 1996a, b). These observations indicate that local release of SP (e. g., in lymph nodes) might be a factor contributing to the immune disorder underlying chronic inflammatory bowel disease, since the presence of specific NK-1 receptors in human peripheral blood T lymphocytes has been demonstrated (Payan et al., 1984).

Chronic inflammation is also clearly correlated with increased risk of developing cancer (Weitzman and Gordon, 1990), since inflammation increases both mitogenesis and mutagenesis (Ames and Gold, 1990). It is also known that a dividing cell has a greater risk of mutation than a quiescent cell (Tong et al., 1980) and that cell division allows adducts to convert to mutations. The time interval for DNA repair during cell division is short, and endogenous or exogenous damage is therefore generally increased if cells are proliferating. Moreover, it is well established that cancer arises in chronically inflamed tissue, and this is particularly notable in the gastrointestinal tract (Macarthur et al., 2004). Inflammation may become chronic either because an inflammatory stimulus persists or because of dysregulation in the control mechanisms that normally turn the process off. It has also been reported that many of the cells, cytokines and processes (e.g., leukocyte migration, dilatation of the local vasculature and angiogenesis...) involved in the inflammation are also found in a variety of tumours. It is important to note that in the vast majority of tumours investigated, NK-1 receptors have been found in intratumoural and peritumoural blood vessels (Hennig et al., 1995) and that NK-1 receptors were highly expressed in blood vessels not only within the tumour mass but also in the peritumoural tissue (Friess et al., 2003).

Chronic inflammation caused by intestinal flora leading to the inflammatory bowel disease, ulcerative colitis and Crohn's disease, is clearly linked with a higher incidence of colon cancer. Moreover, elevated levels of SP and upregulated NK-1 receptor expression have been reported in the rectum and colon of patients with inflammatory bowel disease (O'Connor et al., 2004). It is also known that dietary intake of proinflammatory carcinogens has been associated with prostate cancer and that chronic inflammation resulting from esophageal reflux gives rise to gastroesophageal reflux disease (GERD) and Barrett's esophagus, also linked with a higher incidence of cancer. Chronic *Helicobacter pylori* infection produces chronic inflammation and is therefore a potential risk factor for stomach cancer (Blaser et al., 1995). It is also known that the risk of pancreatic cancer is significantly elevated in subjects with chronic pancreatitis and appears to be independent of sex, country, or type of pancreatitis (Lowenfels et al., 1993). Moreover, the up-regulation of NK-1 receptor mRNA expression in chronic pancreatitis has been reported, there being a strong relationship to the pain syndrome in these patients (Shrikhande et al., 2001). All these observations suggest that chronic inflammation could produce cancer through the SP/NK-1 receptor system, which is up-regulated in the process of chronic inflammation, since it has been demonstrated that SP stimulates mitogenesis in human tumour cell lines (Muñoz et al., 2004b, 2005a, b, 2006a, b, 2007) and that NK-1 receptor expression is increased in cancer cell lines in comparison with normal controls (Friess et al., 2003). Unlike tumour progression, where the role of inflammation in promoting cancer cell proliferation and stromal/matrix degradation is reasonably well understood, the role of inflammation in metastasis is less well defined, although appears to be important. In this sense, an increase in the levels of SP occurs during neurogenic inflammation and, as we have previously mentioned, it is known that SP induces tumour cell migration (metastasis). These data suggest that inflammation could also be involved in metastasis through the SP/NK-1 receptor system.

Substance P, Emotional Behaviour and Cancer: Possible Relationship between the Progression of Cancer and Cerebral Mechanisms

The possibility of links between phsychosocial factors and cancer incidence and progression has generated considerable scientific and public interest. Tachykinins, including substance P, neurokinin A and B, act through three types of transmembrane G-protein coupled receptors denoted NK-1, NK-2 and NK-3. Tachykinins act as neurotransmitters in the peripheral and the central nervous system, but tachykinins and their receptors are also expressed in several non neuronal cells contributing to the fine connections between the nervous system and the peripheral organ system such as, respiratory, cardiovascular, immune, endocrine, gastrointestinal and genitourinary. As tachykinins are involved in regulating numerous physiological functions, they can also concur to pathological conditions including cancer (Palma, 2006).

There are data indicating that psychological factors could be involved in the development and progression of cancer (Hilakivi-Clarke et al., 1994). Thus, mammary tumourgenesis has been associated to the life-style and to the exposure to various forms of stressors. The possible role of life-style factors in breast cancer is important in light of the fact that mortality to this disease is increasing in most countries and that development of curative therapies for breast cancer has not been forthcoming. It seems that the crucial factor affecting tumour growth is the interaction among stress, an individual's personality, and available psychosocial support, and the effect of this interaction on an individual's ability to cope with stress (see Hilakivi-Clarke et al., 1994; Okamura et al., 2005).

It is known that depression and cancer commonly co-occur. Thus, it has been suggested that chronic and severe depression may be associated with elevated cancer risk; that the prevalence of depression among cancer patients increases with disease severity and symptoms such as pain and fatigue; and that depression predicts cancer progression and mortality. Moreover, there are data suggesting that psychosocial support reduces depression, anxiety, and pain, and may increase survival time with cancer. Thus, there is evidence of a bi-directional relationship between cancer and depression, offering new opportunities for therapeutic intervention (see Spiegel and Giese-Davis, 2003). It has been reported that chronic treatment with antidepressant drugs decreased SP concentrations in the striatum, the substantia nigra, and the amygdala and that the expression of the genes encoding the synthesis of tachykinins and the expression of NK-1 receptors in selected brain areas is modified by the action of the psychotropic drugs (Sivam et al., 1989; Humpel et al., 1990; Shibata et al., 1990). These data indicate that a reduction in SP levels in certain brain regions could contribute to a common therapeutic effect of the antidepressant drugs in affective disorders (Shirayama et al., 1996). Moreover, the NK-1 receptor antagonist L-733,060 has been used as an antidepressive agent (Kramer et al., 1998; Varty et al., 2003); it has also been demonstrated that this antagonist has antitumoural activity against human cancer cell lines (Muñoz et al., 2004b, 2005b); and it has been reported that other NK-1 receptor antagonist named Aprepitant (MK-869) (an agent structurally unrelated to L-733,060 with antidepressant activity) is as efficient as the antidepressant drug paroxetine in the treatment of depression. It is also known that Aprepitant is well tolerated and no statistically significant difference in the frequency of adverse events compared to placebo is observed.

As we have previously indicated, both SP and NK-1 receptors are widely distributed in the mammalian central nervous system, including the limbic system (hypothalamus, amygdala...). SP may be involved in the integration of emotional responses to stress, suggesting the possibility that the pathogenesis of depression could be due to an alteration of the SP/NK-1 receptor system, since increased SP levels have been observed in depression (Kramer et al., 1998). Taking all these data together, it seems that depression could induce tumour cell proliferation by activating the SP/NK-1 receptor system and that treatment with NK-1 receptor antagonists could be useful not only for the depression, but also for the treatment of tumour cells. Thus, emotional behaviour (e.g., depression...), cancer and metastasis might be related through alterations in the SP/NK-1 receptor system (Fehder, 1999; De Vane, 2001; Lang et al., 2004). In sum, it seems that through this system an interrelationship between the progression of cancer and cerebral mechanisms could occur, since SP after binding to NK-1 receptors located in the limbic system produces anxiety and depression, whereas SP after binding to the same receptors located in the tumour cells induces cell proliferation. In this sense, NK-1 receptor antagonists in the limbic system exert antidepressant and anxiolytic actions, while such antagonists have an antitumoural action on tumour cells.

Future Perspectives: The Nk-1 Receptor as a New Target in the Treatment of Cancer

NK-1 Receptor Antagonists

As we have previously mentioned, NK-1 receptor antagonists can be used in the treatment of several diseases. Thus, these antagonists could be used in several inflammatory diseases including arthritis, inflammatory bowel disease and cystitis. It is also known that NK-1 receptor antagonists produce analgesia, antidepressive, anxiolytic and antiemetic effects (see Quartara and Maggi, 1998). These antagonists can be divided into peptide antagonists and non-peptide antagonists. The former (Snider et al., 1990; Almeida et al., 2004) suffer from a number of drawbacks, such as partial residual agonist activity, mast cell degranulating activity, poor potency and ability to discriminate between tachykinin receptors, and neurotoxicity after administration in the central nervous system (Lee et al., 1986; see Quartara and Maggi, 1997). Moreover, metabolic instability and the inability to gain access to the central nervous system through the blood-brain barrier of peptide NK-1 receptor antagonists limits their usefulness for in vivo studies

However, the non-peptide antagonists are used in the treatment of a broad range of disorders (Maggi et al., 1993; Harrison et al., 1994; Regoli et al., 1994; see Quartara and Maggi, 1997; Rupniak et al., 2000). Benzylether piperidines (L-733,060, L-741,671, L-742,694...), perhydroisoindolones (RP-67,580, RP-73,467, RPR-100,893...), steroids (WIN-51,708...), benzylamino piperidines (CP-99,994, GR-203,040, GR-205,171, CP-122,721...), benzylamino and benzylether quinuclidine (CP-96,345, L-709,210...), and tryptophan-based (L-732,138, L-737,488...) belong to the non-peptide NK-1 receptor antagonists family (see Quartara and Maggi, 1997). At present, there are more than thirty compounds that act as NK-1 receptor antagonists (Giardina et al., 2003). For example, L-733,060 shows a high affinity for the human NK-1 receptor in vitro and produces analgesia and antidepressive effects; it has been used in the treatment of a broad range of anxiety and mood disorders and inflammatory liver disease (L-733,060 has been successfully used in preclinical assays at 10-20 mg/kg doses (approximately 30-60 μM) as anxiolytic and antiinflammatory agent), most likely by inhibiting the effects of SP; it has been described that L-733,060 inhibits the metastatic progression of NK-1 receptor-expressing cancer cells; and it has recently been used as an antitumoural agent against melanoma, glioma, neuroblastoma, retinoblastoma and pancreas carcinoma cell lines (Muñoz et al., 2004a, b; 2005a, b, 2006a).

Other non-peptide NK-1 receptor antagonist is the compound named L-732,138 (N-acetyl-L-tryptophan 3, 5-bis (trifluoromethyl) benzyl ester) (see Cascieri et al., 1994; MacLeod et al., 1994). This NK-1 receptor antagonist produces an attenuation of hyperalgesia (Cahill and Coderre, 2002); antagonizes H_3 antagonist-induced skin vascular permeability (Hossen et al., 2003); and exerts an antitumoural action against glioma and neuroblastoma cell lines (Muñoz et al., 2006 b, 2007).

Finally, the drug Aprepitant is a selective high-affinity antagonist of human NK-1 receptor. In radio-ligand binding assays, Aprepitant was approximately 3,000-fold selective for the human cloned NK-1 receptor (IC_{50} = 0.1 nM) versus the human cloned NK-3 receptor

(IC_{50} = 300 nM) and 45,000-fold selective over the human cloned NK-2 receptor (IC_{50} = 4500 nM) (Tattersall et al., 2000).

Antitumoural Action of NK-1 Receptor Antagonists

As we have above-mentioned, it has been recently demonstrated that L-733,060 displays antitumoural activity against the human SKN-BE(2) neuroblastoma, GAMG glioma, COLO 679, COLO 858, MEL HO melanoma, WERI-Rb-1, Y-79 retinoblastoma and CAPAN-1 and PA-TU 8902 pancreas carcinoma cell lines (Muñoz et al., 2004a, b, 2005a, b, 2006a); that inhibits the metastatic progression of NK-1 receptor-expressed cancer cells, and that L-733,060 binds specifically to the human NK-1 receptor (Lang et al., 2004), whereas the NK-1 receptor antagonist L-732,138 also elicited antitumoural action on human GAMG glioma and KELLY and IMR-32 neuroblastoma cell lines (Muñoz et al., 2006b, 2007). In addition, it has been reported that the use of SP antagonists other than L-733,060/L-732,138 inhibited the growth of small cell lung cancer, glioma, neuroblastoma and breast cancer cells (Woll and Rozengurt, 1988; Langdon et al., 1992; Reeve and Bleehen, 1994; Seckl et al., 1997; Palma et al., 2000; Bigioni et al., 2005; Mukerji, et al., 2005).

Moreover, there are data indicating that the antitumoural action of the NK-1 receptor antagonists is via the NK-1 receptor. Thus, in competition experiments, exogenous SP human cancer cell lines proliferation (glioma, neuroblastoma, melanoma, retinoblastoma, pancreas carcinoma) was partially reversed by the administration of both L-733,060 and L-732,138 (Muñoz et al., 2004a, b, 2005a, b, 2006a, b, 2007). This means that these NK-1 receptor antagonists block the NK-1 receptors expressed by these human cancer cell lines. Moreover, it has been reported that both L-733,060/L-732,138 induce growth inhibition by apoptosis in human A-204 rabdhomyosarcoma and in SKN-BE(2), KELLY and IMR-32 neuroblastoma cell lines (Esteban et al., 2006; Muñoz et al., 2007), being this observation in agreement with other reports, in which an SP antagonist (other than L-733,060 or L-732,138) induces apoptosis in lung cancer and causes a concentration-dependent loss of cell viability (Reeve and Bleehen 1994; Rosati et al., 1998). These data indicate that the cell death observed in the glioma, neuroblastoma, melanoma, retinoblastoma, rabdhomyosarcoma and pancreas carcinoma cell lines after treatment with NK-1 receptor antagonists was not due to a general toxic effect of L-733,060/L-732,138. Moreover, it has been reported that GAMG glioma, SKN-BE(2) neuroblastoma, CAPAN-1 and PA-TU 8902 pancreas carcinoma, KELLY and IMR-32 neuroblastoma tumour cells express the NK-1 receptor (Muñoz et al., 2005a, 2006a, 2007) and that the antitumoural action of the L-733,060/L-732,138 NK-1 receptor antagonists is dose-dependent (Muñoz et al., 2005a, b 2006a, b, 2007). We have to note that the NK-1 receptor antagonists L-733,060 and L-732,138 are very structurally different designer molecules that bind specifically to the NK-1 receptor, with a Ki value of 0.8 mM in the case of L-733,060 and of 2.3 nM for L-732,138.

In sum, these data suggest that the antiproliferative action of the NK-1 receptor antagonists L-733,060/L-732,138 could involve a signal transduction pathway for apoptosis, after the antagonist is bound to the NK-1 receptors located in different tumour cell lines. This means that a possible common mechanism for cancer cell proliferation mediated by SP and the NK-1 receptor occurs and that NK-1 receptor antagonists could inhibit a large number of tumour cell types in which both SP and NK-1 receptors are expressed (Tarkkanen et al., 1983; Hennig et al., 1995; Khare et al., 1998; Palma et al., 2000; Singh et al., 2000; Friess et al., 2003; Muñoz et al., 2005a, 2006 a, 2007; Schultz et al., 2006).

It is well known that SP activates malignant glial cells to induce cytokine release and proliferation, being both responses relevant for tumour progression. The role of SP in supporting glioma progression in vivo was highlighted by the tumour growth inhibition induced by highly specific and selective human tachykinin NK-1 receptor antagonists (MEN 11467 and MEN 11149). The antitumour activity of MEN 11467 was observed in both subcutaneous and intravenous treatments and it was partially reverted by the concomitant administration of exogenous SP. A novel therapeutic approach to treat malignant gliomas with tachykinin NK-1 receptor antagonists is suggested by these findings (see Palma et al., 2000). Furthermore, antitumoural activity of NK-1 and NK-2 receptor antagonists was demonstrated in nude mice, measuring growth inhibition of MDA-MB-231 tumour cells xenografted subcutaneously. A significant inhibition was found when compounds were administered intravenously (5 mg/kg a day for 2 weeks). These results suggest that the in vivo activity of NK-1 and NK-2 antagonists may be a result of a cytostatic effect rather than a cytotoxic effect (Bigioni et al., 2005).

It has also been reported, that SP induces a metastatogenic tumour cell type by directly regulating gene expression and by increasing migratory activity, which can be prevented by established neurotransmitter antagonists. Thus, the migration of the MDA-MB-468 breast cancer cells can be inhibited by using NK-1 receptor antagonists (Lang et al., 2004).

It is known that SP analogue antagonists (SPA) have potential antitumourigenic activities. Thus, SPA antagonists inhibited multiple neuropeptide-induced Ca^{2+} mobilization, DNA synthesis, and growth on ductal pancreatic cancers in vitro. SPA antagonists also significantly attenuated the growth of HPAF-II tumour xenografts in nude mice; increased markedly apoptosis but moderately decreased proliferation marker; reduced markedly tumour-associated angiogenesis in the HPAF-II xenografts in vivo; and attenuate tumour growth in pancreatic cancer via a dual mechanism involving both the antiproliferative and antiangiogenic properties (Guha et al., 2005). Thus, the NK-1 receptor target could bee used for inhibition of angiogenesis, since it is known that SP stimulates vessel growth enhancing endothelial cell proliferation and that in vivo angiogenesis is mimicked by selective NK-1 receptor agonists and inhibited by NK-1 receptor antagonists (Ziche et al., 1990b).

In summary, the NK-1 receptor antagonists exert an antitumoural action by three mechanisms: 1) Antiproliferative effect by inhibiting tumour cell growth and inducing cellular death by apoptosis; 2) Inhibiting migration of tumour cells (metastasis); and 3) Inhibiting angiogenesis in the mass tumour.

Conclusion

In clinical practice, the non specific action of the cytostatic drugs show a very low safety profile, and very severe side effects are seen (anaemia, leukopenia...). Thereby, the research accent has always been on the search for a drug with the same or greater antitumoural potential, but with fewer side effects. This could only be achieved if drugs act specifically against tumour cells through a specific target.

The data mentioned in previous sections indicate that the NK-1 receptor could be a new and promising target in the treatment of human cancer and that NK-1 receptor antagonists could improve cancer treatment, for the following reasons:

1. NK-1 receptors and substance P are overexpressed in numerous human tumours.
2. Human cancer cell lines express several isoforms of the NK-1 receptor.
3. In human cancer cells SP is a mitogen, an angiogenic and a metastatic agent.
4. Non-peptide NK-1 receptor antagonists such as L-733,060 and L-732,138 exert an antitumoural activity on several human cancer cell lines.
5. NK-1 receptor antagonists induce apoptosis in human tumour cell lines.
6. NK-1 receptor antagonists inhibit both the angiogenesis in the mass tumour and the migration of the tumour cells (metastasis).

The action of NK-1 receptor antagonists is dose-dependent, and depending on the concentration has more positive effects: 1) antiemetic action in situations of untractable vomiting by cytostatics, 2) antidepressive and anxiolytic action, 3) central analgesic, 4) potent antiinflammatory action, 5) antitumoural activity (inhibition of cellular proliferation, inhibition of the production of metastasis, inhibition of peritumour inflammatory reaction and angiogenesis). When used at the highest doses (antitumour or antiinflammatory), NK-1 receptor antagonists produce all the positive pharmacological effects described above and, in addition, they bind specifically to NK-1 receptors and they are selective against tumour cells (these cells express more NK-1 receptors than the non tumour cells). NK-1 receptor visualization with an immunohistochemical method will facilitate identification of tumours with a sufficient receptor overexpression for diagnostic or therapeutic intervention using NK-1 receptor antagonists. In addition, the action of the non-peptide NK-1 receptor antagonists is long-lasting (they are not hydrolysed by peptidases), they can be administered orally, they cross the blood-brain barrier and exert a potent antitumoural action even in melanoma metastasis. Thus, all mentionated above strongly support the use of NK-1 receptor antagonists as an innovative approach to antitumour therapy and confirm the potential use of these antagonists as therapeutic tools.

References

Almeida, T; Rojo, J; Nieto, PM. Tachykinins and tachykinin receptor: structure and activity relationships. *Curr Med Chem* 2004; 11, 763-771.

Ames, BN; Gold, LS. Too many rodent carcinogens: mitogenesis increases mutagenesis. *Science* 1990; 249, 970-971.

Bang, R; Sass, G; Kiemer, AK; Vollmar, AM; Neuhuber, WL; Tiegs, G. Neurokinin-1 receptor antagonists CP-96,345 and L-733,060 protect mice from cytokine-mediated liver injury. *J Pharmacol Exp Ther* 2003; 305, 31-39.

Berger, A; Santic, R; Hauser-Kronberger, C; Schilling, FH; Kogner, P; Ratschek, M; Gamper, A; Jones, N; Sperl, W; Kofler, B. Galanin and galanin receptors in human cancers. *Neuropeptides* 2005; 39, 353-359.

Bigioni, M; Benzo, A; Irrissuto, C; Maggi, CA; Goso, C. Role of NK-1 and NK-2 tachykinin receptor antagonism on the growth of human breast carcinoma cell line MDA-MB-231. *Anticancer Drugs* 2005; 16, 1083-1809.

Blaser, MJ; Chyou, PH; Nomura, A. Age at establishment of *Helicobacter pylori* infection and gastric carcinoma, gastric ulcer and duodenal ulcer risk. *Cancer Res* 1995; 55, 562-565.

Bordey, A; Feltz, P; Trouslard, J. Mobilization of intracellular calcium by substance P in a human astrocytoma cell line (U-373 MG). *Glia* 1994; 11, 277-283.

Burgos, C; Aguirre, JA; Alonso, JR; Coveñas, R. Immunocytochemical study of substance P-like fibres and cell bodies in the cat diencephalon. *J Hirnforsch* 1988; 29, 651-657

Caberlotto, L; Hurd, YL; Murdock, P; Wahlin, JP; Melotto, S; Corsi, M; Carletti, R. Neurokinin 1 receptor and relative abundance of the short and long isoforms in the human brain. *Eur J Neurosci* 2003; 17, 1736-1746.

Cahill, CM; Coderre, TJ. Attenuation of hyperalgesia in a rat model of neuropathic pain after intrathecal pre- or post-treatment with a neurokinin-1 antagonist. *Pain* 2002; 95, 277-285.

Cascieri, MA; MacLeod, AM; Underwood, D; Shiao, LL; Ber, E; Sadowski, S; Yu, H; Merchant, KJ; Swain, CJ; Strader, CD; Fong, TM. Characterization of the interaction of N-acyl-L-tryptophan benzyl ester neurokinin antagonists with the human neurokinin-1 receptor. *J Biol Chem* 1994; 269, 6587-6591.

Chahl, LA; Ladd, RJ. Local oedema and general excitation of cutaneous sensory receptors produced by electrical stimulation of the saphenous nerve in the rat. *Pain* 1976; 2, 25-34.

Cook, GA; Elliott, D; Metwali, A; Blu, AM; Sandor, M; Lynch, R; Weinstock, JV. Molecular evidence that granuloma T lymphocytes in murine *Schistosomiasis mansoni* express an authentic SP (NK1) receptor. *J Immunol* 1994; 152, 1830-1835.

Cuello, AC; Kanazawa, I. The distribution of substance P immunoreactive fibres in the rat central nervous system. *J Comp Neurol* 1978; 178, 129-156.

Dam, TV; Quirion, R. Pharmacological characterization and autoradiographic characterization of substance P receptors in guinea-pig brain. *Peptides* 1986; 7, 855-864.

Danks, JA; Rothman, RB; Cascieri, MA; Chicchi, GG; Liang, T; Herkenham, M. A comparative autoradiographic study of the distributions of substance P and eledoisin binding sites in rat brain. *Brain Res* 1986; 385, 273-281.

De Vane, CL. Substance P: a new era, a new role. *Pharmacotherapy.* 2001; 21, 1061-1069.

Eistetter, HR; Mills, A; Brewster, R; Alouani, S; Rambosson, C; Kawashima, E. Functional characterization of neurokinin-1 receptors on human U373MG astrocytoma cells. *Glia* 1992; 6, 89-95.

Entschladen, F; Lang, K; Drell, TL; Joseph, J; Zaenker, KS. Neurotransmitters are regulators for the migration of tumor cells and leukocytes. *Cancer Immunol Immunother* 2002; 51, 467-482.

Esteban, F; Muñoz, M; González-Moles, MA; Rosso, M. A role for substance P in cancer promotion and progression: a mechanism to counteract intracellular death signals following oncogene activation or DNA damage. *Cancer Metast Rev* 2006; 25, 137-145.

Fehder, WP. Alterations in immune response associated with anxiety in surgical patients. *CNRA* 1999; 10, 124-129.

Feistritzer, C; Clausen, J; Sturn, DH; Djanani, A; Gunsilius, E; Wiedermann, CJ; Kahler, CM. Natural killer cell functions mediated by the neuropeptide substance P. *Regul Pep* 2003; 116, 119-126.

Fowler, CJ; Brannstrom, G. Substance P enhances forskolin-stimulated cyclic AMP production in human UC11MG astrocytoma cells. *Methods Find Exp Clin Pharmacol* 1994; 16, 21-28.

Friess, H; Zhu, Z; Liard, V; Shi, X; Shrikhande, SV; Wang, L; Lieb, K; Korc, M; Palma, C; Zimmermann, A; Reubi, JC; Buchler, MW. Neurokinin-1 receptor expression and its potential effects on tumor growth in human pancreatic cancer. *Lab Invest* 2003; 83, 731-742.

Furness, J; Papka, RE; Della, NG; Costa, M; Eskay, RL. Substance P-like immunoreactivity in nerves associated with the vascular system of guinea-pigs. *Neuroscience* 1982; 7, 447-459.

García, M; Sakamoto, K; Shigekawa, M; Nakanishi, S; Ito, S. Multiple mechanisms of arachidonic acid release in chinese hamster ovary cells transfected with cDNA of SP receptor. *Biochem Pharmacol* 1994; 48, 1735-1741.

Giardina, GAM; Gagliardi, S; Martinelli, M. Antagonists at the neurokinin receptors-Recent patent literature. *Drugs* 2003; 6, 758-772.

Guha, S; Eibl, G; Kisfalvi, K; Fan, RS; Burdick, M; Reber, H; Hines, OJ; Strieter, R; Rozengurt, E. Broad-spectrum G protein–coupled receptor antagonist, [D-Arg1, D-Trp5,7,9,Leu11]SP: a dual inhibitor of growth and angiogenesis in pancreatic cancer. *Cancer Res* 2005; 65, 2738-2745.

Harrison, S; Geppetti, P. Substance P. *Int J Biochem Cell Biol* 2001; 33, 555-576.

Harrison, T; Williams, BJ; Swain, CJ; Ball, RG. Piperidine-ether based hNK1 antagonists 1: determination of the relative and absolute stereochemical requirements. *Biomed Chem Lett* 1994; 4, 2545-2550.

Hennig, IM; Laissue, JA; Horisberger, U; Reubi, JC. Substance-P receptors in human primary neoplasms: tumoral and vascular localization. *Int J Cancer* 1995; 61, 786-792.

Hershey, AD; Polenzani, L; Woodward, RM; Miledi, R; Krause, JE. Molecular and genetic characterization, functional expression, and mRNA expression patterns of a rat substance P receptor. *Ann NY Acad Sci* 1991; 632, 63-78.

Hilakivi-Clarke, L; Rowland, J; Clarke, R; Lippman, ME. Psychosocial factors in the development and progression of breast cancer. *Breast Cancer Res Treat* 1994; 29, 141-160.

Hökfelt, T; Pernow, B; Wahren, J. Substance P: a pioneer amongst neuropeptides. *J Intern Med* 2001; 249, 27-40.

Hökfelt, T; Pernow, B; Nilsson, G; Wetterberg, L; Goldstein, M; Jeffcoate, SL. Dense plexus of substance P-immunoreactive nerve terminals in eminentia medialis of the primate hypothalamus. *Proc Natl Acad Sci USA* 1978; 74, 1013-1015.

Holzer, P. Local effector functions of capsaicin-sensitive sensory nerve endings: Involvement of tachykinins, calcitonin gene-related peptide and other neuropeptides. *Neuroscience* 1988; 24, 739- 768.

Hossen, MA; Fujii, Y; Sugimoto, Y; Kayasuga, R; Kamei, C. Histamine H_3 receptors regulate vascular permeability changes in the skin of mast cell-deficient mice. *Int Immunopharmacol* 2003; 3, 1563-1568.

Humpel, C; Knaus, GA; Auer, B; Knaus, HG; Haring, C; Theodorsson, E; Saria, A. Effects of haloperidol and clozapine on preprotachykinin mRNA tachykinin tissue level release and NK1 receptors in the striatonigral system. *Synapse* 1990; 6, 1-9.

Johnson, CL; Johnson, CG. Substance P regulation of glutamate and cystine transport in human astrocytoma cells. *Receptor Channel* 1993; 1, 53-59.

Johnson, CL; Johnson, CG; Stauderman, KA; Buck, SH. Characterization of substance P receptors in human astrocytoma cells. *Ann NY Acad Sci* 1991; 632, 410-412.

Keranen, U; Jarvinen, H; Karkkaine, P; Kiviluoto, T; Kivilaakso, E; Soinila, S. Substance P – an underlying factor for pouchitis?. Prospective study of substance P and VIP-immunoreactive innervation and mast cells. *Dig Dis Sci* 1996a; 41, 1665-1671.

Keranen, U; Jarvinen, H; Kiviluoto, T; Kivilaakso, E; Soinila, S. Substance P and VIP-immunoreactive innervation in normal and inflamed pouches after restorative proctocolectomy for ulcerative colitis. *Dig Dis Sci* 1996b; 41, 1658-1664.

Khare, VK; Albino, AP; Reed, JA. The neuropeptide/mast cell secretagogue substance P is expressed in cutaneus melanocytic lesions. *J Cutan Pathol* 1998; 25, 2-10.

Kramer, MS; Cutler, N; Feighner, J; Shrivastava, R; Carman, J; Sramek, JJ; Reines, SA; Liu, G; Snavely, D; Wyatt-knowles, E; Halle, JJ; Mills, SG; MacCoss, M; Swain, CJ; Harrison, T; Hill, RG; Hefti, F; Scolnick, EM; Cascieri, MA; Chicchi, GG; Sadowski, S; Williams, AR; Hewson, L; Smith, D; Carlsson, EJ; Hargreaves, RJ; Rupniak, NM. Distinct mechanism for antidepressant activity by blockade of central substance P receptors. *Science* 1998; 281, 1640-1645.

Lang, K; Drell, TL; Lindecke, A; Niggemann, B; Kaltschmidt, C; Zaenker, KS; Entschladen, F. Induction of a metastatogenic tumor cell type by neurotransmitters and its pharmacological inhibition by established drugs. *Int J Cancer* 2004; 112, 231–238.

Langdon, S; Sethi, T; Richie, A; Muir, M; Smyth, J; Rozengurt, E. Broad spectrum neuropeptide antagonists inhibit the growth of small cell lung cancer in vivo. *Cancer Res* 1992; 52, 4554-4557.

Lee, CM; Campbell, NJ; Williams, BJ; Iversen, LL. Molecular characterization of a functional cDNA for rat receptor. *Eur J Pharmacol* 1986; 130, 209-217.

Lee, CM; Tung, WL; Young, JD. Tachykinin-stimulated inositol phospholipid hydrolysis and taurine release from human astrocytoma cells. *J Neurochem* 1992; 59, 406-414.

Lembeck, F. The 1988 Ulf Euler Lecture. Substance P: from extract to excitement. *Acta Physiol Scand* 1988; 133, 435-454.

Lembeck, F; Holzer, P. Substance P as neurogenic mediator of antidromic vasodilation and neurogenic plasma extravasation. Naunyn-Schmiedeberg's *Arch Pharmacol* 1979; 310, 175-183.

Lieb, K; Fiebich, BL; Berger, M; Bauer, J; Schulze-Osthoff, K. The neuropeptide substance P activates transcription factors NF.kB and kB-dependent gene expression in human astrocytoma cells. *J Immunol* 1997; 159, 4952-4958.

Lotz, M; Carson, DA; Vaughan, JH. Substance P activation of rheumatoid synoviocytes: neural pathway in pathogenesis of arthritis. *Science* 1987; 235, 893-895.

Lowenfels, AB; Maisonneuve, P; Cavallini, G; Ammann, RW; Lankisch, PG; Andersen, JR; Dimagno, EP; Andren-Sandberg, A; Domellof, L. Pancreatitis and the risk of pancreatic cancer. *N Engl J Med* 1993; 328, 1433-1437.

Lundberg, JM; Brodin, E; Hua, X; Saria, A. Vascular permeability changes and smooth muscle contraction in relation to capsaicin-sensitive substance P afferents in the guinea-pig. *Acta Physiol Scand* 1984; 120, 217-227.

Luo, W; Sharif, TR; Sharif, M. Substance P-induced mitogenesis in human astrocytoma cells correlates with activation of the mitogen-activated protein kinase signaling pathway. *Cancer Res* 1996; 56, 4983-4991.

Macarthur, M; Hold, GL; El-Omar, EM. Inflammation and cancer II. Role of chronic inflammation and cytokine gene polymorphisms in the pathogenesis of gastrointestinal malignancy. *Am. J. Physiol. Gastrointest Liver Physiol* 2004; 286, G515-G520.

MacDonald, SG; Boyd, ND. Regulation of substance P receptor affinity by guanine nucleotide binding proteins. *J Neurochem* 1989; 53, 264-272.

MacLeod, AM; Merchant, KJ; Brookfield, F; Kelleher, F; Stevenson, G; Owens, AP; Swain, CJ; Cascieri, MA; Sadowski, S; Ber, E; Strader, CD; Fong, TM. Identification of L-tryptophan derivatives with potent and selective antagonist activity at the NK1 receptor. *J Med Chem* 1994; 37, 1269- 1274.

Maeno, H; Kiyama, H; Tohyama, M. Distribution of the substance P (NK1) receptor in the central nervous system. *Mol Brain Res* 1993; 18, 43-58.

Maggi, CA; Meli, A. The sensory-efferent function of capsaicin-sensitive sensory neurons. *Gen Pharmacol* 1988; 19, 1-43.

Maggi, CA; Patacchini, R; Rovero, P; Giachetti, A. Tachykinin receptors and receptor antagonists. *J Auton Pharmacol* 1993; 13, 958-962.

Mai, JK; Stephens, PH; Hopf, A; Cuello, AC. Substance P in the human brain. *Neuroscience* 1986; 17, 709-739.

Mantyh, PW. Substance P and the inflammatory and immune response. *Ann NY Acad Sci* 1991; 632, 263-271.

McGillis, JP; Mitsuhashi, M; Payan, DG. Immunomodulation by tachykinin neuropeptides. *Ann NY Acad Sci* 1990 ; 594, 85-94.

Medrano, S; Gruenstein, E; Dimlich, RV. Sunstance P receptor on human astrocytoma cells are linked to glycogen breakdown. *Neurosci Lett* 1994; 167, 14-18.

Mitsuhashi, M; Ohashi, Y; Schichijo, S; Christian, C; Sudduth-Klinger, J; Harrowe, G; Payan, DG. Multiple intracellular signalling pathways of the neuropeptide SP receptor. *J Neurosci Res* 1992; 32, 437-443.

Mochizuki-Oda, N; Nakajima, Y; Nakanishi, S; Ito, S. Characterization of the SP receptor-mediated Ca influx in cDNA transfected CHO cells. *J Biol Chem* 1994; 269, 9651-9658.

Mukerji, I; Ramkissoon, SH; Reddy, KK; Rameshwar, P. Autocrine proliferation of neuroblastoma cells is partly mediated through neurokinin receptors: relevance to bone marrow metastasis. *J Neurooncol* 2005; 71, 91-98.

Muñoz, M; Pérez, A; Coveñas, R; Rosso, M; Castro E. Antitumoural action of L-733,060 on neuroblastoma and glioma cell lines. *Arch Ital Biol* 2004a; 142, 105- 112.

Muñoz, M; Pérez, A; Rosso, M; Zamarriego, C; Rosso, R. Antitumoural action of NK1 receptor antagonist L-733,060 on human melanoma cell lines. *Melanoma Res* 2004b; 14, 183-188.

Muñoz, M; Rosso, M; Pérez, A; Coveñas, R; Rosso, R; Zamarriego, C; Piruat, JI. The NK1 receptor is involved in the antitumoural action of L-733,060 and in the mitogenic action of substance P on neuroblastoma and glioma cell lines. *Neuropeptides* 2005a; 39, 427-432.

Muñoz, M; Rosso, M; Pérez, A; Coveñas, R; Rosso, R; Zamarriego, C; Soult, JA; Montero, I. Antitumoural action of the neurokinin-1- receptor antagonist L-733,060 and mitogenic action of substance P on human retinoblastoma cell lines. *Invest Ophthalmol Vis Sci* 2005b; 46, 2567-2570.

Muñoz, M; Rosso, M; Coveñas, R. The NK-1 receptor is involved in the antitumoural action of L-733,060 and in the mitogenic action of substance P on human pancreatic cancer cell lines. *Lett Drug Des Discovery* 2006a; 3, 323-329.

Muñoz M; Rosso M; Soult JA; Coveñas R. Antitumoural action of neurokinin-1 receptor antagonists on human brain cancer cell lines. In: Yang AV, editor. *Brain Cancer: Therapy and surgical intervention.* New York: Nova Science Publishers; 2006b; pp. 45-75.

Muñoz, M; Rosso, M; Coveñas, R. NK-1 receptor antagonists as new anti-tumoural agents: action on human neuroblastoma cell lines. In: Fernandes JA, editor. *Focus on Neuroblastoma Research.* New York: Nova Science Publishers; 2007; (in press).

Nakajima, Y; Tsuchida, K; Negishi, M; Ito, S; Nakanishi, S. Direct linkage of three tachykinin receptors to stimulation of both phosphatidylinositol hydrolysis and cAMP cascades in transfected chinese hamster ovary cells. *J Biol Chem* 1992; 267, 2437-2442.

Nakata, Y; Hiraoka, C; Segawa, T. Apparent molecular weight of the substance P binding site in rat brain. *Eur J Pharmacol* 1988; 152, 171-174.

Nakaya, Y; Kaneko, T; Shigemoto, R; Nakanishi, S; Mizuno, N. Immunohistochemical localization of substance P receptor in the central nervous system of the adult rat. *J Comp Neurol* 1994; 347, 249-274.

Nilsson, J; von Euler, AM; Dalsgaard, CJ. Stimulation of connective tissue cell growth by substance P and substance K. *Nature* 1985; 315, 61-63.

Nowicki, M; Miskowiak, B. Comparison of the cell immunophenotype of metastatic and primary foci in stage IV-S neuroblastoma. *Folia Histochem Cytobiol* 2002; 40, 297-303.

O'Connor, TM; O'Connell, J; O'Brien, DI; Goode, T; Bredin, CP; Shanahan, F. The role of substance P in inflammatory disease. *J Cell Physiol* 2004; 201, 167-180.

Okamura, M; Yamawaki, S; Akechi, T; Taniguchi, K; Uchitomi, Y. Psychiatric disorders following first breast cancer recurrence: prevalence, associated factors and relationship to quality of life. *Jpn J Clin Oncol* 2005; 35, 302-309.

Organist, ML; Harvey, JP; McGillis, JP; Payan, DG. Processing of the human IM-9 lymphoblast substance P receptor. Biosynthetic and degradation studies using a monoclonal anti-receptor antibody. *Biochem Biophys Res Commun* 1988; 151, 535-541.

Oury-Donat, F; Lefevre, IA; Thurneyssen, O; Gauthier, T; Bordey, A; Feltz, P; Edmonds-Alt, X; Le Fur, G; Soubrie, P. SR 140333, a novel, selective, and potent nonpeptide antagonist of the NK1 tachykinin receptor: characterization on the U373 MG cell line. *J Neurochem* 1994; 62, 1399-1407.

Palma, C. Tachykinins and their receptors in human malignancies. *Curr Drug Targets* 2006; 7, 1043-1052.

Palma, C; Manzini, S. Substance P induces secretion of immunomodulatory cytokines by human astrocytoma cells. *J Neuroimmunol* 1998; 81, 127-137.

Palma, C; Bigioni, M; Irrissuto, C; Nardelli, F; Maggi, CA; Manzini, S. Anti-tumour activity of tachykinin NK1 receptor antagonists on human glioma U373 MG xenograft. *Br J Cancer* 2000; 82, 480-487.

Palma, C; Nardelli, F; Manzini, S; Maggi, CA. Substance P activates responses correlated with tumour growth in human glioma cells line bearing tachykinin NK1 receptors. *Br J Cancer* 1999; 79, 236-243.

Pascual, DW; Xu-Amano, J; Kiyono H; McGhee, JR; Bost, KL. Substance P acts directly upon cloned B Lymphoma cells to enhace IgA and IgM production. *J Immunol* 1991; 146, 2130-2136.

Payan, DG; Brewster, DR; Missirian-Bastian, A; Goetzl, EJ. Substance P recognition of a subset of human T lymphocytes. *J Clin Investigation* 1984; 74, 1532-1539.

Payan, DG; Brewster, DR; Goetzl, EJ. Specific stimulation of human T lymphocytes by substance P. *J Immunol* 1983; 131, 1613-1615.

Payan, DG; McGillis, JP; Organist, ML. Binding characteristics and affinity labeling of protein constituents of the human IM-9 lymphoblast receptor for substance P. *J Biol Chem* 1986; 261, 14321-14329.

Pernow, B. Substance P. *Pharmacol Rev* 1983; 35, 85-141.

Quartara, L; Maggi, CA. The tachykinin NK1 receptor. Part I: ligands and mechanisms of cellular activation. *Neuropeptides* 1997; 31, 537-563.

Quartara, L; Maggi, CA. The tachykinin NK1 receptor. Part II: distribution and pathophysiological roles. *Neuropeptides* 1998; 32, 1-49.

Reeve, JG; Bleehen, NM. [D-Arg^{1}-D-Phe^{5}, D-Trp^{7,9}, Leu^{11}] Substance P induces apoptosis in lung cancer cell lines in vitro. *Biochem Bioph Res Com* 1994; 199, 1313-1319.

Regoli, D; Boudon, A; Fauchere, J. Receptors and antagonists for substance P and related peptides. *Pharmacol Rev* 1994; 46, 551-599.

Reubi, JC. Peptide receptors as molecular targets for cancer diagnosis and therapy. *Endocr Rev* 2003; 24, 389-427.

Rosati, R; Adil, MR; Ali, MA; Eliason, J; Orosz, A; Sebestye, NF; Kalemkerian, GP. Induction of apoptosis by a short-chain neuropeptide analog in small cell lung cancer-stimulation of multiple neuropeptides and inhibition by broad spectrum antagonists in vitro and in vivo. *Peptides* 1998; 19, 1519-1523.

Rupniak, NM; Carlson, E; Boyce, S; Webb, JK; Hill, RG. Enantioselective inhibition of the formalin paw late phase by the NK1 receptor antagonist L-733,060 in gerbils. *Pain* 1996; 67, 189-195.

Rupniak, NM; Carlson, EC; Harrison, T; Oates, B; Seward, E; Owen, S; de Felipe, C; Hunt, S; Wheeldon, A. Pharmacological blockade or genetic deletion of substance P (NK(1)) receptors attenuates neonatal vocalisation in guinea-pigs and mice. *Neuropharmacology* 2000; 39, 1413-1421.

Saffroy, M; Beaujouan, JC; Torrens, Y; Besseyre, J; Bergstrom, L; Glowinski, J. Localization of tachykinin binding sites (NK1, NK2, NK3 ligands) in the rat brain. *Peptides* 1988; 9, 227-241.

Samsam, M; Coveñas, R; Ahangari, R; Yajeya, J; Narváez, JA; Tramu, G. Simultaneous depletion of neurokinin A, substance P and calcitonin gene-related peptide from the caudal trigeminal nucleus of the rat during electrical stimulation of the trigeminal ganglion. *Pain* 2000; 84, 389-395.

Schulz, S; Stumm, R; Röcken, C; Mawrin, C; Schulz, S. Immunolocalization of Full-length NK1 Tachykinin Receptors in Human Tumors. *J Histochem Cytochem* 2006; 54, 1015-1020.

Seabrook, GR; Fong, TM. Thapsigargin blocks the mobilization of intracellular Ca caused by activation of human NK1 (long) receptors expressed in CHO cells. *Neurosci Lett* 1993; 152, 9-12.

Seckl, MJ; Higgins, T; Wildmer, F; Rozengurt, E. [D-Arg^{1}, D-Trp^{5,7,9}, Leu^{11}] Substance P: a novel potent inhibitor of signal transduction and growth in vitro and in vivo in small cell lung cancer cells. *Cancer Res* 1997; 57, 51-54.

Sharif, TR; Luo, W; Houghron, PJ; Sharif, M. Substance K peptide induces mitogenesis by activating the mitogen-activated protein kinase signalling pathway through the substance P receptor (NK-1 subtype) in human astrocytoma. *Cell Pharmacol* 1996; 3, 441-449.

Shibata, K; Haverstick, DM; Bannon, MJ. Tachykinin gene expression in rat limbic nuclei: modulation by dopamine antagonists. *J Pharmacol Exp Ther* 1990; 255, 388-392.

Shirayama, Y; Mitsushio, H; Takashima, M; Ichikawa, H; Takayashi, K. Reduction of substance P after chronic antidepressant treatment in the striatum, substantia nigra and amygdala of the rat. *Brain Res* 1996; 739, 70-78.

Shrestha, P; Yang, L; Liu, B; Namba, M; Takagi, H; Hosaka, M; Mori, M. Neuroblastoma of parotid gland: report of a case and immunohistochemical characteristics. *Eur J Cancer B Oral Oncol* 1994; 30, 356-361.

Shrikhande, SV; Friess, H; di Mola, FF; Tempia-Caliera, A; Conejo-García, JR; Zhu, ZW; Zimmermann, A; Büchler, MW. NK-1 receptor gene expression is related to pain in chronic pancreatitis. *Pain* 2001; 91, 209-217.

Singh, D; Joshi, DD; Hameed, M; Qian, J; Gascon, P; Maloof, PB; Mosenthal, A; Rameshwar, P. Increased expression of preprotachykinin-I and neurokinin receptors in human breast cancer cells: implications for bone marrow metastasis. *Proc Natl Acad Sci USA* 2000; 97, 388-393.

Sivam, SP; Krause, JE; Takeuchi, K; Li, S; McGinty, J; Hong, JS. Lithium increases rat striatal β- and γ-preprotachykinin messenger RNAs. *J Pharmacol Exp Ther* 1989; 248, 1297-1301.

Snider, RM; Constantine, JW; Lowwe, IJ. A potent nonpeptide antagonist of the substance P (NK1) receptor. *Science* 1991; 251, 435-437.

Spiegel, D; Giese-Davis, J. Depression and cancer: mechanisms and disease progression. *Biol Psychiatry* 2003; 54, 269-282.

Sporn, MB. The war on cancer. *Lancet* 1996; 347, 1377-1381.

Takeda, Y; Blount, P; Sachais, BS; Hershey, AD; Raddatz, R; Krause, JE. Ligand binding kinetics of substance P and neurokinin A receptors stably expressed in CHO cells and evidence for differential stimulation of inositol 1, 4, 5-triphosphate and camp second messenger responses. *J Neurochem* 1992; 59, 740-745.

Tarkkanen, A; Tervo, T; Tervo, K; Eranko, L; Eranko, O; Cuello, AC. Substance P immunoreactivity in normal human retina and in retinoblastoma. *Ophthalmic Res* 1983; 15, 300-306.

Tattersall, FD; Rycroft, W; Cumberbatch, M; Mason, G; Tye, S; Williamson, DJ; Hale, JJ; Mills, SG; Finke, PE; MacCoss, M; Sadowski, S; Ber, E; Cascieri, M; Hill, RG; MacIntyre, DE; Hargreaves RJ. The novel NK-1 receptor antagonist MK-0869 (L-754,030) and its water soluble phosphoryl prodrug, L-758,298, inhibit acute and delayed cisplatin-induced emesis in ferrets. *Neuropharmacology* 2000; 39, 652-663.

Tong, C; Fazio, M; Williams, GM. Cell cycle-specific mutagenesis at the hypoxanthine phosphoribosyltransferase locus in adult rat liver epithelial cells. *Proc Natl Acad Sci USA* 1980; 77, 7377-7379.

Torrens, Y; Beaujouan, JC; Saffroy, M; Glowinski, J; Tence, M. Functional coupling of the NK1 tachykinin receptor to phospholipase D in Chinese hamster ovary cells ans astrocytoma cells. *J Neurochem* 1998; 70, 2091-2098.

Turner, AJ. Endopeptidase-24.11 and neuropeptide metabolism. In: Turner AJ, editor. *Neuropeptides and Their Peptidases*. Chichester: Ellis Horwood; 1987; pp. 183-201.

Unger, T; Rascher, W; Schuster, C; Pavlovitch, R; Schömig, A; Dietz, R; Ganten, D. Central blood pressure effects of substance P and angiotensin II: role of the sympathetic nervous system and vasopressin. *Eur J Pharmacol* 1981; 71, 33-42.

van Ginkel, FW; Pascual, DW. Recognition of neurokinin 1 receptor (NK1-R): an antibody to a peptide sequence from the third extracellular region binds to brain NK1-R. *J Neuroimmunol* 1996; 67, 49-58.

Varty, GB; Cohen-Williams, ME; Hunter, JC. The antidepressant-like effects of neurokinin NK1 receptor antagonists in a gerbil tail suspension test. *Behav Pharmacol* 2003; 14, 87-95.

von Euler, US ; Gaddum, JH. An unidentified depressor substance in certain tissue extracts. *J Physiol* 1931; 72, 74-87.

Walsh, DT; Weg, VB; Williams, TJ; Nourshargh, S. Substance P induced inflammatory responses in guinea-pig skin: the effect of specific NK1 receptor antagonists and the role of endogenous mediators. *Br J Pharmacol* 1995; 114, 1343-1350.

Weinstock, JV; Blum, A; Walden, J; Walden, T. Eosinophils from granulomas in murine *Schistosomiasis mansoni* produce SP. *J Immunol* 1988; 141, 961-966.

Weitzman, SA; Gordon, LJ. Inflammation and cancer: role of phagocyte-generated oxidants in carcinogenesis. *Blood* 1990; 76, 655-663.

Whitty, CJ; Walker, PD; Goebel, DJ; Poosch, MS; Bannon, MJ. Quantitation, cellular localization and regulation of neurokinin receptor gene expression whithin the rat substantia nigra. *Neuroscience* 1995; 64, 419-425.

Wolf, SS; Moody, TW; Quirion, R; O'Donahue, TL. Biochemical characterization and autoradiographic localization of central SP receptor using [^{125}I] physalaemin. *Brain Res* 1985; 332, 299-307.

Woll, PJ; Rozengurt E. [D-Arg^{1}-D-Phe^{5}, D-Trp^{7,9}, Leu^{11}] Substance P, a potent bombesin antagonist in murine Swiss 3T3 cells, inhibits the growth of human small cell lung cancer cells in vitro. *Proc Natl Acad Sci USA* 1988; 35, 1859-1863.

Ziche, M; Morbidelli, L; Pacini, M; Dolara, P; Maggi, CA. NK1-receptors mediate the proliferative response of human fibroblasts to tachykinins. *Br J Pharmacol* 1990a; 100,11-4.

Ziche, M; Morbidelli, L; Pacini, M; Gepetti, P; Alessandri, G; Maggi CA. Substance P stimulates neovascularization in vivo and proliferation of cultured endothelial cells. *Microvasc Res* 1990b; 40, 264-278.

In: Neuropeptide Research Trends
Editor: Bernice A. Levine, pp. 121-141
ISBN: 978-1-60021-640-4

Chapter IV

Possible Role of Orexigenic Neuropeptides in the Reservoir Function of the Stomach

Motoi Kobashi[*]
Department of Oral Physiology, Okayama University Graduate School of Medicine, Dentistry and Pharmaceutical Sciences, Okayama 700-8525, Japan

Abstract

Recently, several neuropeptides have been discovered as the ligand for orphan receptors in the brain, some of which are appetite-regulating peptides that induce satiety or hunger. Since feeding is closely related to digestive function, the effects of these peptides on gastric acid secretion, bile secretion and gastric motility are well investigated. Orexigenic (appetite-enhancing) peptides generally facilitate gastric acid secretion and also facilitate phasic contractions of the distal stomach, which enhances the ability to churn gastric contents to quicken gastric emptying. These gastric functions are particularly important for digestion. In addition, accommodation of food is also important for smooth digestion. The proximal region of the stomach, which serves as a reservoir, relaxes to accommodate swallowed food. Reflex relaxations, induced by the activation of esophageal and/or pharyngo-laryngeal receptors before food reaches the stomach, are well known as so-called "receptive relaxation"; however, the relationship between feeding-related peptides and gastric relaxation of the proximal stomach has never been clarified. Gastric relaxation is achieved by the intramural nervous system and extrinsic nervous system. The vagus nerves, whose cell bodies are in the dorsal motor nucleus of the vagus (DMV), have a major role in the extrinsic control of gastric motility. Neural circuits in the dorsal vagal complex (DVC), which consists of the area postrema, the nucleus of the solitary tract and the DMV, form output signals for regulating gastric motor activities; therefore, the DVC must be the main target of feeding-related peptides

[*] Corresponding author: Motoi Kobashi Ph.D.; Department of Oral Physiology, Okayama University Graduate School of Medicine, Dentistry and Pharmaceutical Sciences, Shikata-cho 2-5-1, Okayama 700-8525, Japan; Tel: +81-86-235-6641; Fax: +81-86-235-6641; E-mail: mkobashi@md.okayama-u.ac.jp

to regulate gastric motility. Recently, our studies revealed that some orexigenic neuropeptides induced gastric relaxation of the proximal stomach by way of the DVC. The administration of orexin-A into the fourth ventricle induced gastric relaxation of the proximal stomach, as well as enhanced gastric contractility of the distal stomach. Neuropeptide Y (NPY), which is a classical but substantial orexigenic neuropeptide, also induced relaxation in the proximal stomach via Y1 receptors (Y1R) situated in the DVC. Since several studies revealed that the Y1R subtype plays a prominent role in mediating feeding induced by NPY, gastric relaxation induced by NPY might be closely associated with food intake. Ghrelin, which was originally isolated from the stomach and has orexigenic effects, induced gastric relaxation of the proximal stomach. These observations strongly suggest that central orexigenic neuropeptides facilitate accommodation by the stomach of a meal without an increase in intragastric pressure. Fasting is usually followed by feeding, therefore, gastric relaxation achieved by the actions of orexigenic neuropeptides, which are released by hunger, to enhance the easy accommodation of food is rational. In this report it is hypothesizes that orexigenic neuropeptides induce anticipatory gastric relaxation to compensate for hyperphagia. Centrally released neuropeptides might cooperate with the acute reflex relaxation just after swallowing to facilitate reservoir function of the stomach.

1. Introduction

Food intake is essential for animals to live. Energy homeostasis is achieved by using peripheral and central nervous functions. It is well established that the hypothalamus plays a principal role in the regulation of food intake. Peripheral and/or central neural information and humoral information reach the nuclei in the hypothalamus. The regulatory mechanism of food intake by both the ventromedial hypothalamic nucleus (VMH) and the lateral hypothalamic area (LHA) has long been accepted. According to the classical dual-center model, the VMH and LHA have a reciprocal role for food intake based on the gluco-static theory (Anand et al., 1961; Oomura et al., 1964; Oomura et al., 1969). Electrical stimulation of the VMH and those of the LHA resulted in the cessation and acceleration of eating, respectively, whereas lesions of these nuclei induced opposite responses. Therefore, the VMH and LHA are known as the satiety center and feeding center, respectively. The VMH has glucose receptor neurons which excite in response to the administration of glucose. Conversely, the LHA has glucose-sensitive neurons which are suppressed in response to the administration of glucose; therefore, glucose is considered the most important mediator to transmit feeding-related information to the brain.

Recently, a new concept of the hypothalamic control system of feeding has been considered after identification of a number of feeding-related neuropeptides in the hypothalamus (Kawano et al., 2002). Orexigenic neuropeptides, which enhance food intake, include melanin-concentrating hormone (MCH), orexin, neuropeptide Y (NPY), agouti-related peptide (AgRP) and ghrelin. The anorectic neuropeptides, which reduce food intake, include alfa-melanocyte-stimulating hormone (alpha-MSH), cocaine- and amphetamine-regulated transcript (CART), thyrotropin-releasing hormone (TRH), corticotropin-releasing hormone (CRH) and cholecystokinin (CCK). The key peptide, which transmits peripheral energy balance information to the brain, is leptin. Leptin is an anorectic hormone and was discovered in white adipose tissue. Circulating leptin relays the condition of peripheral energy balance to NPY- and alpha-MSH-containing neurons in the arcuate nucleus of the hypothalamus (ARC), as well as orexin- and MCH-containing neurons in the LHA (Schwartz et al., 2000; Williams et al., 2001). Thus, the classical dual-center model consisting of VMH and LHA is now replaced by a new scheme showing a complex network consisting of plural feeding-related neuropeptides.

Ingested food is considered as a disturbance of homeostasis. Food intake changes the internal environment such as the digestive tract, chemical composition of blood, metabolism and so on. Animals use several strategies to adapt themselves to these changes. Roughly speaking, there are two types of regulatory mechanisms: feedback regulation and feed-forward (or anticipatory) regulation. Feedback regulation is precise, whereas feed-forward regulation is rough but rapid. When the magnitude of disturbance is known beforehand, feed-forward regulation is very useful to reduce the fluctuation of physiological values. To elucidate these two types of mechanisms, regulation of the blood glucose level is often mentioned. Food intake increases blood glucose concentration. Glucose facilitates insulin release from the pancreas, and then the glucose concentration returns to the normal level. This type of regulation is considered the feedback type. Oral sensation also induced insulin release. Rapid insulin release after food intake was observed using free-moving rats (Steffens, 1976). The insulin level rose in the first minute after the start of food intake, although the glucose level began to increase only in the third minute after food ingestion. Since the insulin response was triggered before the increase in blood glucose level, this type of regulation is considered the feed-forward type. Oral sensation must play an important role in regulating insulin release. Feed-forward regulation is very useful to avoid a rapid increase in glucose concentration. Animals skillfully use two types of regulation to adapt to various changes in the internal environment caused by food intake. The gastric motor responses caused by feeding-related neuropeptides are possibly included in feed-forward regulation.

The motor function of the alimentary tract is closely related with food intake, and is controlled by humoral and/or neural factors. The extrinsic nervous system, in addition to the intramural nervous system, is an important factor to regulate alimentary tract motility. The central nervous system strongly contributes to gastric motor function, since the stomach has rich innervation of extrinsic nerves. The stomach is, to consider motor function, divided into two parts, the proximal and distal stomach. Phasic contractions appearing in the distal stomach after ingesting food are to churn gastric contents to mix with gastric juice and to facilitate gastric emptying. The proximal stomach usually relaxes, associated with swallowing, to facilitate reservoir function. These gastric motor responses are achieved by both the intramural and/or extrinsic nervous system. The reflex relaxation observed after food intake is divided into two types. The first is adaptive relaxation, caused when swallowed food reaches the stomach then mechanically stimulates the gastric wall (Abrahamsson, 1973); therefore, adaptive relaxation is classified as feedback regulation. The other is so-called "receptive relaxation" (Cannon and Lieb, 1911), which is elicited by swallowing. Animals with a gastric fistula also show this relaxation just after swallowing. Mechanoreceptive afferents from the pharynx and esophagus, and chemoreceptive afferents from the pharyngo-laryngeal region induced this reflex relaxation (Abrahamsson and Jansson, 1969; Rogers et al., 1999; Kobashi et al., 2000). Thus, receptive relaxation occurrs before food reaches the stomach; therefore, receptive relaxation is classified into feed-forward regulation.

It is well-known that orexigenic (appetite-enhancing) neuropeptides contribute to gastric motor functions. The effects on gastric acid secretion and peristaltic movement of the upper digestive tract have been well investigated just after the discovery of new orexigenic neuropeptides. Orexigenic neuropeptides usually enhance gastric acid secretion and contractile activity. These responses are rational to accelerate digestion; however, few studies have investigated the relationship between orexigenic neuropeptides and gastric relaxation. We are now investigating how orexigenic neuropeptides participate in gastric relaxation and have found that central orexigenic neuropeptides seem to enhance gastric relaxation. Central neuropeptides might cooperate with acute reflex relaxation just after swallowing to facilitate the accommodation of a meal by the stomach.

This chapter deals with gastric motor response in relation to central feeding-related neuropeptides. Although many feeding related-peptides are present in both the periphery and central nervous system, only the central effect of neuropeptides on gastric motility is included in this chapter. Background knowledge of the extrinsic neural mechanism to achieve gastric motor response was initially presented, now continuing to how central orexigenic neuropeptides work to regulate gastric motor activity, especially for relaxation responses, showing our recent data and those of other studies. Furthermore, the relationship between anorectic substances and gastric motor activity is mentioned.

2. Extrinsic Control of Gastric Motility and the Neural Mechanism

Gastric motility is controlled by the autonomic nervous system in addition to the intramural nervous system. Generally speaking, the parasympathetic nerve facilitates gastric motility and the sympathetic nerve suppresses it. The parasympathetic nerve predominantly works to control gastric motility. The suppressive effects of the sympathetic nerve on gastric motility diminished after dissecting parasympathetic nerve, suggesting that the sympathetic nerve inhibits parasympathetic neural activity to suppress gastric motor activity (Roman and Gonella, 1987). Thus, the parasympathetic nerves play an essential role in the extrinsic control of gastric motility. The cell bodies of parasympathetic preganglionic nerves (vagus nerve) are situated in the dorsal motor nucleus of the vagus (DMV) in the dorsal medulla. The DMV and their vicinity have an essential role in regulating gastric motility.

The dorsal vagal complex (DVC), which consists of the DMV, the nucleus of the solitary tract (NTS) and the area postrema (AP), processes peripheral afferent information, descending signals from higher center and humoral factors. The NTS receives visceral afferent inputs such as gastric baroreceptor, duodenal chemoreceptor, hepatic chemoreceptor and so on (Adachi et al., 1984; Kobashi and Adachi, 1985), and their afferent signals regulate gastric motility (Kobashi et al., 1998; Sakaguchi and Shimojyo, 1984). NTS neurons relay vagal afferent to a higher center, including the hypothalamus, directly and indirectly (Kobashi and Adachi, 1986, 1988). Conversely, the DVC receive inputs from the hypothalamus (Nishimura and Oomura, 1987); therefore, it is possible that hypothalamic neurons that contain orexigenic neuropeptides affect the neural activities of DVC neurons. DVC neurons are closely associated with food intake itself and the feeding-related responses. Microinjection into the DVC of ghrelin, an orexinergic neuropeptide, enhanced food intake (Faulconbridge et al., 2003). The AP, which has a loose blood-brain barrier, is called the chemoreceptor trigger zone to induce vomiting (Borison, 1989), and is also related with the intake of palatable food (Edwards and Ritter, 1981). Thus, the DVC is closely related to feeding behavior; therefore, it is plausible that the DVC is suitable for regulating the gastric motor response induced by food intake.

Phasic contraction in the distal stomach is enhanced by the excitation of parasympathetic preganglionic neurons. The consensus is that neuroeffector transmission from the excitatory vagal pathway is cholinergic via muscarinic type 3 receptors (Kajimura et al., 1997), and is consequently blocked by atropine or hyoscine. On the other hand, the neural mechanism of relaxation is rather complex. Early work, using cats, revealed two types of vagal preganglionic neurons, classified based on the response to esophageal inflation, which caused gastric relaxation (Miolan and Roman, 1978): one (VEF; vagal excitatory neuron) showed inhibition and the other (VIF; vagal inhibitory neurons) showed excitation during esophageal inflation (figure 1). As much of the tonic input to the stomach is excitatory, the ablation of tonic inputs to the stomach causes the inhibition of gastric tone. Indeed, the reduction of gastric tone induced by the antral (or esophageal) stretch receptor or superior laryngeal afferents appears to be caused by the inhibition of vagal preganglionic neurons (McCann and Rogers, 1992; Rogers et al., 1999; Kobashi, 2002). In addition, nonadrenergic and noncholinergic (NANC) myenteric neurons, activated by the excitation of vagal preganglionic neurons, induce the inhibition of gastric motility (Abrahamsson and Jansson, 1973). Thus, the inhibition of gastric tone can occur because of the inhibition of VEF connected to cholinergic myenteric neurons and/or the excitation of VIF connected to NANC myenteric neurons (figure 1).

The candidate neurotransmitters for NANC transmission are vasointestinal polypeptides (VIP) and/or nitric oxide (NO) (Goyal and Ratttan. 1980; Krowicki and Hornby, 1996). After the identification of NO in endothelial cells, the relationship between NO and gastric relaxation were well investigated. An inhibitor of nitric oxide synthase disrupted adaptive relaxation in the isolated rat stomach (Desai et al., 1991). The gastric relaxation induced by electrical stimulation of vagal efferent nerves was also inhibited by the administration of an inhibitor of NO synthase in intact rats (Lefebvre et al., 1992). Thus, NO certainly has an important role in rapid reflex relaxation, that is, adaptive and receptive relaxation. Furthermore, NO might mediate transmission in the parasympathetic ganglia. Although the transmission between pre- and postganglionic neurons ordinarily mediates nicotinic receptors, anatomical studies revealed that DMV neurons which project to the proximal stomach have a nitric oxide synthase (NOS). It is possible that VIFs transmit information using NO in the autonomic ganglia (Krowicki et al., 1997; Zheng et al., 1999). Anatomical and physiological studies also suggested that the caudal DMV (posterior to the obex) is closely related with the relaxation of the proximal stomach (Krowicki et al., 1997; Zheng et al., 1999; Rogers et al., 1999). Thus, the neural circuits in the caudal DVC have a prominent role in gastric relaxation. A recent study revealed that prolactin-releasing neuropeptide affected this sited to induce gastric relaxation (Grabauskas et al., 2004); therefore, orexigenic neuropeptides might affect the caudal DVC to induce gastric relaxation. A possible neural wiring diagram to induce gastric relaxation is shown in figure 1.

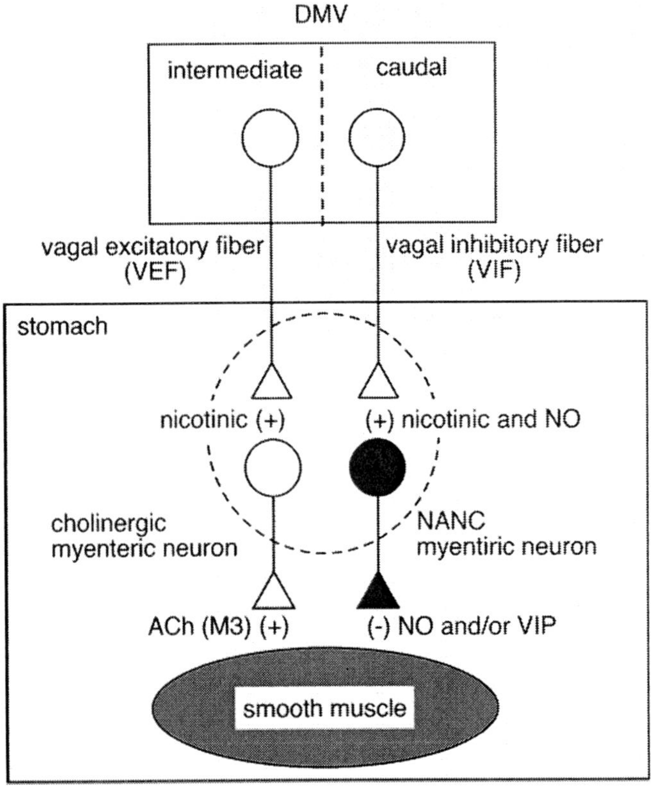

Figure 1. Neural wiring diagram to induce gastric relaxation. Two types of parasympathetic (vagal) preganglionic neurons, whose cell bodies are situated in the dorsal motor nucleus of the vagus (DMV), participate in gastric relaxation: vagal excitatory fibers (VEF), which connect with excitatory cholinergic myenteric neurons, and vagal inhibitory fibers (VIF), which connect with inhibitory nonadrenergic-noncholinergic (NANC) myenteric neurons (Miolan & Roman, 1978). Inhibition of the VEF reduces vagal tone, thereby causing a reduction in intragastric pressure (IGP). Excitation of the VIF facilitates NANC transmission between myenteric neurons and smooth muscle. The cell body of the VEF is situated in the DMV just rostral to the obex (intermediate part). The cell body of the VIF is situated in the DMV caudal to the obex (caudal part) (Krowicki et al., 1997; Zheng et al., 1999; Rogers et al., 1999).

3. Orexigenic Neuropeptides and Gastric Motility

3.1. Orexin-A

Orexins (orexin-A and orexin-B) are neuropeptides that were identified as endogenous ligands for an orphan G-protein-coupled receptor (Sakurai et al., 1998; de Leccea et al., 1998). Orexin-containing neurons are specifically localized in the LHA and their axons project widely throughout the brain to suggest their participation in several body regulations. Since the LHA is feeding center, the relationship between orexin and food intake are considered at their identification (Mondal et al., 2000). A certain part of the orexigenic effect of orexin is thought to be mediated by the release of NPY, which is a classical potent orexigenic neuropeptide produced in the arcuate nucleus of the hypothalamus (ARC) (Yamanaka et al., 2000). Retrograde tracing experiments revealed that as many as 20% of hypothalamic orexin neurons project to the DVC (Zheng et al., 2005). Two orexin receptors, orexin receptor 1 and orexin receptor 2, have been identified. These receptors display different affinities for the two orexin peptides. Specifically, orexin receptor 1 binds orexin-A and orexin-B with similar affinity, whereas orexin receptor 1 is more responsive to orexin-A than orexin-B (Sakurai et al., 1998; Okumura et al., 2001). Orexin-A is more potent to induce feeding, drinking and gastric acid secretion than orexin-B (Edward, 1999; Kunii et al., 1999; Takahashi et al., 1999), suggesting that orexin-A is closely associated with the regulation of food intake and gastric function.

We examined the effects of orexin-A on gastric motility using urethane-chloralose anaesthetized rats. We separately recorded the intragastric pressure (IGP) of the proximal and distal stomach through a balloon inserted into each region (figure 4A). Orexin-A was administered into the fourth ventricle to affect neurons in the DVC. The dose (3 nmol, 3 μl) used in this study is comparable with that to induce feeding and gastric acid secretion (Edward, 1999; Kunii et al., 1999; Takahashi et al., 1999). Consequently, the administration of orexin-A but not a vehicle induced relaxation of the proximal stomach lasting for more than 30 min (figure 4Ca). Phasic contractions in the distal stomach were facilitated in response to the administration of orexin-A but not a vehicle (figure 4Cb). Facilitation in the distal stomach was not observed in animals which underwent sectioning of the bilateral vagi at the subdiaphragmatic level. Relaxation of the proximal stomach was observed in vagotomized animals but the magnitude of relaxation was significantly smaller than that in intact animals. These results suggest that central orexin facilitates distal stomach motility and relaxation of the proximal stomach via vagal preganglionic neurons (Kobashi et al, 2002).

Krowicki et al. also investigated the gastric motor response in response to the adminisration of orexins (Krowicki et al., 2002). They used a rather precise method to administer orexins into the dorsal medulla, applying a small amount of orexins (1 - 10 pmol, 10 nl) to the DMV using multi-barrel micropipettes. Excitatory amino acid, L-glutamate was injected beforehand to identify the regional specificity to the gastric motor response, and then orexins were injected into the same site. When orexin-A was administered into the DMV rostral to the obex, where L-glutamate increased contractility, the motility of the antrum showed facilitation. This observation is coincident with our result. In contrast, the microinjection of orexin-A into the DMV caudal to the obex, where L-glutamate evokes gastric relaxation, did not significantly alter gastric motor function, that is, gastric relaxation is not observed in response to the administration of orexin-A. This result seems to differ from that obtained in our study. We cannot clarify why microinjection of orexin-A did not induce gastric relaxation. Considering that orexin receptors are found in the NTS (Marcus et al., 2001), and orexin-containing fibers terminate in the NTS (Harrison et al., 1999), it is possible that orexin-A may act on NTS neurons, which in turn affect DMV neurons; therefore, direct injection of orexin-A into the DMV did not induce gastric relaxation.

The gastric motor responses induced by orexin-administration are peculiar compared with receptive relaxation induced by swallowing. Receptive relaxation is transient (duration of response is a few minutes). The response appeared in the proximal stomach and that appearing in the distal stomach showed a similar tendency during receptive relaxation, that is, the proximal stomach relaxed (decreased in IGP) and distal stomach contractility is suppressed (figure 4Ba and 4Bb). Receptive relaxation works to permit a bolus of food to fall into the stomach smoothly. On the other hand, the orexin response is long-lasting. The responses in the proximal and distal stomach are opposite to receptive relaxation, but are reasonable. The role of distal stomach contractility is to churn gastric contents to mix with gastric juice and to facilitate gastric emptying. In fact, intracerebroventricular administration of orexin-A facilitates gastric emptying (Furudono et al., 2006). Relaxation of the proximal stomach facilitates reservoir function, that is, the proximal stomach contains food without increasing IGP. It is inferred that these responses are common effects of orexingenic neuropeptides on gastric motility.

3.2. Neuropeptide Y (NPY)

NPY is a peptide of 36 amino acids isolated from porcine brain (Takemoto, 1982). The potent orexigenic effects of NPY are well known (Clark et al., 1984, 1987). NPYergic transmissions are considered an essential component of the common final pathway in the hypothalamic integration of energy homeostasis. Kalra et al., citing a large number of studies, described NPY as the only messenger molecule for a physiological appetite transducer in the brain (Kalra et al., 1999). Several studies revealed that the Y1 receptor subtype (Y1R) is the most prominent factor in mediating feeding induced by NPY (Kalra SP et al., 1991a; Kalra PS et al., 1998; Xu et al., 1998). Food intake induced by MCH was attenuated by the antagonistic effects of Y1R (Chaffer and Morris, 2002). Neurons containing other orexigenic neuropeptides, such as orexin or ghrelin, send information to NPY-containing neurons (Funahashi et al., 2003). Thus, NPY is a key peptide in the regulation of body energy homeostasis.

In addition to the role of NPY in feeding behavior, NPY is involved in several abdominal functions. Centrally administered NPY regulated gastric acid secretion (Matsuda et al., 1991; Humphreys et al., 1992; Penner et al., 1993; Geoghegan et al., 1994), and stimulated bile secretion via the Y1R (Yoneda et al., 2001). Microinjection of NPY into the dorsal vagal complex (DVC) enhanced basal motility of the antrum via Y1R (Chen et al., 1997). Central administered NPY delayed gastric emptying via Y2 receptors (Ishiguchi et al., 2001a), and inhibited gastric distension-induced pyloric relaxation via Y1R (Ishiguchi et al., 2001b); thus, NPY contributes to the control of digestion via upper abdominal functions. However, the relationship between NPY and gastric relaxation of the proximal stomach, which serves as a reservoir, had never been clarified. Since NPY affects the DVC neurons, which receive afferent information from the stomach (Schwartz and Moran, 2002) and send efferent information through axons to the gastrointestinal tract (Browning and Travagli, 2003), it is possible that NPY induces gastric relaxation to accommodate ingested food; therefore, we conducted research to demonstrate the role of NPY in gastric relaxation of the proximal stomach to facilitate reservoir function (Kobashi et al., 2006).

The results of the research showed that the administration of NPY into the fourth ventricle induced relaxation of the proximal stomach in a dose-dependent manner (0.3, 1.0 and 3.0 nmol). YIR must mediate the NPY response if the gastric response is closely associated with food intake. In fact, the administration of an Y1R agonist [Leu^{31}, Pro^{34}] NPY induced a greater relaxation than NPY. On the other hand, the administration of a Y2 receptor agonist (NPY 13 - 36) did not induce significant changes in motility. These results suggest that gastric relaxation in the proximal stomach induced by NPY administration correlates with food intake.

As described in section 2, activation of caudal DMV neurons induced gastric relaxation (Krowicki et al., 1997; Zheng et al., 1999). In our study, microinjections of [Leu^{31}, Pro^{34}] NPY (6 pmol, 60 nL) into the caudal part of the DVC induced relaxation of the proximal stomach (figure 2). In addition, the administration of NPY into the fourth ventricle did not induce relaxation after bilateral injections of a Y1R antagonist (1229U91) into the caudal DVC. These results indicate that NPY induces relaxation in the proximal stomach via Y1R situated in the caudal DVC. In contrast, similar injections into the intermediate part of the DVC (rostral to the obex) increased the IGP of the proximal stomach (figure 2). Previous studies showed that the injection of L-glutamate, orexins or NPY facilitated atral motility and increased IGP (Chen et al., 1997; Krowicki et al., 2002); therefore, a similar tendency was observed in the proximal stomach. We are currently unable to give an adequate explanation for the meaning of the regional differences of the DVC. It is possible that proximal stomach relaxation induced by the fourth ventricular administration of NPY is caused by the activation of caudal DVC neurons.

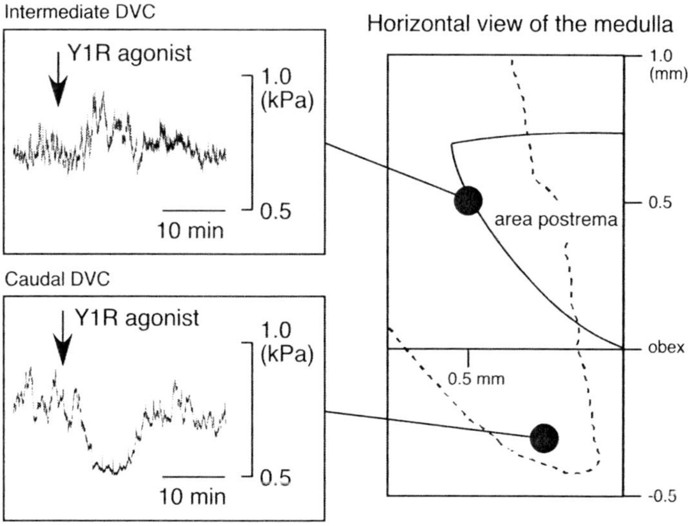

Figure 2. Regional specificity of the response induced by microinjection of the Y1 receptor agonist, [Leu^{31}, Pro^{34}] NPY, into the dorsal vagal complex (DVC) (Kobashi, et al., 2006). The administration of [Leu^{31}, Pro^{34}] NPY into the caudal part of the DVC induced relaxation of the proximal stomach (lower left). The administration of [Leu^{31}, Pro^{34}] NPY into the intermediate part of the DVC, conversely, increased intragastric pressure of the proximal stomach (upper left). The right panel show the horizontal view of the caudal medulla. Filled circles show injection sites of [Leu^{31}, Pro^{34}] NPY. The broken line shows the border of the DMV. These results revealed that fourth ventricular administration of NPY induced proximal stomach relaxation via Y1 receptors situated in the caudal DVC.

The NST sends a descending projection to the sympathetic preganglionic neurons in the spinal cord, and also projects into several sympathetic premotor nuclei including the rostral and caudal ventrolateral areas (Saper, 1995); therefore, a sympathetic contribution to the NPY response in the present research should be considered. However, bilateral sectioning of the subdiaphragmatic vagi completely abolished the NPY response of the stomach in the present study. The proximal stomach relaxation observed in our study was probably achieved by parasympathetic preganglionic neurons. Furthermore, atropine methyl nitrate, a peripherally acting muscarinic cholinergic receptor antagonist (Rauhofer et al., 1996), attenuated the magnitude of relaxation induced by the administration of NPY. The activation of Y1R decreased IGP of the proximal stomach, inhibiting cholinergic vagal efferent outflow. An electrophysiological study of *in vitro* slice preparations showed that excitatory postsynaptic currents (EPSCs) of DMV neurons induced by electrical stimulation of the NST was partly inhibited by presynaptic actions of NPY via the Y1R (Browning and Travagli, 2003). This result corresponds with our observations, that is, activation of the Y1R reduced the activity of vagal excitatory preganglionic neurons that connect with cholinergic myenteric neurons. Figure 3 shows the possible neural circuit to achieve NPY-induced gastric relaxation.

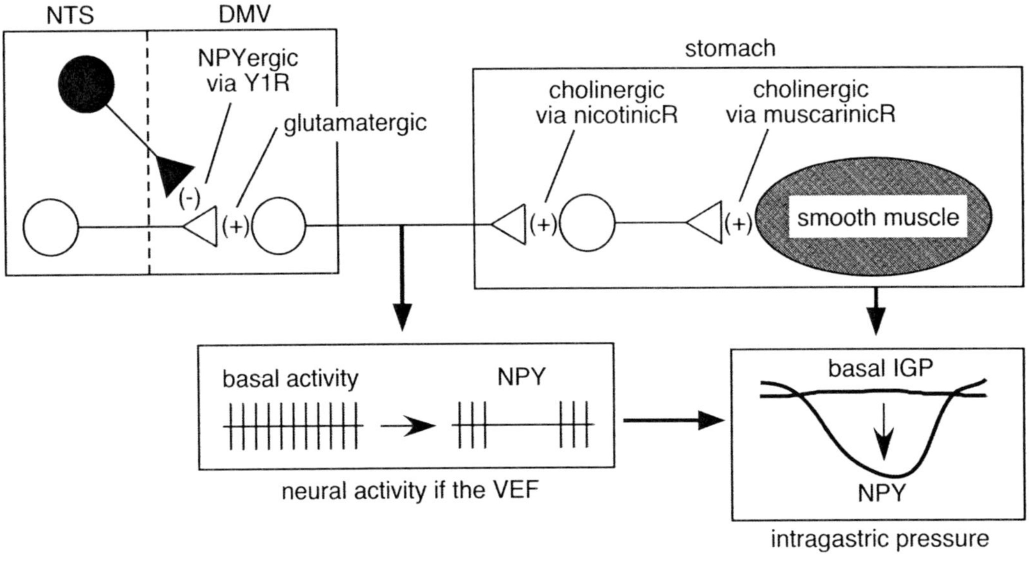

Figure 3.Possible neural mechanisms underlying the gastric relaxation induced by NPY. The main effect of NPY in the DVC is to inhibit glutamatergic synaptic transmission between the nucleus of the solitary tract (NTS) and the DMV. NPY exerts its presynaptic inhibitory action on glutamate currents via mainly the Y1 receptor (Browing & Travagli, 2003). The administration of atropine blocks the relaxation of the proximal stomach induced by the fourth ventricular administration of NPY (Kobashi et al., 2006). NPY might reduce the neural activity of VEF to induce gastric relaxation.

It is well known that orexigenic neuropeptides in the hypothalamus mediate NPY release to achieve food intake (Funahashi et al., 2003). A recent behavioral study revealed that the food intake facilitated by the fourth ventricular administration of ghrelin was attenuated by pretreatment with an NPY antagonist (Faulconbridge et al., 2005); therefore, NPY might mediate the effect caused by other orexigenic neuropeptides in the DVC.

3.3. Ghrelin

Ghrelin is a recently identified endogenous ligand for growth hormone secretagogue receptor (GHS-R), and was originally isolated from the stomach (Kojima et al., 1999). In addition to GH-releasing activity, ghrelin has an orexigenic effect. Ghrelin is also shown to be present in the hypothalamus, and participates in the regulation of food intake through the peripheral and central nervous system (Nakazato et al., 2001; Date et al., 2002). Ghrelin, which is secreted from the stomach, inhibits the gastric vagal afferent neurons, and then facilitates food intake (Date et al., 2002). The central effect of ghrelin seems to mediate glucose receptor neurons and glucose-sensitive neurons (Chen et al., 2005). Ghrelin excites glucose sensitive neurons in the LHA, which is known as a feeding center, and inhibits glucose receptor neurons in the VMH and the parvocellular region of the paraventricular nucleus of the hypothalamus (pPVN), which is known as a satiety center. Central ghrelin also stimulates NPY-containing neurons and orexin-containing neurons (Funahashi et al., 2003). A rather complex relationship between feeding-related neuropeptides and the neural substrate is considered.

It is interesting whether ghrelin induces gastric relaxation as well as orexin and NPY. Intravenous administration of ghrelin facilitates gastric acid secretion and gastric motility in rats (Masuda et al., 2000). Central and peripheral administration of ghrelin induced fasted motor activity of the gastrointestinal tract in rats (Fujino et al., 2003). Histological study confirmed GHS-R mRNA expression in several hypothalamic nuclei, many of which have long been recognized as playing roles in body weight and food intake in rats and mice (Zigman et al., 2006). In addition, GHS-R mRNA has also been found in the DVC, including the AP, NTS and DMV. Immunohistochemical study revealed that GHS-R was expressed in the NTS and DMV in rats (Lin et al., 2004). Microinjection of ghrelin into the DVC facilitated food intake (Faulconbridge et al., 2003). It is plausible that ghrelin induces gastric relaxation as well as orexin and NPY.

We therefore examined the effect of ghrelin on motility of the proximal stomach in anaesthetized rats (Kobashi et al., 2007 abstract). The administration of ghrelin into the fourth ventricle induced relaxation of the proximal stomach in a dose-dependent manner. Significant reduction in IGP was observed at doses of 3, 10 or 30 pmol. Simultaneous administration of ghrelin and GHS-R antagonist ([D-Lys^3] GHRP-6) did not induce a significant change in IGP. The administration of [D-Lys^3] GHRP-6 did not induce significant changes in IGP. Microinjections of ghrelin into the DVC also induced relaxation of the proximal stomach. These results indicate that ghrelin induces relaxation of the proximal stomach via GHS-R situated in the DVC. Since both bilateral sectioning of the vagi at the cervical level abolished the relaxation induced by the administration of ghrelin into the fourth ventricle, the relaxation induced by ghrelin is mediated by vagal preganglionic neurons. These results indicated that central ghrelin facilitates reservoir function of the proximal stomach as well as orexin and NPY. Furthermore, we confirmed that the fourth ventricular administration of ghrelin inhibits reflex swallowing elicited by electrical stimulation of the superior laryngeal nerve (Xuan et al., 2007, abstract). Ghrelin might regulate extensive autonomic phenomena associated with food intake.

3.4. Melanin-Concentrating Hormone (MCH)

It is described above that orexin, NPY and ghrelin induced relaxation of the proximal stomach to facilitate reservoir function of the stomach. MCH, as well as orexins, is mainly produced in LHA neurons (Bittencourt et al., 1992) and has an orexigenic effect (Qu et al., 1996; Rossi et al., 1997). MCH-containing fibers, as well as orexin-containing fibers, are widely distributed in rats (Kawano et al., 2002; Elias et al., 1998). Since there are few MCH fibers within the hypothalamus but they are rich in the extrahypothalamic regions, MCH neurons seem to integrate signals on food intake arising from hypothalamic neurons and relay the signals to extrahypothalamic regions (Kawano et al., 2002). The axons of MCH-containing neurons project into the DVC, which have MCH receptors; therefore, it is expected that MCH affects gastric motility as shown by other orexigenic neuropeptides. We also examine the effect of the fourth ventricular administration of MCH on gastric motor response. Although enough MCH to induce food intake (3 nmol) was administered, no gastric responses were observed. The administration of MCH did not induce either phasic contractions in the distal stomach or relaxation of the proximal stomach (Furudono et al., 2006). This could not be addressed in this paper, except to mention that orexigenic neuropeptides, which affect gastric motor function, induced both enhancement of phasic contractions in the distal stomach and relaxation of the proximal stomach.

4. Reverse Effects of Anorexigenic Agents on Gastric Motility

I have already mentioned that the feeding-related neuropeptides produced in the central nervous system are divided into two groups based upon their effect on feeding, that is, orexigenic and anorectic neuropeptides. Leptin, which is secreted from white adipose tissue, has an anorectic effect on feeding. Leptin works for negative energy homeostasis by way of central feeding-related peptides (Schwartz et al., 2000; Williams et al., 2001). Leptin stimulates glucose receptor neurons in the VMH, and POMC neurons in the ARC. Conversely, leptin inhibits NPY neurons in the ARC, and orexin neurons in the LHA (Funahashi et al., 2003); thus, leptin is a key peptide of energy homeostasis including feeding behavior. Since the fourth ventricular and intracerebroventricular administration of leptin inhibit gastric emptying in rats (Smedh et al., 1998; Martinez et al., 1999), leptin and/or the other anorectic peptides might suppress phasic contraction of the stomach. In fact, anorectic peptide CCK, which was administered intraperitoneally, attenuated phasic contractions of the stomach in freely moving rats (Flanagan et al., 1989); however, the role of leptin (or other anorectic peptides) on gastric reservoir function has not been clarified. The activation of gastric stretch receptors excites vagal afferent neurons. Their neural information transmits to the NTS then elicits reflex relaxation of the stomach. In the rat NTS, the neural responses induced by gastric inflation disappeared after microinjection of TRH into the DVC (Rogers and McCann, 1989), suggesting that TRH disrupts reflex relaxation of the stomach. Thus, anorectic neuropeptides seem to have opposite effects from orexigenic neuropeptides on the gastric motor response.

We have not yet examined the effect of anorectic neuropeptides on gastric motor responses. Apomorphine, a non-specific dopamine agonist, is a potent emetic agent. Rodents do not vomit, but show anorexia in response to emetic agents. We examined the effect of the intravenous (or fourth ventricular) administration of apomorphine on the gastric motor response in rats (Koga et al., 2003). Apomorphine inhibited phasic contractions in the distal stomach accompanied by increasing IGP (figure 4Db). In the proximal stomach, IGP increased in response to the administration of apomorphine (figure 4Da). These responses are omitted by bilateral sectioning of the vagi, lesions of the AP or pre-treatment with a D2 receptor antagonist. These results indicated that distinctive motor responses were induced by D2 receptors situated in the AP. In dogs and cats, which show vomiting in response to emetic stimuli, gastric relaxation of the proximal stomach was observed in response to the administration of apomorphine (Abrahamsson et al., 1973; Blancquaert et al., 1985). Conversely, motility of the distal stomach was promoted during fictive vomiting. Relaxation of the proximal stomach is explained as accommodating the returned gastrointestinal contents, and the promotion of distal stomach motility is thought to mix contents in the stomach prior to their expulsion in carnivores. Since rats do not vomit, gastric motor responses appearing after the administration of apomorphine may delay gastric emptying to prevent toxic contents moving into the intestine. Lowered gastric contractions and increased gastric tone seem to delay gastric emptying to raise the probability of survival. In rats, anorectic agents induce an opposite gastric response from orexigenic neuropeptides; therefore, I predict that anorectic neuropeptides increased IGP in the proximal stomach, at least in rats.

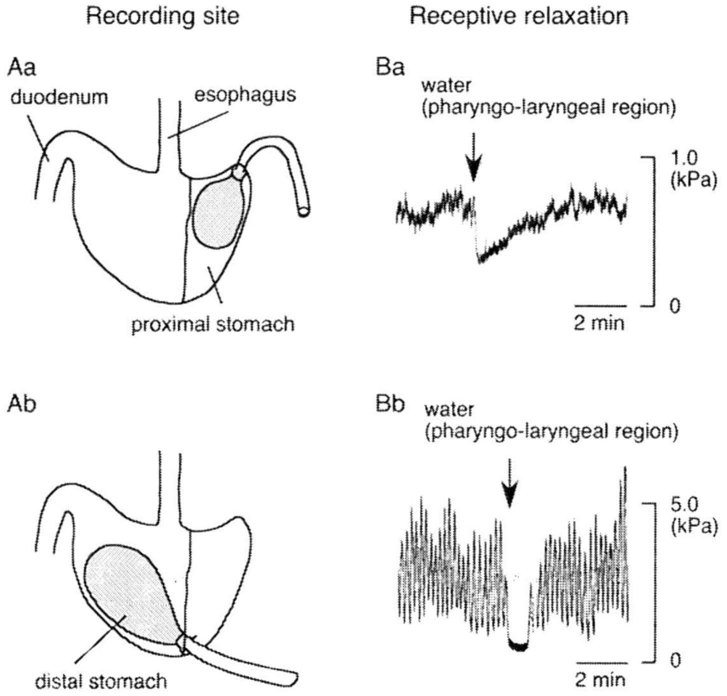

Figure 4 Continued.

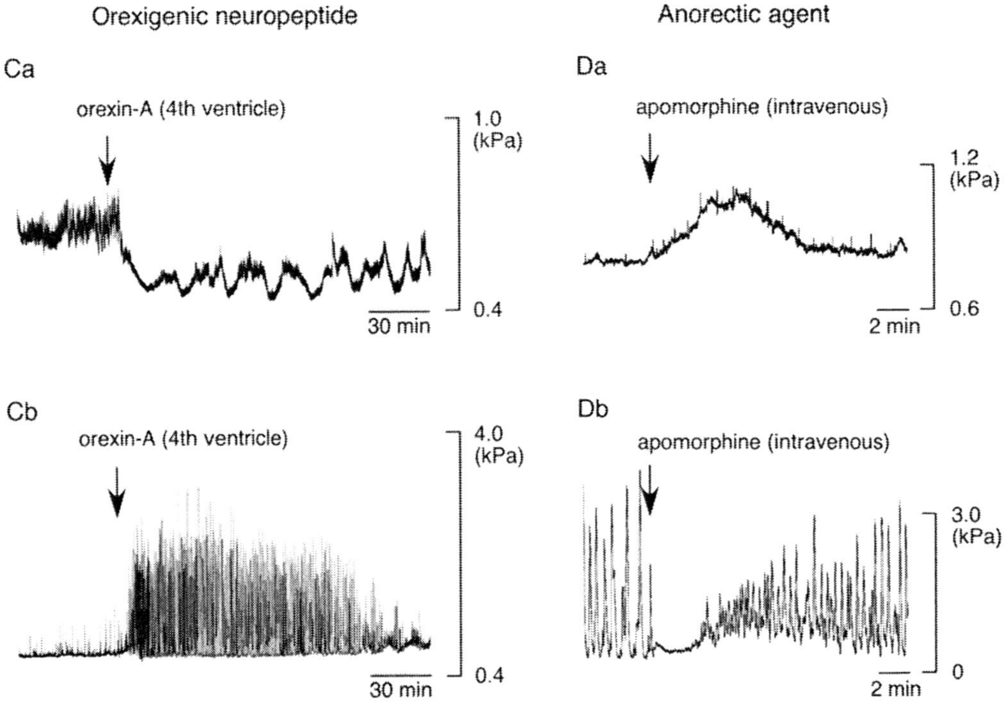

Figure 4. Summary of the 3 types of stimuli which induce different responses of the proximal and distal stomach. A: The balloon for recording proximal stomach motility (a) and that for recording distal stomach motility (b) is shown. B: Reflex relaxation induced by water fibers in the superior laryngeal nerve. Gastric relaxation of the proximal stomach (a) and the inhibition of phasic contractions of the distal stomach (b) induced by the administration of water (0.2 ml) around the epiglottis (Kobashi et al., 2000). Response tendency is the same as the proximal and distal stomach. B: The effects of orexigenic neuropeptides on gastric motility administered into the fourth ventricle are shown. The administration of orexin-A (3 nmol) induced gastric relaxation of the proximal stomach (a) and conversely, facilitated phasic contractions in the distal stomach (b) (Kobashi et al., 2002). Note that the durations of responses are drastically longer than those appearing in B and D. D: the effects of intravenously administered anorectic agents on gastric motility are shown. The administration of apomorphine (0.1 mg / kg body weight) increased intragastric pressure in the proximal stomach (a) and conversely, suppressed phasic contractions of the distal stomach (b) (Koga, et al., 2003). Note that these responses provide a marked contrast to those induced by orexigenic neuropeptides.

5. Conclusion

Our recent studies strongly suggest that central orexigenic neuropeptides facilitate accommodation by the stomach of a meal without an increase in IGP. Fasting is usually followed by feeding; therefore, gastric relaxation achieved by the actions of orexigenic neuropeptides, which are released by fasting, to enhance reservoir function of the stomach, is rational. It is hypothesized that orexigenic neuropeptides induce anticipatory gastric relaxation to compensate for hyperphagia. Centrally released neuropeptides might cooperate with acute reflex relaxation just after swallowing to facilitate accommodation by the stomach of a meal.

6. Acknowledgements

This work was partly supported by a Grant-in-Aid for Scientific Research from the Japanese Society for the Promotion of Science (No: 17591936 to Motoi Kobashi).

7. References

Abrahamsson H. (1973). Vagal relaxation of the stomach induced from the gastric antrum. *Acta Physiologica Scandinavica,* 89, 406-414.

Abrahamsson H. and Jansson G. (1969). Elicitation of reflex vagal relaxation of the stomach from pharynx and esophagus in the cat. *Acta Physiologica Scandinavica,* 77, 172-178.

Abrahamsson H., Jansson G. and Martinson J. (1973). Vagal relaxation of the stomach induced by apomorphine in the cat. *Acta Physiologica Scandinavica,* 88, 296-302.

Abrahamsson H. and Thoren P. (1973). Vomiting and reflex vagal relaxation of the stomach elicited from heart receptors in the cat. *Acta Physiologica Scandinavica,* 88, 433-439.

Adachi A., Shimizu N., Oomura Y. and Kobashi M. (1984). Convergence of hepatoportal glucose-sensitive afferent signals to glucose-sensitive units within the nucleus of the solitary tract. *Neurosci. Lett.,* 46, 215-218.

Anand B. K. (1961). Nervous regulation of food intake. *Physiol. Rev.,* 41, 677-708.

Bernstein I. L., Chavez M., Allen D. and Taylor E. M. (1992). Area postrema mediation of physiological and behavioral effects of lithium chloride in the rat. *Brain Res.,* 575, 132-137.

Bittencourt J. C., Presse F., Arias C., Peto C., Vaughan J., Nahon J. L., Vale W. and Sawchenko P. E. (1992). The melanin-concentrating hormone system of the rat brain: an immuno- and hybridization histochemical characterization. *J. Comp. Neurol.,* 319, 218-245.

Borison H. L. (1989). Area postrema: chemoreceptor circumventricular organ of the medulla oblongata. *Progress in Neurobiol.,* 32, 351-390.

Browning K. N. and Travagli R. A. (2003). Neuropeptide Y and peptide YY inhibit excitatory synaptic transmission in the rat dorsal motor nucleus of the vagus. *J. Physiol.* (Lond.), 549, 775-785.

Blancquaert J. P., Lefebvre, R. A. and Willems J. L., (1985). Influence of peripheral and central administration of apomorphine and morphine on intragastric pressure in the dog, *Eur. J. Pharmacol.,* 107, 133-140.

Browning K. N. and Travagli R. A. (2003). Neuropcptide Y and peptide YY inhibit excitatory synaptic transmission in the rat dorsal motor nucleus of the vagus. *J. Physiol.* (Lond.), 549, 775-785.

Cannon W. B. and Lieb C. W. (1911). The receptive relaxation of the stomach. *Am. J. Physiol.,* 29, 267-273.

Chaffer C. L. and Morris M. J. (2002). The feeding response to melanin-concentrating hormone is attenuated by antagonism of the NPY Y1-Receptor in the rat. *Endocrinology,* 143, 191-197.

Chen CH, Stephens RL, Jr. and Rogers RC. (1997). PYY and NPY: control of gastric motility via action on Y1 and Y2 receptors in the DVC. *Neurogastroenterol. Motil.,* 9, 109-111,.

Chen X., Ge Y. L., Jiang Z. Y., Liu C. Q., Depoortere I. and Peeters T. L., Effects of ghrelin on hypothalamic glucose responding neurons in rats. *Brain Res.* 1055, 131-136, 2005.

Clark J. T., Kalra P. S., Crowley W. R. and Karla S. P. (1984). Neuropeptide Y and human pancreatic polypeptide stimulate feeding behavior in rats. *Endocrinology,* 115, 427-429.

Date Y., Murakami N., Toshinai K., Matsukura S., Niijima A., Matsuo H., Kangawa K. and Nakazato M. (2002). The role of the gastric afferent vagal nerve in ghrelin-induced feeding and growth hormone secretion in rats. *Gastroenterology,* 123, 1120-1128.

de Lecea L., Kilduff T. S., Peyron C., Gao X., Foye P. E., Danielson P. E., Fukuhara C., Battenberg E. L., Gautvik V. T., Bartlett F. S., Frankel W. N., van den Pol A. N., Bloom F. E., Gautvik K. M. and Sutcliffe J. G. (1998). The hypocretins: hypothalamus-specific peptides with neuroexcitatory activity. *Proc. Nat. Acad. Sci. USA,* 95, 322-327.

Desai K. M., Sessa W. C. and Vane J. R. (1991). Involvement of nitric oxide in the reflex relaxation of the stomach to accommodate food or fluid. *Nature,* 351, 477-9.

Edwards, C. M., Abusnana, S., Sunter, D., Murphy, K. G., Ghatei, M. A. and Bloom, S. R. (1999). The effect of the orexins on food intake: comparison with neuropeptide Y, melanin-concentrating hormone and galanin, *J. Endocrinol.,* 160, R7-R12.

Edwards G. L. and Ritter R. C. (1981). Ablation of the area postrema causes exaggerated consumption of preferred foods in the rat. *Brain Res.,* 216, 265-276.

Elias C. F., Saper C. B., Maratos-Flier E., Tritos N. A., Lee C., Kelly J., Tatro J. B., Faulconbridge L. F., Grill H. J. and Kaplan J. M. (2005). Distinct forebrain and caudal brainstem contributions to the neuropeptide Y mediation of ghrelin hyperphagia. *Diabetes,* 2005 54,1985-1993.

Faulconbridge L. F., Cummings D. E., Kaplan J. M. and Grill H. J. (2003). Hyperphagic effects of brainstem ghrelin administration. *Diabetes,* 52, 2260-2265.

Faulconbridge L. F., Grill H. J. and Kaplan J. M. (2005). Distinct forebrain and caudal brainstem contributions to the neuropeptide Y mediation of ghrelin hyperphagia. *Diabetes,* 54, 1985-1993.

Flanagan L. M., Verbalis J. G. and Stricker E. M. (1989). Effects of anorexigenic treatments on gastric motility in rats. *Am. J. Physiol.,* 256, R955-R961.

Fujino K., Inui A., Asakawa A., Kihara N., Fujimura M. and Fujimiya M. (2003). Ghrelin induces fasted motor activity of the gastrointestinal tract in conscious fed rats. *J. Physiol.* (Lond.), 550, 227-240.

Funahashi H., Takenoya F., Guan J. L., Kageyama H., Yada T. and Shioda S. (2003). Hypothalamic neuronal networks and feeding-related peptides involved in the regulation of feeding. *Anat. Sci. Int.,* 78, 123-138.

Furudono Y., Ando C., Yamamoto C., Kobashi M. and Yamamoto T. (2006). Involvement of specific orexigenic neuropeptides in sweetener-induced overconsumption in rats. *Behav. Brain Res.,* in press (doi:10.1016/j.bbr.2006.08.031).

Gaige S., Abysique A. and Bouvier M. (2003). Effects of leptin on cat intestinal motility. *J. Physiol.* (Lond.)., 546, 267-277.

Geoghegan J. G., Lawson D. C., Cheng C. A. and Pappas T. N. (1994). Intracerebroventricular neuropeptide Y increases gastric acid secretion by decreasing tonic adrenergic inhibition of acid in dogs. *Brain Res.*, 635, 118-124.

Goyal, RK and Ratttan, S. (1980). VIP as a possible neurotransmitter of non-cholinergic, non-adrenergic inhibitory neurons. *Nature.* (Lond.), 288, 378-380.

Grabauskas G., Zhou S. Y., Das S., Lu Y., Owyang C. and Moises H. C. (2004). Prolactin-releasing peptide affects gastric motor function in rat by modulating synaptic transmission in the dorsal vagal complex. *J. Physiol.* (Lond.), 561, 821-39.

Harrison T. A., Chen C. T., Dun N. J. and Chang J. K. (1999). Hypothalamic orexin A-immunoreactive neurons project to the rat dorsal medulla. *Neurosci. Lett.,* 273, 17-20.

Hoffman G. E., Ollmann M. M., Barsh G. S., Sakurai T., Yanagisawa M. and Elmquist J. K. (1998). Chemically defined projections linking the mediobasal hypothalamus and the lateral hypothalamic area. *J. Comp. Neurol.,* 402, 442-459.

Humphreys G. A., Davison J. S. and Veale W. L. (19920. Hypothalamic neuropeptide Y inhibits gastric acid output in rat: role of the autonomic nervous system. Am. J. Physiol. Gastrointest. *Liver Physiol.,* 263, G726-G732.

Ishiguchi T., Amano T., Matsubayashi H., Tada H., Fujita M. and Takahashi T. (2001a). Centrally administered neuropeptide Y delays gastric emptying via Y2 receptors in rats. *Am. J. Physiol. Regul. Integr. Comp. Physiol.,* 281, R1522-R1530.

Ishiguchi T., Nakajima M., Sone H., Tada H., Kumagai A. K. and Takahashi T. (2001b)Gastric distension-induced pyloric relaxation: central nervous system regulation and effects of acute hyperglycaemia in the rat. *J. Physiol.* (Lond.), 533, 801-813.

Kajimura M., Takeuchi K., Kodaira M., Hanai H. and Kaneko E. (1997). Expression of muscarinic receptor subtypes in rat gastric smooth muscle: effect of M3 selective antagonist on gastric motility and emptying. *Dig. Dis. Sci.,* 1997 42, 907-914.

Kalra P. S., Dube M. G., Xu B., Farmerie W. G. and Kalra S. P. (1998). Neuropeptide Y (NPY) Y1 receptor mRNA is upregulated in association with transient hyperphagia and body weight gain: evidence for a hypothalamic site for concurrent development of leptin resistance. *J. Neuroendocrinol.,* 10, 43-49.

Kalra S. P., Dube M. G., Fournier A. and Kalra P. S. (1991). Structure-function analysis of stimulation of food intake by neuropeptide Y: effects of receptor agonists. *Physiol. Behav.,* 50, 5-9.

Kalra S. P., Dube M. G., Pu S., Xu B. and Horvath T. L. (1999). Interacting appetite-regulating pathways in the hypothalamic regulation of body weight. *Endocr. Rev.,* 20, 68-100.

Kawano H., Honma S., Honma A., Horie M., Kawano Y. and Hayashi S. (2002). Melanin-concentrating hormone neuron system: the Wide Web that controls the feeding. *Anat. Sci. Int.,* 77, 149-160.

Kunii, K., Yamanaka, A., Nambu, T., Matsuzaki, I., Goto, K. and Sakurai, T. (1999). Orexins/hypocretins regulate drinking behaviour, *Brain Res.,* 842, 256-261.

Kobashi M. and Adachi A. (1988). A direct hepatic osmoreceptive afferent projection from nucleus tractus solitarius to dorsal hypothalamus. *Brain Res. Bull.,* 20, 487-492.

Kobashi M. and Adachi A. (1985). Convergence of hepatic osmoreceptive inputs on sodium-responsive units within the nucleus of the solitary tract of the rat. *J. Neurophysiol.*, 54, 212-219.

Kobashi M. and Adachi A. (1986). Projection of nucleus tractus solitarius units influenced by hepatoportal afferent signal to parabrachial nucleus. *J. Auton. Nerv. Syst.*, 16, 153-158.

Kobashi M., Furudono Y., Matsuo R. and Yamamoto T. (2002). Central orexin facilitates gastric relaxation and contractility in rats. *Neurosci. Lett.*, 332, 171-174.

Kobashi M., Mizutani M. and Adachi, A. (1998). Facilitation of gastric motility induced by portal infusion of hyper- and hypotonic solution in rats. *J. Auton. Nerv. Syst.*, 73, 156-162.

Kobashi M., Mizutani M. and Matsuo R. (2000). Water stimulation of the posterior oral cavity induces inhibition of gastric motility. *Am. J. Physiol. Regulatory Integrative Comp. Physiol.*, 279, R778-R786.

Kobashi M., Shimatani Y., Shirota K., Xuan S.-Y., Mitoh Y. and Matsuo R. (2006). Central neuropeptide Y induces proximal stomach relaxation via Y1 receptors in the dorsal vagal complex of the rat. *Am. J. Physiol. Regul. Integr. Comp. Physiol.*, 290, R290-R297.

Kobahsi M., Xuan S.-Y., Mitoh Y. and Matsuo R. (2007). Central ghrelin induces relaxation of the proximal stomach in rats. *J. Physiol. Sci.*, 57, abstract.

Koga T., Kobashi M., Mizutani M., Tsukamoto G. and Matsuo R. (2003). Area postrema mediates gastric motor response induced by apomorphine in rats. *Brain Res.*, 960, 122-131.

Kojima M., Hosoda H., Date Y., Nakazato M., Matsuo H., and Kangawa K. (1999). Ghrelin is a growth-hormone-releasing acylated peptide from stomach. *Nature.* (Lond.)., 402, 656-660.

Krowicki Z. K. and Hornby P. J. (1996). Contribution of acetylcholine, vasoactive intestinal polypeptide and nitric oxide to CNS-evoked vagal gastric relaxation in the rat. *Neurogastroenterol. Motil.*, 8, 307-317.

Krowicki Z. K., Burmeister M. A., Berthoud H. R., Scullion R. T., Fuchs K. and Hornby P. J. (2002). Orexins in rat dorsal motor nucleus of the vagus potently stimulate gastric motor function. *Am. J. Physiol. Gastrointest. Liver Physiol.*, 283, G465-G672.

Krowicki Z. K., Sharkey K. A., Serron S. C., Nathan N. A. and Hornby P. J. (1997). Distribution of nitric oxide synthase in rat dorsal vagal complex and effects of microinjection of nitric oxide compounds upon gastric motor function. *J. Comp. Neurol.*, 377, 49-69.

Lefebvre R. A., Hasrat J. and Gobert A. (1992). Influence of NG-nitro-L-arginine methyl ester on vagally induced gastric relaxation in the anaesthetized rat. *Br. J. Pharmacol.*, 313-320.

Lin Y., Matsumura K., Fukuhara M., Kagiyama S., Fujii K. and Iida M. (2004). Ghrelin acts at the nucleus of the solitary tract to decrease arterial pressure in rats. *Hypertension,* 43, 977-982.

Marcus J. N., Aschkenasi C. J., Lee C. E., Chemelli R. M., Saper C. B., Yanagisawa M. and Elmquist J. K. (2001). Differential expression of orexin receptors 1 and 2 in the rat brain. *J. Comp. Neurol.*, 435, 6-25.

Martinez V., Barrachina M. D., Wang L. and Tache Y. (1999). Intracerebroventricular leptin inhibits gastric emptying of a solid nutrient meal in rats. *Neuroreport,* 10, 3217-3221.

Masuda Y., Tanaka T., Inomata N., Ohnuma N., Tanaka S., Itoh Z., Hosoda H., Kojima M. and Kangawa K. (2000). Ghrelin stimulates gastric acid secretion and motility in rats. *Biochem. Biophys. Res. Commun.,* 276, 905-908.

Matsuda M., Aono M., moriga M. and Okuma M. (1991). Centrally administered NPY stimulated gastric acid and pepsin secretion by a vagally mediated mechanism. *Regul. Pept.,* 35, 31-41.

McCann M. J. and Rogers R. C. (1992). Impact of antral mechanoreceptor activation on the vago-vagal reflex in the rat - functional zonation of responses. *J. Physiol.* (Lond.) , 453, 401-411.

Miolan J. P. and Roman C. (1978). Activite des fibres vagales efferentes destinees a la musculature lisse du cardia du chien. *J. Physiol.* (Paris), 74, 709-723.

Mondal M. S., Nakazato M. and Matsukura S. (2000). Orexins (hypocretins): novel hypothalamic peptides with divergent functions. *Biochem. Cell Biol.,* 78, 299-305.

Nakazato M., Murakami N., Date Y., Kojima M., Matsuo H., Kangawa K. and Matsukura S. (2001). A role for ghrelin in the central regulation of feeding. *Nature.* (Lond.), 409, 194-198.

Nishimura H. and Oomura Y. (1987). Effects of hypothalamic stimulation on activity of dorsomedial medulla neurons that respond to subdiaphragmatic vagal stimulation. *J. Neurophysiol.,* 58, 655-675.

Okumura T., Takeuchi S., Motomura W., Yamada H., Egashira S., Asahi S., Kanatani A., Ihara M. and Kohgo Y. (2001). Requirement of intact disulfide bonds in orexin-A-induced stimulation of gastric acid secretion that is mediated by OX1 receptor activation. *Biochem. Biophys. Res. Commun.,* 280, 976-981.

Oomura Y., Kimura K., Ooyama H., Maeno T., Iki M. and Kuniyoshi M. (1964). Reciprocal activities of the ventromedial and lateral hypothalamic areas of cats. *Science,* 143, 484-485.

Oomura Y., Ono T., Ooyama H., and Wayner M. J. (1969). Glucose and osmosensitive neurones of the rat hypothalamus. *Nature.* (Lond.) , 222, 282-284.

Penner S. B, Smyth D. D. and Glavin G. B. (1993). Effects of neuropeptide Y and [Leu31,Pro34] neuropeptide Y on experimental gastric lesion formation and gastric secretion in the rat. *J. Pharmacol. Exp. Ther.,* 266, 339-343.

Qu D., Ludwig D. S., Gammeltoft S., Piper M., Pelleymounter M. A., Cullen M. J., Mathes W. F., Przypek R., Kanarek R. and Maratos-Flier E. (1996). A role for melanin-concentrating hormone in the central regulation of feeding behaviour. *Nature.* (Lond), 380, 243-247.

Rauhofer E. A., Greenberg D. and Smith G. P. (1996). Peripheral atropinization does not change meal size controlled by prepyloric mechanisms. *Physiol. Behav.,* 59, 237-240.

Rogers R. C., Hermann G. E. and Travagli R. A. (1999). Brainstem pathways responsible for esophageal control of gastric motility and tone in the rat. *J. Physiol.* (Lond.), 514, 369-383.

Rogers R. C., Hermann GE and Travagli R. A. (1999). Brainstem pathways responsible for esophageal control of gastric motility and tone in the rat. *J. Physiol.* (Lond.), 514, 369-383.

Rogers R. C. and McCann M. J. (1989). Effects of TRH on the activity of gastric inflation-related neurons in the solitary nucleus in the rat. *Neurosci. Lett.,* 104, 71-76.

Roman C. and Gonella J. (1987). Extrinsic control of digestive tract motility. In L. R. (Eds.), *Physiology of the gastrointestinal tract.* (Second Edition, pp. 507-553). New York: Raven Press.

Rossi M., Choi S. J., O'Shea D., Miyoshi T., Ghatei M. A. and Bloom S. R. (1997). Melanin-concentrating hormone acutely stimulates feeding, but chronic administration has no effect on body weight. *Endocrinology,* 138, 351-355.

Sakaguchi T. and Shimojo E. (1984). Inhibition of gastric motility induced by hepatic portal injections of D-glucose and its anomers. *J. Physiol.* (Lond.), 351, 573-581.

Sakurai T., Amemiya A., Ishii M., Matsuzaki I., Chemelli R. M., Tanaka H., Williams S. C., Richarson J. A., Kozlowski G. P., Wilson S., Arch J. R., Buckingham R. E., Haynes A. C., Carr S. A., Annan R. S., McNulty D. E., Liu W. S., Terrett J. A., Elshourbagy N. A., Bergsma D. J. and Yanagisawa M. (1998). Orexins and orexin receptors: a family of hypothalamic neuropeptides and G protein-coupled receptors that regulate feeding behavior. *Cell,* 92, 573-585.

Saper B. C. (1995). Central autonomic system. In: G. PAXINOS (EDS.), *The rat nervous system.* (PP. 107-135). San Diego: Academic Press.

Schwartz G. J. and Moran T. H. (2002). Leptin and neuropeptide Y have opposing modulatory effects on nucleus of the solitary tract neurophysiological responses to gastric loads: implications for the control of food intake. *Endocrinology,* 143, 3779-3784.

Schwartz M. W, Woods S. C., Porte D. Jr., Seeley R. J. and Baskin D. G. (2000). Central nervous system control of food intake. *Nature.* (Lond.), 404, 661-671.

Smedh U., Hakansson M. L., Meister B. and Uvnas-Moberg K. (1998). Leptin injected into the fourth ventricle inhibits gastric emptying. *Neuroreport,* 9, 297-301.

Steffens A. B. (1976). Influence of the oral cavity on insulin release in the rat. *Am. J. Physiol.,* 230, 1411-1415.

Takahashi N., Okumura T., Yamada H. and Kohgo Y. (1999). Stimulation of gastric acid secretion by centrally administered orexin-A in conscious rats. *Biochem. Biophys. Res. Commun,.* 254, 623-627.

Takemoto K. (1982). Neuropeptide Y: complete amino acid sequence of the brain peptide. *Proc. Natl. Acad Sci. U S A,* 79, 5485-5489.

Williams G., Bing C., Cai X. J., Harrold J. A., King P. J. and Liu X. H. (2001). The hypothalamus and the control of energy homeostasis: different circuits, different purposes. *Physiol. Behav.,* 74, 683-701.

Xu B., Kalra P. S., Moldawer L. L. and Kalra S. P. (1998). Increased appetite augments hypothalamic NPY Y1 receptor gene expression: effects of anorexigenic ciliary neurotropic factor. *Regul. Pept.,* 76, 391-395.

Xuan S. Y., Kobahsi M., ,Mitoh Y. and Matsuo R. (2007). Central ghrelin induces relaxation of the proximal stomach in rats. *J. Physiol. Sci. J. Physiol. Sci.,* 57, abstract.

Yamanaka A., Kunii K., Nambu T., Tsujino N., Sakai A., Matsuzaki I., Miwa Y., Goto K., and Sakurai T. (2000). Orexin-induced food intake involves neuropeptide Y pathway. *Brain Res.,* 859, 404-409.

Zheng H., Patterson L. M., and Berthoud H. R. (2005). Orexin-A projections to the caudal medulla and orexin-induced c-Fos expression, food intake, and autonomic function. *J. Comp. Neurol.,* 485, 127-142.

Zheng Z. L., Rogers R. C. and Travagli R. A. (1999). Selective gastric projections of nitric oxide synthase-containing vagal brainstem neurons. *Neuroscience,* 90, 685-694.

Zigman J. M., Jones J. E, Lee C. E., Saper C. B. and Elmquist J. K. (2006). Expression of ghrelin receptor mRNA in the rat and the mouse brain. *J. Comp. Neurol.,* 494, 528-48.

In: Neuropeptide Research Trends
Editor: Bernice A. Levine, pp. 143-167
ISBN: 978-1-60021-640-4

Chapter V

Gender Differences in CGRP Concentrations in Cardiovascular Tissues in Type 2 Diabetic Rats

Jitka Kuncová, Zbyněk Tonar, Jitka Švíglerová and Jana Slavíková
Department of Physiology and Department of Histology and Embryology, Medical School Plzeň, Charles University, Czech Republic

Abstract

Calcitonin gene-related peptide (CGRP) is a potent vasodilatory and cardiotonic neuropeptide with widespread distribution in the central and peripheral nervous systems. CGRP-positive nerve terminals have been detected in the heart and vasculature, particularly in arteries. The perivascular sensory nerve endings have been suggested to be the major source of circulating CGRP. In the present study, we investigated the impact of type 2 diabetes on CGRP concentrations in the heart, selected arteries and plasma in relation to basal cardiovascular parameters, i.e. resting heart rate, systolic and diastolic blood pressures. Type 2 diabetes was induced in 2-day-old male and female rats by a single intraperitoneal streptozotocin injection (STZ; 100 mg/kg; n2-STZ model). Diabetic state closely resembling human type 2 diabetes was verified by determinations of blood glucose levels, glucose tolerance tests, plasma and pancreatic insulin concentrations, and immunohistochemical analysis of the pancreatic tissue. CGRP concentrations were measured in the heart compartments, thoracic and abdominal aortae, mesenteric and femoral arteries, and in the plasma using radioimmunoassay diagnostic kits. At the age of 3 months, diabetic animals displayed significantly higher blood glucose levels, lower plasma insulin concentrations and impaired glucose tolerance. At this stage, no differences were noted in the resting heart rate, blood pressure, and tissue and plasma CGRP concentrations between n2-STZ and control rats. At the age of 18 months, diabetic rats of both sexes had blood glucose levels not differing from the age-matched control values. However, impaired glucose tolerance associated with decreased insulin response to the glucose challenge and elevated pancreatic insulin concentrations suggested defective insulin secretion in rats after neonatal STZ administration. Although systolic pressure did not differ between the control and n2-STZ rats, diastolic pressure was significantly higher in diabetic animals of both sexes. CGRP concentrations were significantly lower in the plasma, arteries and heart compartments of diabetic rats, reaching 40-70% of the control values. Interestingly, female diabetic hearts displayed significantly lower peptide concentrations than diabetic males. With regard to the well-documented cardioprotective and vasodilatory effects of CGRP, changes in CGRP synthesis and release may be associated with the development of cardiovascular complications of type 2 diabetes, i.e. decreased protection of the myocardium against ischemic injury and increased total peripheral resistance. In addition, profound deficit in myocardial CGRP levels revealed in female diabetic rats might be implicated in the loss of "female advantage" in the prevalence of coronary heart disease frequently described in diabetic patients.

1. Introduction

Alpha calcitonin gene-related peptide (CGRP) is a 37-amino-acid neuropeptide encoded in the calcitonin/CGRP gene, which is localized on chromosome 11 [1, 2]. Production of calcitonin has been demonstrated in the parafollicular cells of the thyroid gland and in the bronchial Kulchitsky cells of the lungs [3, 4]. However, in both central and peripheral nervous systems, alternative splicing of the primary RNA transcript of the same gene leads to the formation of α-CGRP [5]. In addition, a second CGRP gene, coding peptide with nearly identical biological effects as α-CGRP has been discovered on chromosome 11. This peptide was called β-CGRP and it differs from the human α-CGRP by three amino acids and from the homologous rat α-CGRP by one amino acid only [6, 7].

The primary site of CGRP synthesis in the peripheral nervous system is the dorsal root ganglia that send their processes to the peripheral organs including the heart and blood vessels [8]. In the sensory ganglia, CGRP is often co-stored with substance P and related tachykinins and is transported to the target tissues by axonal transport [9]. In the heart, CGRP-positive nerve fibres have been demonstrated in all regions, particularly close to the coronary vasculature, within the papillary muscles and in the sinoatrial and atrioventricular nodes [10]. In the vasculature, CGRP-positive nerve fibre terminals have been documented at the junction of the adventitia and the media and in the muscle layer with higher density in the arteries than in the veins [11].

CGRP contributes to the regulation of cardiac performance by acting on specific myocardial CGRP receptors composed of two proteins: calcitonin receptor-like receptor (CRLR), which is the binding component of the receptor complex, and receptor activity modifying protein (RAMP1) [12]. It has been well documented that CGRP exerts positive inotropic and chronotropic effects in various mammalian species including humans [13-15]. In the vasculature, the peptide influences microvascular permeability and is the most potent endogenous vasodilator exerting its effect through both endothelium-dependent and endothelium-independent mechanisms [16-18].

In addition to its immediate actions, CGRP seems to be a trophic factor in the mammalian cardiovascular system as evidenced by its hypertrophic effect on the myocardial cells and CGRP-induced angiogenesis [19, 20]. CGRP may also play an important role in the protection of the heart provided by ischemic preconditioning, i.e. brief exposures to ischemia resulting in an increased resistance to severe ischemia/reperfusion [21].

The role of CGRP has been suggested in numerous pathological conditions: coronary heart disease, heart failure, hypertension [22], i.e. complications frequently associated with chronic diabetes. Diabetic patients are particularly prone to develop cardiovascular autonomic neuropathy (CAN) resulting from damage to both efferent and afferent innervation of the heart and blood vessels. CAN is typically manifested by the reduced heart rate variability, exercise intolerance, orthostatic hypotension and silent myocardial ischemia [23]. The prevalence of CAN has been reported to be higher in type 2 than type 1 diabetics ranging from 22 to 34% depending on the criteria used [24]. Coronary heart disease, the leading cause of mortality in diabetic patients, is more frequent in patients with CAN compared to those without CAN [25, 26]. In addition, diabetic women have been reported to have significantly higher risk of coronary disease than diabetic men [27, 28].

Multiple role of CGRP in the regulation of cardiovascular functions, particularly in the coronary perfusion and protection against ischemia and reperfusion injury, along with differential regulation of the peptide expression in males and females [29], give rise to hypothesis that changes in CGRP signalling could be involved in the development of major cardiovascular complications of diabetes. The data reporting the impact of type 1 diabetes on the sensory innervation of the heart and vessels are limited and those dealing with the relationship between type 2 diabetes and CGRP signalling system are sparse. Thus, in the present study, we have investigated the changes in CGRP concentrations in the plasma, heart compartments and selected arteries in male and female type 2 diabetic young and aged rats (i.e. at the ages of 3 and 18 months, respectively). The chosen model of type 2 diabetes was carefully evaluated by measurements of blood glucose concentrations, plasma and pancreatic insulin levels, performance of glucose tolerance tests and immunohistochemical analysis of the pancreatic tissue.

2. Materials and Methods

2.1. Materials

Streptozotocin (STZ) and other chemicals were from Sigma (St. Louis, MO, USA) or Lachema (Brno, Czech Republic), if not stated otherwise. All chemicals were of analytical grade.

2.2. Animals

Wistar rats bred in our laboratory were used. Pregnant rats were housed individually with free access to food and water. After birth, litters were made up of 8-10 pups each. To induce the n2-STZ model of type 2 diabetes, 2-day-old rats were injected with STZ (100 mg/kg body weight, intraperitoneally) dissolved in citrate buffer (pH 4.5). Control animals were injected with vehicle only. The pups were weaned at the age of four weeks. All experiments were conducted in accordance with European Directive for the Protection of Vertebrate Animals Used for Experimental and Other Scientific Purposes (86/609/EU), the relevant Guidelines of the Czech Ministry of Agriculture for scientific experimentation on animals and were approved by the University Committee for Experiments on Laboratory Animals. At the age of three months, animals were checked for blood glucose levels by glucose oxidase method (Bio-La-Test, Lachema, Czech Republic). STZ-injected rats having non-fasting blood glucose concentrations lower than 10 mmol/l were excluded from the study. The animals under study were used for further experiments at the ages of 3 and 18 months.

2.3. Histology and Immunohistochemistry

The pancreata of 5 control and 5 n2-STZ 18-month-old rats of both sexes were fixed by immersion in 4% buffered formaldehyde, embedded in paraffin and cut into 0.5 mm thick slabs, one of which was randomly selected for further histological sectioning and then cut exhaustively into a series of 5 μm tissue sections. The sections were mounted on slides, deparaffinised and rehydrated. Endogenous peroxidase activity was blocked by a solution composed of hydrogen peroxide (1 volume) and methanol (50 volumes). For insulin detection, sections were incubated with polyclonal guinea pig anti-insulin (with swine insulin used as immunogen) (1:150; Dako, CA, USA) for 12 hours at 4°C. The secondary antibody (45 min, 37°C) and avidin-biotin peroxidase complex (45 min, 37°C) were applied, using the Novostain Super ABC Universal Kit (Novocastra Laboratories Ltd., GB) with diaminobenzidin as chromogen. Following immunohistochemistry, the sections were counterstained with Gill's haematoxylin (30 s; Bio-Optica, Italy), dehydrated and mounted with DPX (Fluka, Buchs, Switzerland). According to the manufacturer's recommendation, unspecific rabbit IgG were used for negative control instead of the primary antibody (due to extensive homology between guinea pig and rabbit immunoglobulins, secondary antibodies directed against rabbit immunoglobulins are well-suited for the detection of this antibody to insulin).

2.4. Quantitative Assessment of Insulin Positivity within the Pancreatic Islets

For a quantitative analysis we used the stereological point-grid method [30] allowing a reliable estimation of volume fraction of insulin-positive cells within pancreatic islets *Vv(insulin, islets)*, see Equation 1:

$$Vv(insulin, islets) = \frac{estV(insulin)}{estV(islets)}, \qquad (1)$$

where *estV(insulin)* was the estimated volume of cytoplasm of cells stained with the insulin antibody, and *estV(islets)* was the total volume of the islets within the specimen.

In order to estimate both the *estV(insulin)* and *estV(islets)* in each sample, sampled histological sections underwent stereological analysis with the use of the PointGrid module (figure 1-2) of the Ellipse software (ViDiTo, Kosice, Slovakia). Within the reference volume, 5 µm-thick equidistant sections were selected through systematic uniform random sampling. The position of the first tissue section in the volume was random, i.e. equal to a product of *(72*n)*, where *n* was a random number between 0 and 1. Starting with this section, every eighth section was captured with two constant magnifications, so that the distance between the two neighbouring calibrated photomicrographs (sampling period) was 40 µm. We assessed the area of the profiles of insulin-positive cells and whole islets, respectively, in each tissue section according to Equation 2:

$$estA = a \cdot P\,, \qquad (2)$$

where *estA* was the estimated area of profiles of insulin-positive cells (or of whole islets), grid parameter *a* was the area corresponding to one test point and *P* was the number of test points hitting the structure under study. The total number of points counted was at least 200 in all sections of each series. The repeatability of quantification, i.e. the intraobserver variability, was assessed with the intraclass correlation coefficient (ICC) [31] for two observations of 36 micrographs of the sample C2. In order to affect the second measurement by the knowledge of the first one as little as possible, the interval between the first and the second observation was seven days.

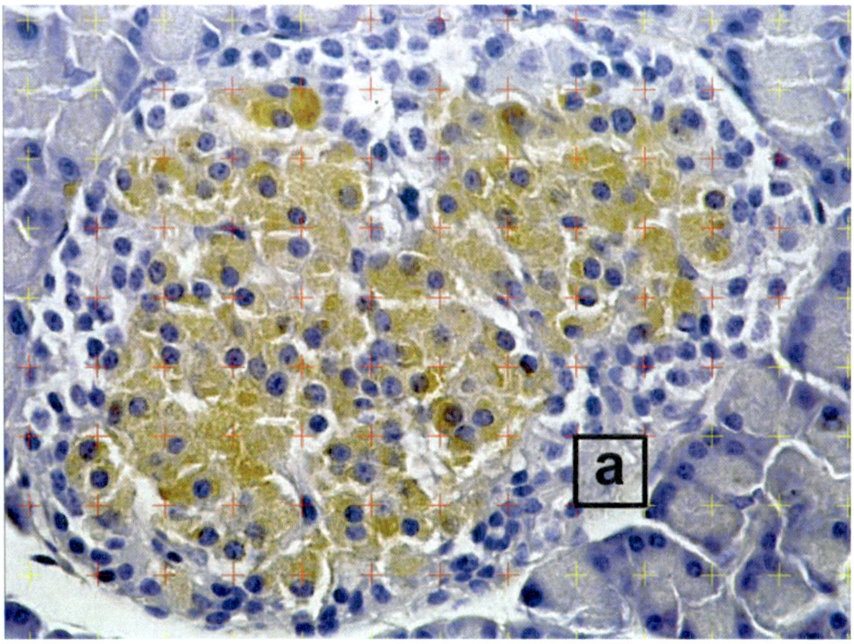

Figure 1. Counting points. Assessment of the total volume of the islets within the specimen *estV(islets)* was based on counting the test points hitting the islet (highlighted as red). The grid parameter *a* corresponds to the area of one test point.

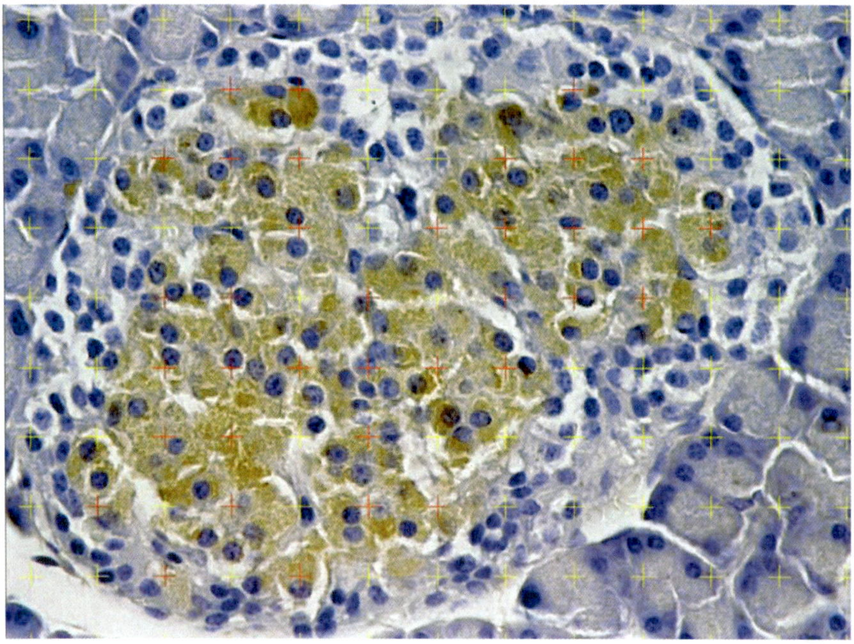

Figure 2. Counting points. Assessment of the volume of cells stained with the insulin antibody *estV(insulin)* was based on counting the test points hitting the cells (highlighted as red).

The Cavalieri principle was used for the estimation of the volume *estV*, see Equation 3:

$$estV = T \cdot (A_1 + A_2 + \ldots + A_m), \qquad (3)$$

where *estV* was the Cavalieri volume estimator, $T=0.040\ mm$ was the distance between the two following selected sections, A_i was the area of the insulin-positive cells or of the islets in the i-th section, and m stood for the total number of sections selected from the series. To estimate the value of m, i.e. the variation caused by sampling the serial sections, we calculated the coefficient of error *(CE)* according to Gundersen and Jensen [32] from a series of 72 sections, see Equation 4.

$$estCE = \frac{\sqrt{3A + C - 4B}}{\sum_{i=1}^{n} a_i},$$

where $A = \sum_{i=1}^{n} a_i.a_i$, $B = \sum_{i=1}^{n-1} a_i.a_{i+1}$, $C = \sum_{i=1}^{n-2} a_i.a_{i+2}$ (4)

Preliminary analysis proved that the number of sections sampled within each tissue block had to be $m=9$ in order to keep the $CE \approx 0.05$ (figure 3).

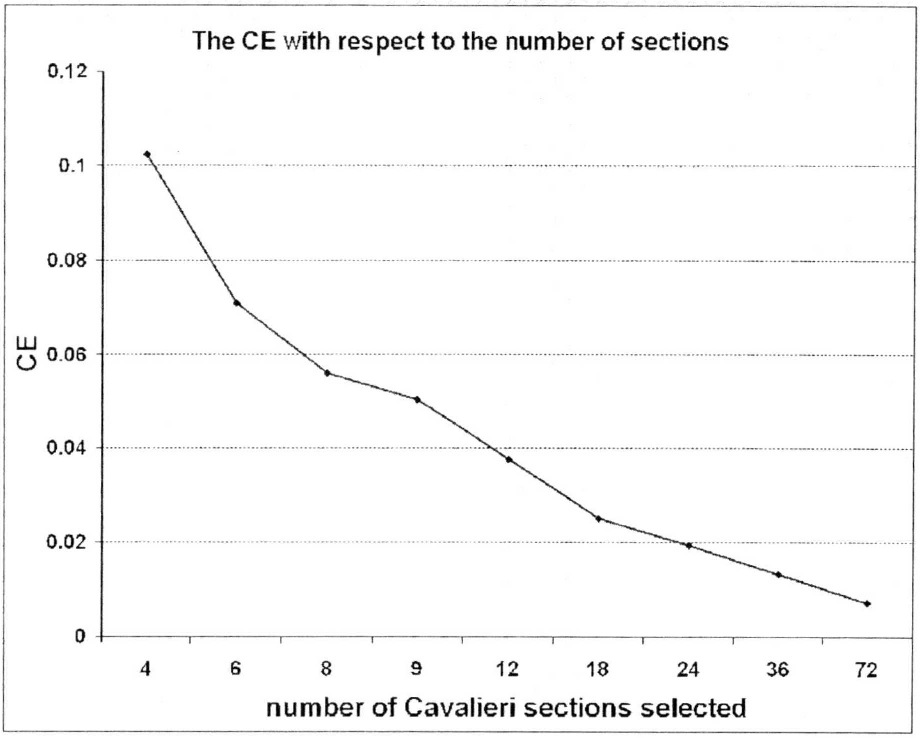

Figure 3. Analysis of sampling error. Average values of the Gundersen and Jensen's coefficient of error (*CE*) for all sampling combinations of corresponding values of the number of sections. These are the results from a pilot study performed for the parameter *Vv(insulin, islets)* in pancreas of a control rat. To keep the CE ≈ 0.05, the suitable value of *m=9* was found.

2.5. Glucose Tolerance Tests

Glucose tolerance tests were performed in male and female control and n2-STZ rats at the age of 3 and 18 months (n=4-6 per group). The rats were fasted from evening before and anaesthetized with urethane 1.5 g/kg of body weight (b.w.). Glucose (2 g/kg b.w.) was administered intraperitoneally and blood samples were collected from the orbital sinus at 15, 30, 60, 90, 120, 150, and 180 min after the glucose challenge.

Blood glucose concentrations in the collected samples were determined by glucose oxidase method (Bio-La-Test, Lachema, Czech Republic).

2.6. Measurement of the Blood Pressure and Heart Rate

One day prior to dissection, animals were placed in a small chamber with electrodes in the floor that were connected to electrocardiograph (EKG Seiva Praktik, Czech Republic). Animals were let to adapt for 20 min and then the resting heart rate was measured five times in one-minute intervals. The values mentioned in the results represent means of these measurements.

Blood pressure (systolic and diastolic) was measured non-invasively by tail-cuff method using RTBP apparatus (Kent Scientific Co., CT, USA). Blood pressure values were derived from an average of 10-15 measurements per animal at each time point. Two preliminary training sessions were performed during 2 weeks before final blood pressure determinations.

2.7. Extraction of CGRP and Insulin

Rats were anaesthetized with ether and killed by decapitation. Blood was collected on EDTA and centrifuged (1000×*g*, 4°C, 15 min). The plasma was removed and stored at -20°C for determination of insulin and CGRP concentrations. Heart, thoracic and abdominal aorta, mesenteric and femoral arteries, and pancreas were rapidly excised. Tissues were rinsed with ice-cold saline, freed of connective tissue and fat and the heart was divided into the left atrium with the interatrial septum, right atrium, and free walls of both ventricles. Immediately after dissection, the tissues were frozen on dry ice and weighed. The CGRP-like material was extracted from pulverized heart compartments and vessels by boiling in 10 volumes of 0.1 mol/ l HCl containing 100 mmol/ l EDTA and 0.01% $Na_2S_2O_5$ for 15 min, rapid cooling on ice and homogenization for 30 s using an Ultra-Turrax homogenizer. The homogenates were centrifuged (10,000×*g*, 4°C, 20 min); the supernatants were neutralized with 1 mol/l Tris-base and centrifuged again (5,000×*g*, 4°C, 10 min). The supernatants were stored at -70°C until radioimmunoassay (RIA) for CGRP concentration measurements.

2.8. Biochemical Assays

Insulin concentrations in the plasma and pancreatic tissue were measured using commercial RIA kits (LINCO Research, MO, USA) with rat insulin as the standard. CGRP concentrations were assayed in tissue extracts and plasma by RIA using commercial kits (Phoenix Pharmaceuticals, CA, USA).

2.9. Data Analysis

Results are presented as mean ± SEM. Statistical analysis was performed by means of analysis of variance (ANOVA) followed by post hoc Fisher's Least Significant Difference test, using software package STATISTICA Cz, version 7 (StatSoft, Inc., 2004). Normality of populations and homogeneity of variances were tested before each ANOVA. The results were considered significantly different when $P < 0.05$.

3. Results

3.1. General

Table 1 summarizes the body weights, blood glucose levels and plasma and pancreatic insulin levels in male and female control and n2-STZ rats at the ages of 3 and 18 months.

At 3 months of age, body weights of n2-STZ rats of both sexes were similar to their respective age-matched controls. Blood glucose levels were significantly higher and plasma insulin levels significantly lower in 3-month-old diabetic animals than in controls; however, no significant difference was detected in insulin concentrations in the pancreas between the control and n2-STZ rats. At the age of 18 months, body weights of n2-STZ rats of both sexes were significantly lower than in the respective control animals. Blood glucose concentrations and plasma insulin levels did not significantly differ among the groups of 18-month-old animals. Interestingly, insulin concentrations in the pancreatic tissues of 18-month-old n2-STZ rats were significantly elevated when compared to their respective controls.

Table 1. Body weights (BW), blood glucose concentrations, plasma and pancreatic insulin levels in 3- and 18-month-old control and diabetic (n2-STZ) male (M) and female (F) rats. Data are means ± SEM. *P< 0.05 compared with the respective control value

	Age	Control M	n2-STZ M	Control F	n2-STZ F	n
BW [g]	3M	345 ± 9	361 ± 6	226 ± 7	245 ± 5	10
	18M	610 ± 9	538 ± 21*	340 ± 4	293 ± 12*	10
Blood glucose [mmol/l]	3M	5.5 ± 0.4	14.8 ± 0.8*	5.8 ± 0.6	13.4 ± 0.8*	10
	18M	5.2 ± 0.5	5.7 ± 0.5	6.0 ± 0.7	6.1 ± 0.7	10
Plasma insulin [ng/ml]	3M	1.48 ± 0.20	0.19 ± 0.04*	1.16 ± 0.12	0.17 ± 0.02*	6
	18M	1.37 ± 0.17	0.71 ± 0.13	1.35 ± 0.18	0.94 ± 0.06	6
Pancreatic insulin [ng/g]	3M	17.7 ± 2.2	11.6 ± 2.1	17.4 ± 1.9	9.4 ± 2.3	6
	18M	15.2 ± 2.2	33.7 ± 4.7*	16.2 ± 2.7	29.6 ± 3.7*	5

3.2. Immunohistochemistry of the Pancreas

In all animals under study, the pancreas was well preserved, without autolysis, and islets showed no signs of either florid inflammation or lymphocytes/macrophages infiltration. In both the control and n2-STZ group, the pancreas contained insulin-positive cells. In n2-STZ rats, both the volume fraction of insulin-positive cells and intensity of reaction signals during immunohistochemistry were comparable to controls (table 2). In these animals the islets were of comparable size to those in control animals, but of highly irregular form, and consisted of both insulin-positive and insulin-negative cells. The latter finding was quantified with the *Vv(insulin,islet)* parameter. The average volume proportion of insulin-positive cells within islets was 0.4008 in n2-STZ female animals and 0.4566 in control female animals. In males *Vv(insulin,islet)* reached 0.5467 in n2-STZ animals and 0.5763 in control males. The overall shrinkage of islet mass and depopulation of β-cells were present in n2-STZ group compared to control animals. This was accompanied by islet reorganization, impairing the natural structure of endocrine trabecular epithelium. Next to the typical islets, small clusters of β-cells were scattered throughout the whole pancreas. Morphology of islets and distribution of insulin-positive cells are presented in figure 4. Estimated volume fraction of insulin-positive cells within pancreatic islets *Vv(insulin, islets)* assessed in each sample are summarized together with blood glucose levels in table 2. The intraclass coefficient for two estimations of the sample C2 was 0.949 with a 95% confidence interval ranging between 0.890-0.988.

Table 2. Glycaemia and volume fraction of insulin-positive cells within pancreatic islets *Vv(insulin, islets)* in the 18-month-old diabetic (n2-STZ) and control (C) rats. The results of the *Vv(insulin,islets)* parameter are based on evaluation of a set of 9 micrographs in each tissue sample. *M* stays for male, *F* for female

Sample	Sex	Glycaemia [mmol/l]	*Vv(insulin,islets)*
C1	F	2.8	0.4538
C2	F	5.8	0.4593
C3	M	5.2	0.4851
C4	M	4.9	0.6493
C5	M	5.1	0.5944
n2-STZ1	F	4.3	0.4024
n2-STZ2	F	4.6	0.4419
n2-STZ3	F	4.3	0.3581
n2-STZ4	M	4.0	0.5037
n2-STZ5	M	4.7	0.5897

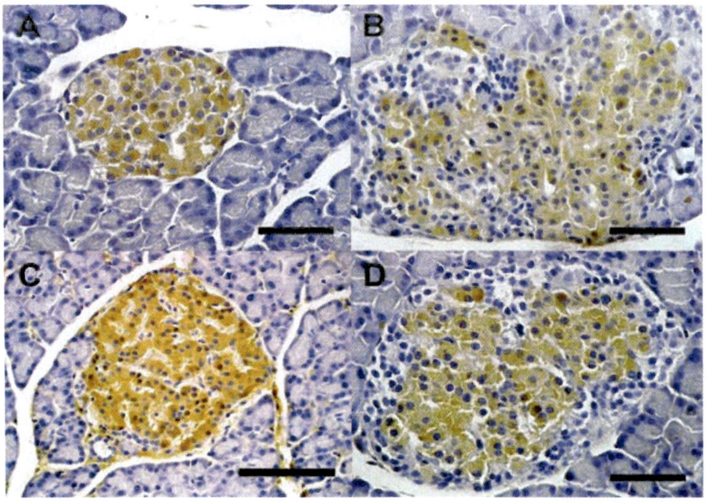

Figure 4. Immunohistochemistry of the pancreatic islets of 18-month-old rats showing insulin-positive cells (brownish). Counterstained with Gill's haematoxylin. A – sample C1 (control female), glycaemia 2.8 mmol/l, *Vv*=0.4538, bar 50 μm. B – sample n2-STZ3 (n2-STZ female), glycaemia 4.3 mmol/l, *Vv*=0.3581, bar 60 μm. C – sample C3 (control male), glycaemia 5.2 mmol/l, *Vv*=0.4851, bar 100 μm. D – sample n2-STZ5 (n2-STZ male), glycaemia 4.7 mmol/l, *Vv*=0.5897, bar 60 μm. Normal architecture of endocrine trabecular epithelium is present in figure 5A and 5C only. *Vv* stood for *Vv(insulin,islets)*.

3.3. Glucose Tolerance Tests

At the age of 3 months, blood glucose levels of n2-STZ rats of both sexes were significantly elevated and also their glucose tolerance tests indicated marked glucose intolerance (figure 5). Although blood glucose levels and plasma insulin concentrations were not significantly different between 18-month-old n2-STZ rat and their respective controls, their glucose tolerance was markedly different as shown in figure 5 for male rats. After intraperitoneal glucose challenge, n2-STZ rats of both sexes had blood glucose concentrations significantly higher than their respective controls (P < 0.01) and even after 3 hours they did not return to the fasting values. No gender-related differences were noted in the glucose tolerance at both ages studied.

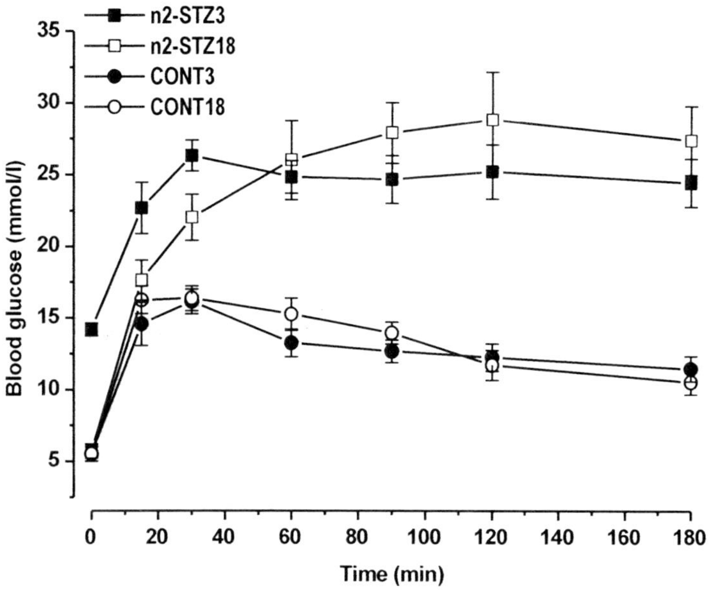

Figure 5. Blood glucose concentrations in given time intervals after intraperitoneal administration of glucose (2g/kg b.w.) to type 2 diabetic male rats at the ages of 3 and 18 months (n2-STZ3 and n2-STZ18) and their age-matched controls (CONT3 and CONT18). Values are means ± SEM (n=5).

3.4. Resting Heart Rate, Systolic and Diastolic Blood Pressures

As shown in table 3, there were no significant differences in the values of the resting heart rate and systolic blood pressure among all control and n2-STZ groups. In contrast, both male and female n2-STZ rats at the age of 18 months displayed diastolic and pulse pressures significantly higher and lower, respectively, when compared to the respective controls.

Table 3. Systolic pressure, diastolic pressure, pulse pressure and resting heart rate in 3- and 18-month-old control and diabetic (n2-STZ) male (M) and female (F) rats. Data are means ± SEM. *P< 0.05 compared with the respective control value (n=8 per group)

	Age	Control M	n2-STZ M	Control F	n2-STZ F
Systolic pressure [mmHg]	3M	123 ± 2	118 ± 2	119 ± 3	116 ± 3
	18M	127 ± 2	121 ± 2	126 ± 2	118 ± 2
Diastolic pressure [mmHg]	3M	68 ± 2	63 ± 2	69 ± 3	61 ± 3
	18M	67 ± 3	85 ± 4*	66 ± 2	81 ± 3*
Pulse pressure [mmHg]	3M	54 ± 3	56 ± 2	49 ± 2	55 ± 3
	18M	59 ± 3	37 ± 3*	60 ± 2	38 ± 2*
Resting heart rate [min^{-1}]	3M	370 ± 8	366 ± 8	367 ± 12	373 ± 8
	18M	359 ± 8	358 ± 12	371 ± 9	364 ± 9

3.5. CGRP Concentrations in the Arteries and Plasma of Control Rats and N2-STZ Rats

As shown in figure 6, CGRP concentrations in the selected arteries of control rats of both sexes reached the highest values in the mesenteric artery and the lowest ones in the thoracic aorta. No differences in the peptide levels were revealed between the respective specimens obtained from male and female rats. In addition, age did not have any impact on CGRP concentrations in the vessels studied.

Compared to age-matched control rats of both sexes, no differences in CGRP arterial concentrations were noted in 3-month-old diabetic animals. In contrast, the peptide levels were significantly decreased in the selected arteries of diabetic rats at the age of 18 months ($P < 0.05$; figure 7). No gender-related differences were revealed in the CGRP arterial tissue levels at both ages studied.

Diabetes lasting 3 months had no impact on CGRP concentrations in the plasma (table 4). However, at the age of 18 months, n2-STZ animals of both sexes displayed significantly lower CGRP plasma levels ($P < 0.05$).

Table 4. CGRP concentrations in the plasma of 3- and 18-month-old control and diabetic (n2-STZ) male (M) and female (F) rats. Data are means ± SEM. *P< 0.05 compared with the respective control value (n=6 per group)

	Age	Control M	n2-STZ M	Control F	n2-STZ F
CGRP in the plasma [pg/ml]	3M	319 ± 16	380 ± 25	353 ± 21	397 ± 25
	18M	360 ± 30	237 ± 36*	386 ± 40	207 ± 17*

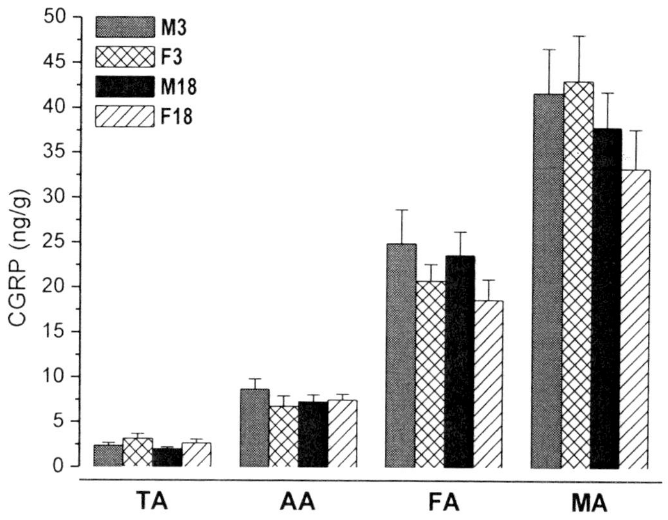

Figure 6. CGRP concentrations expressed in ng/g wet weight in the thoracic aorta (TA), abdominal aorta (AA), femoral artery (FA) and mesenteric artery (MA) of 3- and 18-month-old male (M3 and M18, respectively) and female (F3 and F18) control rats. Values are means ± SEM (n=7).

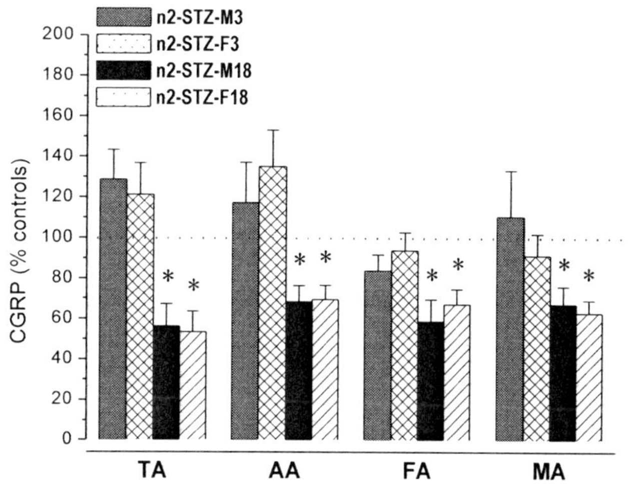

Figure 7. The impact of n2-STZ diabetes on CGRP concentrations in the thoracic aorta (TA), abdominal aorta (AA), femoral artery (FA) and mesenteric artery (MA) of male and female rats at the ages of 3 and 18 months (M3, F3, M18, F18, respectively) expressed in % of the respective control values (dotted line). Values are means ± SEM (n=6). *$P < 0.05$ compared with the respective control rats.

3.6. CGRP Concentrations in the Hearts of Control and Diabetic Rats

In 3-month-old control male rats (n=7), the mean values of CGRP concentrations in the heart compartments expressed in ng/g tissue wet weight were 10.6 ± 0.6 in the right atria, 11.1 ± 0.5 in the left atria, 6.6 ± 0.4 in the right ventricles, and 2.8 ± 0.3 in the left ventricles and they did not significantly differ from the respective values in the heart compartments of female rats of the same age (figure 8). CGRP levels in the atria were significantly higher than those in the ventricles in both males and females ($P < 0.05$). The values of cardiac CGRP concentrations slightly declined with age in both sexes. When compared to 3-month-old animals, in 18-month-old male rats the peptide levels reached by ~22%, 24% and 32% lower values in the atria, right and left ventricles, respectively. In the heart compartments of the female 18-month-old rats, the decline in the peptide concentrations associated with age seemed to be more expressed than in male rats, since the peptide cardiac concentrations reached ~60% of values obtained from rats at the age of 3 months. However, the absolute values of CGRP content in the heart did not significantly differ between aged males and females.

Compared to age-matched control rats of both sexes, no differences in CGRP cardiac concentrations were noted in 3-month-old diabetic animals. In contrast, the peptide levels were significantly decreased in all heart compartments of diabetic rats at the age of 18 months (figure 9). Interestingly, the absolute values of CGRP concentrations were significantly lower in female diabetic atria and left ventricles than in the male hearts, reaching 3.2 ± 0.3, 3.9 ± 0.3, 2.1 ± 0.2, and 0.7 ± 0.1 ng/g tisue wet weight in the female right atria, left atria, right ventricles, and left ventricles, respectively, and 5.3 ± 0.8, 5.4 ± 0.3, 2.9 ± 0.6, and 1.3 ± 0.2 ng/g in the respective male heart compartments.

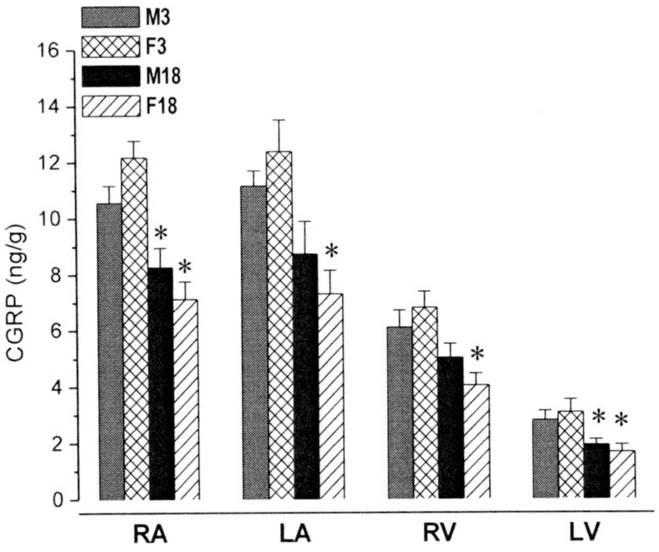

Figure 8. CGRP concentrations expressed in ng/g wet weight in the right (RA) and left atria (LA) and the free walls of the right (RV) and left ventricles (LV) of 3- and 18-month-old male (M3 and M18, respectively) and female (F3 and F18) control rats. Values are means ± SEM (n=7). *$P < 0.05$, compared with the respective value obtained from 3-month-old rats.

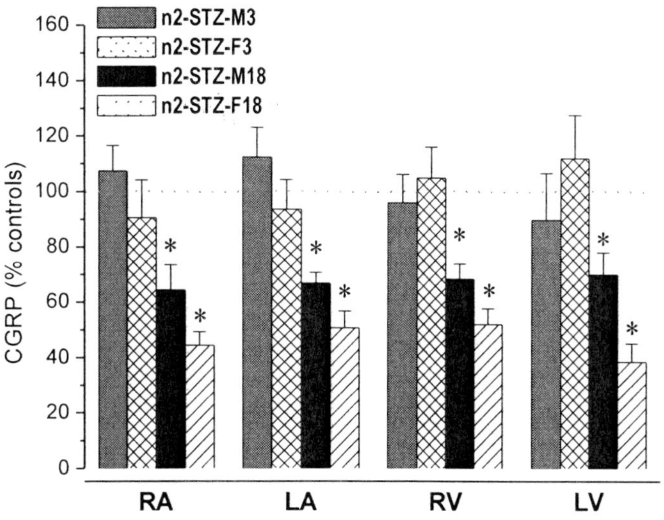

Figure 9. The impact of n2-STZ diabetes on CGRP concentrations in the right (RA) and left atria (LA) and the free walls of the right (RV) and left ventricles (LV) of male and female rats at the ages of 3 and 18 months (M3, F3, M18, F18, respectively) expressed in % of the respective control values (dotted line). Values are means ± SEM (n=6). *$P < 0.05$ compared with the respective control rats.

4. Discussion

The present study demonstrates significant changes in the cardiovascular and plasma concentrations of the sensory neuropeptide CGRP in male and female rats with experimental diabetes closely resembling type 2 diabetes in humans. The main findings reported in this chapter are: 1. A reliable model of type 2 diabetes in aged rats of both sexes was induced by STZ administration to 2-day-old animals (n2-STZ model). 2. Long-term type 2 diabetes led to significant increase in the diastolic blood pressure in both male and female rats. 3. CGRP concentrations in the aortae, femoral and mesenteric arteries were significantly decreased to the same extent in both male and female diabetic rats at the age of 18 months. 4. Long-term diabetes led to significant decline in the cardiac CGRP content that was more expressed in diabetic females compared to diabetic males.

4.1. Model of Type 2 Diabetes

The first step of this work was to evaluate the chosen model of experimental diabetes. Diabetes is a complex metabolic disorder characterized by either absolute insulin deficit due to the destruction of pancreatic β-cells and fasting hyperglycaemia (type 1) or rather glucose intolerance associated with defective function of pancreatic β-cells and insulin resistance (type 2). Whereas stable and reproducible model of type 1 diabetes can be induced in adult rats by a single STZ administration [33], reports on metabolic parameters in chemically induced rat models of type 2 diabetes differ in relation to the age and strain of experimental animals and to the dose of STZ used [34, 35]. In various models of type 2 diabetes induced in rats by STZ administration at birth or shortly after birth, (n0-, n2-STZ diabetes), STZ injection led to destruction of pancreatic β-cells that was followed by signs of regeneration of insulin-positive cells from the postnatal day 4 onwards. However, the regeneration process was still incomplete in 4-month-old animals [35]. To our best knowledge, morphological analysis of the endocrine pancreas of aged rats with regard to possible sexual dimorphism has not been published yet.

The results presented in this chapter show that Wistar rats injected at the age of 2 days with 100 mg/kg STZ displayed after 3 months fasting hyperglycaemia, low insulin plasma and pancreatic concentrations and marked intolerance to the glucose challenge, i.e. conditions rather resembling the symptoms of the human type 1 diabetes. However, at the age of 18 months, blood glucose and plasma insulin levels did not differ from the control values. The response to the glucose load indicated marked glucose intolerance in spite of significantly increased pancreatic insulin content, suggesting a secretory deficit and/or insensitivity of the target tissues to insulin. Immunohistochemical evaluation of the pancreatic tissue indicated that at the age of 18 months, n2-STZ pancreatic specimens displayed overall shrinkage of islet mass associated with islet reorganization and depopulation of β-cells despite comparable size of the islets in diabetic and control animals of both sexes.

Islets of Langerhans are far from only clusters of endocrine cells. Islets are complex structures with a vascular network, nerve supply, cell-to-cell contacts that are believed to be important for normal islet function [36]. As β-cells are the most abundant cells in the endocrine pancreas, changes in their biological dynamics are the most important factors that determine islets morphology [37]. The total volume (or mass) of pancreatic β-cells is a critical factor in the regulation of glucose homeostasis. Endocrine cells secreting insulin are dispersed through the entire gland in structural units ranging from single cells up to islets with many thousands cells. Spatial distribution of pancreatic endocrine cells is not random. Large islets are often located close to ducts and blood vessels whereas the smaller ones seem to be more randomly embedded within the endocrine tissue [38]. Thus, when counting the islets, an extreme variation in islet volume has to be considered.

In the present study, we estimated neither the volume fraction of islets nor their total volume within the whole pancreas. Both these parameters would have required much more extensive sampling of the whole pancreas; as the distribution of the islets within the pancreas is uneven. As the total insulin secretion correlates rather with total volume of the β-cells than with the volume fraction of β-cells within the islets, we are aware of the fact that the results presented in this chapter do not describe the functional capability of the whole endocrine pancreas to synthesize and release insulin. On that account, the *Vv(insulin,islets)* parameter might not be always straightforwardly correlated with glycaemia or total content of insulin in the pancreas. Provided that the distribution of the damage caused to β-cells by STZ was even in all parts of the pancreas, it could be still considered as one of quantitative parameters in the animals under study, where the quantitative differences between STZ-treated and control group would have been hardly diagnosed without stereological estimation.

In several juvenile rodents, there is a considerable dynamical development of the endocrine pancreas continuous with the fetal period. In mice, generation of new β-cells from replication of existing β-cells has been reported in the period from postnatal week 6 until 1 year later [39]. During physiological growth in Sprague-Dawley rats there was a linear correlation between the volume-weighted mean islet volume and the total islet volume (for review see [36]). Another study in Sprague-Dawley rats [40] demonstrated that the total β-cell volume increased from postnatal day 4 until day 16. This was followed by a plateau until day 24, after which the total β-cell volume increased further. This plateau was caused by β-cell hypotrophy as well as decreased β-cell formation. The β-cell volume (or mass) per unit body weight (the relative β-cell mass) was five times higher at birth compared with the adults constant level that was reached at approximately 24 days of age. In post-weaning young Wistar rat pancreas, the number of islets increased steadily with age from 3 to 122 weeks of age, whereas the volume-weighted mean volume of individual islets was not significantly different among all age groups of rats. There was a strong positive correlation between total islet number and islet mass, and between volume fraction and islet mass [37]. With use of stereological methods, the impact of genetic background on the size and structure of the endocrine pancreas has been investigated in mice [41]. Mouse strain, body weight and pancreas mass reached statistical significance as explanatory parameters for the islet volume, β-cell volume, and α-cell volume. Unfortunately, the study did not evaluate the possible differences between male and female mice.

Taken together, neonatal damage to pancreatic β-cells induced by STZ led to a transient phase of hyperglycaemia and hypoinsulinaemia that was followed by long-term regeneration of pancreatic endocrine tissue that was structurally and functionally impaired.

4.2. The Impact of Type 2 Diabetes on CGRP-Ergic Neurotransmission in the Cardiovascular System

Evaluation of the basal cardiovascular parameters, i.e. the resting heart rate, systolic and diastolic pressures in 3- and 18-month-old type 2 diabetic rats showed some interesting differences in comparison with similar studies dealing with type 1 diabetes induced by STZ in adult rats. Whereas in type 1 diabetic rats, bradycardia is a frequent finding [42], the resting heart rate in our type 2 diabetic rats did not differ from the control values at any time point studied. Data on blood pressure in type 1 diabetic rats are inconsistent, showing unchanged, increased or even decreased values of both systolic and diastolic pressures [43]. In contrast, studies dealing with regulation of blood pressure in type 2 diabetic rats mostly reported unchanged systolic, diastolic and mean blood pressures in animals aged 6-14 months [44-46]. Our study did not reveal any changes in the systolic pressure in type 2 diabetic animals of both sexes at the ages of 3 and 18 months. However, the diastolic pressure was significantly elevated in 18-month-old diabetic rats and the increase was comparable in both sexes. This increase could be at least partly attributed to the changes in CGRP concentrations found in the aortae, femoral and mesenteric arteries associated with decreased plasma levels of the peptide. This hypothesis is supported by several previously reported findings documenting implication of CGRP in the regulation of blood pressure: 1. Administration of capsaicin, ingredient of hot chilli pepper causing degeneration of unmyelinated sensory fibres [47], leads to a dramatic decrease in CGRP concentrations in the vasculature associated with decreased peptide levels in the plasma and elevated blood pressure [48]. 2. Complete deletion of the α-CGRP gene in mice produces a significant increase in basal mean arterial pressure [49, 50]. 3. In the spontaneously hypertensive rats, genetic model of essential hypertension, CGRP in arteries is down-regulated with age [51]. 4. CGRP synthesis and release are increased in various models of experimental hypertension, like subtotal nephrectomy [52], deoxycorticosterone-salt hypertension [53], two-kidney, one-clip [54], L-NAME-induced hypertension [55]. 5. Plasma CGRP is largely derived from the perivascular nerve terminals and is thought to represent a spillover phenomenon related to the release of the peptide [56, 57].

To our knowledge, changes in CGRP content in the diabetic rat heart have been documented in diabetes induced by STZ administration to young adult animals (type 1). In these reports, CGRP concentrations in the heart ventricles were significantly elevated 4-16 weeks after STZ treatment most probably due to intra-axonal accumulation of the peptide [58] and CGRP-immunoreactive nerve fibres were structurally degenerated in the heart valves 12 and 24 months after the onset of diabetes [59]. The impact of type 2 diabetes on the sensory innervation of the heart has not been documented yet. In our study, CGRP cardiac levels were unchanged in 3-month-old type 2 diabetic animals and they were significantly decreased at the age of 18 months. Interestingly, female diabetic hearts displayed significantly lower peptide concentrations than diabetic male ones.

Decrease in CGRP tissue content could be attributed to: a) greater degree of CGRP degradation, b) increased release of CGRP leading to a depletion of the intracellular pool of the peptide, c) structural degeneration of the sensory nerve endings, and/or d) decreased synthesis of the peptide.

Neutral endopeptidase (NEP) is an endothelial cell surface metalloproteinase, which is involved in the degradation of several regulatory peptides including natriuretic peptides, angiotensin II, bradykinin, substance P and CGRP released from the sensory nerve endings [60]. Although beneficial effects of NEP inhibition have been reported in spontaneously diabetic Goto-Kakizaki rats [61], direct effect of NEP inhibition on the bradykinin levels in the heart effluent showed rather decreased activity of the enzyme in the diabetic heart [62]. The relationship between NEP activity and CGRP concentrations in the heart has not been studied yet.

CGRP release from the human diabetic heart has been documented in a single study reporting no change in the peptide release stimulated by capsaicin [63]. In contrast, studies dealing with the impact of type 1 diabetes induced by STZ administration in rats indicated compromised CGRP release from various tissues including the heart [64-66].

CGRP contained in the heart originates mainly from nerve cell bodies localized in the nodose and dorsal root ganglia C6-T6 [67]. Ganglionic neurones sending their projections to different target organs, i.e. heart, lungs, vessels, upper part of the gastrointestinal tract, skin etc., are located in a very close proximity; target-specific impairment of individual classes of sensory neurons is thus difficult to evaluate. However, general observations indicated decreased CGRP content, lower expression of CGRP precursor mRNA and reduced number of CGRP-positive neurones in dorsal root ganglia of rats suffering from type 1 diabetes [68]. In addition, Murakawa et al. [69] reported functional and structural unmyelinated and small myelinated fibre neuropathy associated with impaired neuropeptide synthesis in the lumbar sensory ganglia of the persistently glucose intolerant and insulinopenic Goto-Kakizaki rats serving as a model for mild type 2 diabetes.

The expression of the calcitonin/CGRP gene in the nervous tissue is regulated by a number of hormones and second messengers including glucocorticoids, estrogens, progesterone, and cAMP [70, 71]. In particular, it has been documented that female sex steroids regulate CGRP synthesis in dorsal root ganglia [29] as well as the sensitivity of the coronary and mesenteric vasculature to the peptide [72]. In addition, numerous data indicate that circulatory CGRP levels are increased in women who use oral contraceptives or hormone replacement therapy [73, 74]. The finding of reduced circulating levels of estradiol in female rats 3 months after the onset of diabetes [75] gives rise the hypothesis of putative gender-specific impact of diabetes on CGRP synthesis. In our study, CGRP concentrations in the heart compartments of diabetic females at the age of 18 months were significantly lower than in the age-matched diabetic males. Thus, long-standing type 2 diabetes might have more prominent effect on CGRP levels in the female heart. This finding might be of clinical importance with regard to the role of CGRP in the protection of the heart against ischemic injury and in the mediation of ischemic preconditioning [76, 77]. It has been documented that the protection afforded by CGRP-mediated ischemic preconditioning is attenuated in aging and diabetic rat hearts [66, 78]. In addition, multiple lines of evidence suggest considerably higher diabetes-related relative risk for a major coronary heart disease event in diabetic women than in men [79, 80]. It is thus tempting to speculate that the differential gender effect on CGRP concentrations in type 2 diabetic heart reported in this chapter might contribute to the elimination of the "female advantage" in the prevalence of coronary heart disease in diabetics.

5. Conlusion

In conclusion, administration of 100 mg/kg STZ to 2-day-old rats led to the development of typical symptoms of type 2 diabetes at the age of 18 months. CGRP concentrations were significantly lower in the plasma, arteries and heart compartments of diabetic rats, reaching 40-70% of the control values. Interestingly, female diabetic hearts displayed significantly lower peptide concentrations than diabetic males. With regard to the well-documented cardioprotective and vasodilatory effects of CGRP, limited CGRP synthesis in neurons supplying cardiovascular tissue might be associated with the development of cardiovascular complications of type 2 diabetes, i.e. decreased protection of the myocardium against ischemic injury and increased total peripheral resistance. In addition, profound deficit in myocardial CGRP levels revealed in female diabetic rats might be implicated in the loss of "female advantage" in the prevalence of coronary heart disease frequently described in diabetic patients.

6. Acknowledgements

The present study was supported by the Research project MSM 0021620819: Replacement of and Support to Some Vital Organs awarded by the Ministry of Education, Youth and Sports of the Czech Republic and by the Grant Agency of the Czech Republic, grant No. 305/04/0144. We thank Šárka Faitová, Jaroslava Beránková and Kateřina Pícková for excellent technical assistance.

7. References

[1] Rosenfeld, MG; Mermod, JJ; Amara, SG; Swanson, LW; Sawchenko, PE; Rivier, J; Vale, WW; Evans, RM. Production of a novel neuropeptide encoded by the calcitonin gene via tissue specific RNA processing. *Nature,* 1983, 304, 129-136.

[2] Kittur, SD; Hoppener, JWM; Antonarakis, SE. Linkage map of the short arm of human chromosome 11: location of the genes for catalase, calcitonin, and insulin-like growth factor II. *Genetics,* 1985, 82, 5064-5067.

[3] Foster, GV; MacIntyre, I; Pearse, AEG. Calcitonin production and the mitochondrial rich cells of the dog thyroid. *Nature*, 1964, 203, 1029-1030.

[4] Becker, KL; Monaghan, KG; Silva, OL. Immunocytochemical localization of calcitonin in Kulchitsky cells of human lung. *Arch. Pathol. Lab. Med,* 1980, 104,196-198.

[5] Amara, SG; Jonas, V; Rosenfeld, MG; Ong, ES; Evans, RM. Alternative RNA processing in calcitonin gene expression generates mRNAs encoding different polypeptide products. *Nature,* 1982, 296, 240-244.

[6] Hoppener, JWM; Steenberg, PH; Zandberg, J; Lips, CJM. The second calcitonin/CGRP gene is located on chromosome 11. *Hum. Genet,* 1985, 70, 259-263.

[7] Wimalawansa, SJ; Morris, HR; Etienne, A; Blench, I; Panico, M; MacIntyre, I. Isolation, purification and characterization of β-hCGRP from human spinal cord. *Biochem. Biophys. Res. Commun,* 1990, 167, 993-1000.

[8] Gibbins, IL; Furness, JB; Costa, M. Pathway-specific patterns of the co-existence of substance P, calcitonin gene-related peptide, cholecystokinin and dynorphin in neurons of the dorsal root ganglia of the guinea-pig. *Cell Tissue Res,* 1987, 248, 417-437.

[9] Kashihara, Y; Sakaguchi, M; Kuno, MJ. Axonal transport and distribution of endogenous calcitonin gene-related peptide in rat peripheral nerve. *J. Neurosci,* 1989, 9, 3796-3802.

[10] Mulderry, PK; Ghatei, MA; Rodrigo, J; Allen, JM; Rosenfeld, MG; Polak, JM; Bloom, SR. Calcitonin gene-related peptide in cardiovascular tissues of the rat. *Neuroscience,* 1985, 14, 947-954.

[11] Holzer, P. Local effector functions of caosaicin-sensitive sensory nerve endings: involvement of tachykinins, CGRP and other neuropeptides. *Neuroscience,* 1988, 24, 739-768.

[12] McLatchie, LM; Fraser, NJ; Main, MJ; Wise, A; Brown, J; Thompson, N; Solari, R; Lee, MG; Foord, SM. RAMPs regulate the transport and ligand specificity of the calcitonin-receptor-like receptor. *Nature,* 1998, 393, 333-339.

[13] Wang, X; Fiscus, RR. Calcitonin gene-related peptide increases cAMP, tension, and rate in rat atria. *Am. J. Physiol,* 1989, 256, R421-R428.

[14] Gennari, C; Fischer, JA. Cardiovascular action of calcitonin gene-related peptide in humans. *Calcif Tissue Int,* 1985, 37, 581-584.

[15] Ishikawa, T; Okamura, N; Saito, A; Masaki, T; Goto, K. Positive inotropic effect of calcitonin gene-related peptide mediated by cyclic AMP in guinea pig heart. *Circ. Res,* 1988, 63, 726-734.

[16] Tam, C; Brain, SD. The assessment of vasoactive properties of CGRP and adrenomedullin in the microvasculature: a study using in vivo and in vitro assays in the mouse. *J. Mol. Neurosci,* 2004, 22, 117-124.

[17] Holzer, P; Wachter, C; Heinemann, A; Jocic, M; Lippe, IT; Herbert, MK. Sensory nerves, nitric oxide and NANC vasodilatation. *Arch. Int. Pharmacodyn. Ther,* 1995, 329, 67-79.

[18] Benrath, J; Eschenfelder, C; Zimmerman, M; Gillardon, F. Calcitonin gene-related peptide, substance P and nitric oxide are involved in cutaneous inflammation following ultraviolet irradiation. *Eur. J. Pharmacol,* 1995, 293, 87-96.

[19] Ito, H; Bell, D; Tamamori, M; Nozato, T; Shimojo, T; Adachi, S; Abe, S; Marumo, F; Hiroe, M. Calcitonin gene-related peptide (CGRP) and hypertrophy of cardiomyocytes. *Heart Vessels,* 1997, Suppl 12, 15-17.

[20] Manek, S; Terenghi, G; Shurey, C; Nishikawa, H; Green, CJ; Polak, JM. Angiogenesis and reinnervation in skin flaps: the effects of ischaemia examined in an animal model. *Int. J. Exp. Pathol,* 1994, 75, 243-255.

[21] Li, YJ; Xiao, ZS; Peng, CF; Deng, HW. Calcitonin gene-related-induced preconditioning protects against ischemia–reperfusion injury in isolated rat hearts. *Eur. J. Pharmacol,* 1996, 311, 163-167.

[22] Wimalawansa, SJ. Calcitonin gene-related peptide and its receptors: molecular genetics, physiology, pathophysiology, and therapeutic potentials. *Endocr. Rev,* 1996, 17, 533-585.

[23] Vinik, AI; Maser, RE; Mitchell, BD; Freeman, R. Diabetic autonomic neuropathy. *Diabetes Care,* 2003, 26, 1553-1579.

[24] Ziegler, D; Gries, FA; Muhlen, H; Rathmann, W; Spuler, M; Lessmann, F. Prevalence and clinical correlates of cardiovascular autonomic and peripheral diabetic neuropathy in patients attending diabetes centers. The Diacan Multicenter Study Group. *Diabete Metab,* 1993; 19, 143-151.

[25] Janka, HU. Increased cardiovascular morbidity and mortality in diabetes mellitus: identification of the high risk patient. *Diabetes Res. Clin. Pract,* 1996, 30 (Suppl), 85-88.

[26] Forsblom, CM; Sane, T; Groop, PH; Totterman, KJ; Kallio, M; Saloranta, C; Laasonen, L; Summanen, P; Lepantalo, M; Laatikainen, L; Matikainen, E; Teppo, AM; Koskimies, S; Groop, L. Risk factors for mortality in Type II (non-insulin-dependent)

diabetes: evidence of a role for neuropathy and a protective effect of HLA-DR4. *Diabetologia,* 1998, 41, 1253-1262.

[27] Colhoun, H. Coronary heart disease in women: why the disproportionate risk? *Curr. Diab. Rep,* 2006, 6, 22-28.

[28] Wilson, PW. Diabetes mellitus and coronary heart disease. *Am. J. Kidney Dis,* 1998, 32, S89-100.

[29] Gangula, PRR; Lanlua, P; Wimalawansa, SJ; Supowit SC; DiPette, DJ; Yallampalli, C. Regulation of calcitonin gene-related peptide expression in dorsal root ganglia of rats by female sex steroid hormones. *Biol. Reprod,* 2000, 62, 1033-1039.

[30] Howard, CV; Reed, MG. Unbiased Stereology: Three Dimensional Measurement in Microscopy. In: Royal Microscopical Society. *Microscopy Handbook Series No. 41.* New York: Springer-Verlag; 1998; 39-54.

[31] Shrout, PE; Fleiss, JL. Intraclass correlations: Uses in assessing rater reliability. *Psychol. Bull,* 1979, 2, 420-428.

[32] Gundersen, HJG; Jensen, EB. The efficiency of systematic sampling in stereology and its prediction. *J. Microsc,* 1987, 147, 229-263.

[33] Rakieten, N; Rakieten, ML; Nadkarni, ML. Studies on the diabetogenic action of streptozotocin (NSC-37917). *Cancer Chemother. Rep,* 1963, 29, 91-98.

[34] Ar'Rajab, A; Ahren, B. Long-term diabetogenic effect of streptozotocin in rats. *Pancreas,* 8, 1993, 50-57.

[35] Arulmozhi, DK; Veeranjaneyulu, A; Bodhankar, SL. Neonatal streptozotocin-induced rat model of Type 2 diabetes mellitus: a glance. *Ind. J. Pharmacol,* 2004, 36, 217-221.

[36] Bock T. The source(s) for new pancreatic beta cells in adult life. *Bioessays,* 2004, 26, 1156-1159.

[37] Inuwa, IM; El Mardi, AS. Correlation between volume fraction and volume-weighted mean volume, and between total number and total mass of islets in post-weaning and young Wistar rats. *J. Anat,* 2005, 206, 185-192.

[38] Skau, M; Pakkenberg, B; Buschard, K; Bock, T. Linear correlation between the total islet mass and the volume-weighted mean islet volume. *Diabetes,* 2001, 50, 1763-1770.

[39] Dor, Y; Brown, J; Martinez, OI; Melton, DA. Adult pancreatic beta-cells are formed by self-duplication rather than stem-cell differentiation. *Nature,* 2004, 429, 41-46.

[40] Svenstrup, K; Skau, M; Pakkenberg, B; Buschard, K; Bock T. Postnatal development of beta-cells in rats. Proposed explanatory model. *APMIS,* 2002, 110, 372-378.

[41] Bock, T; Pakkenberg, B; Buschard, K. Genetic background determines the size and structure of the endocrine pancreas. *Diabetes,* 2005, 54, 133-137.

[42] Hicks, KK; Seifen, E; Stimers, JR; Kennedy, RH: Effects of streptozotocin-induced diabetes on heart rate, blood pressure and cardiac autonomic nervous control. *J. Auton. Nerv. Sys,* 1998, 69, 21-30.

[43] Tomlinson, KC; Gardiner, SM; Hebden, A; Bennett T. Functional consequences of streptozotocin-induced diabetes mellitus, with particular reference to the cardivascular system. *Pharm. Rev,* 1992, 44, 103-150.

[44] Mozaffari, MS; Warren, BK; Russell, CM; Schaffer, SW. Renal function in the noninsulin-dependent diabetic rat: effects of unilateral nephrectomy. *J. Pharmacol. Toxicol. Meth,* 1997, 37, 197-203.

[45] Mozaffari, MS; Patel, C; Warren, BK; Schaffer, SW. NaCl-induced hypertensive rat model of non–insulin-dependent diabetes: role of sympathetic modulation. *Am. J. Hypertens,* 2000, 13, 540-546.

[46] Uriu, K; Kazo, K; Qie, YL; Kai, K; Eto, S. Effect of Acute Thromboxane A2 Inhibition on the Renal Hemodynamics in a Spontaneously Non-Insulin-Dependent Diabetic Rat, Otsuka Long-Evans Tokushima Fatty Rat. *J. Diab. Comp,* 1999, 13,182-186.

[47] Nagy, JI; Iversen, LL; Goedert, M; Chapman, D; Hunt, SP. Dose-dependent effects of capsaicin on primary sensory neurons in the neonatal rat. *J. Neurosci,* 1983, 3, 399-406.

[48] Wimalawansa, SJ. Effects of neonatal capsaicin therapy on plasma contents and tissue levels of immunoreactive-CGRP. *Peptides,* 1993, 14, 247-252.

[49] Gangula, PRR; Zhao, H; Supowit, SC; Wimalawansa, SJ; Dipette, DJ; Westlund, KN; Gagel, RF; Yallampalli, C. Increased blood pressure in alpha-calcitonin gene-related peptide/calcitonin gene knockout mice. *Hypertension,* 2000, 35, 470-475.

[50] Oh-hashi, Y; Shindo, T; Kurihara, Y; Imai, T; Wang, Y; Morita, H; Imai, Y; Kayaba, Y; Nishimatsu, H; Suematsu, Y; Hirata, Y; Yazaki, Y; Nagai, R; Kuwaki, T; Kurihara, H. Elevated sympathetic nervous activity in mice deficient in alphaCGRP. *Circ. Res,* 2001, 89, 983-990.

[51] Kawasaki, H; Saito, A; Takaksaki, K. Age-related decrease of calcitonin gene-related peptide-containing vasodilator innervation in the mesenteric resistance vessel of the spontaneously hypertensive rat. *Circ. Res,* 1990, 67, 733-743.

[52] Supowit, SC; Zhao, H; Hallman, DM; DiPette, DJ. Calcitonin gene-related peptide is a depressor in subtotal nephrectomy hypertension. *Hypertension,* 1998, 31, 391-396.

[53] Supowit, SC; Zhao, H; Hallman, DM; DiPette, DJ. Calcitonin gene-related peptide is a depressor of deoxycorticosterone-salt hypertension in the rat. *Hypertension,* 1997, 29, 945-950.

[54] Deng, PY; Ye, F; Zhu, HQ; Cai, WJ; Deng, HW; Li, YJ. An increase in the synthesis and release of calcitonin gene-related peptide in two-kidney, one-clip hypertensive rats. *Regul. Pept,* 2003, 114, 175-182.

[55] Gangula, PRR; Supowit, SC; Wimalawansa, SJ; Zhao, H; Hallmann, DM; Dipette DJ; Yallampalli, C. Calcitonin gene-related peptide is a depressor in N^G-nitro-L-arginine methyl ester-induced hypertension during pregnancy. *Hypertension,* 1997, 29, 248-253.

[56] Zaidi, M; Bevis, PJR; Girgis, SI; Lynch, C; Stevenson, JC; MacIntyre, I. Circulating CGRP comes from perivascular nerves. *Eur. J. Pharmacol,* 1985, 117, 283-284.

[57] Zaidi, M; Bevis, PJR; Abeyasekera, G; Girgis, SI; Wimalawansa, SJ; Morris, HR; MacIntyre, I. The origin of calcitonin gene-related peptide in the rat. *J. Endocrinol,* 1986, 110, 185-190.

[58] Chottová Dvořáková, M; Kuncová, J; Pfeil, U; Mcgregor, GP; Švíglerová, J; Slavíková, J; Kummer, W. Cardiomyopathy in streptozotocin-induced diabetes involves intra-axonal accumulation of calcitonin gene-related peptide and altered expression of its receptor in rats. *Neuroscience,* 2005, 134, 51-58.

[59] Kumar, SD; Tay, SS. Changes in peptidergic nerves in the atrioventricular valves of streptozotocin-induced diabetic rats: a confocal microscopy study. *Anat. Rec,* 2000, 258, 277-285.

[60] Le Greves, P; Nyberg, F; Hokfelt ,T; Terenius, L. Calcitonin gene-related peptide is metabolized by an endopeptidase hydrolyzing substance P. *Regul. Pept,* 1989, 25, 277-286.

[61] Gronholm, T; Cheng, ZJ; Palojoki, E; Eriksson, A; Backlund, T; Vuolteenaho, O; Finckenberg, P; Laine, M; Mervaala, E; Tikkanen, I. Vasopeptidase inhibition has beneficial cardiac effects. *Eur. J. Pharmacol,* 2005, 519, 267-276.

[62] Koch, M; Wendorf, M; Dendorfer, A; Wolfrum, S; Schulze, K; Spillmann, F; Schultheiss, HP; Tschope, C. Cardiac kinin level in experimental diabetes mellitus: role of kininases. *Am. J. Physiol. Heart Circ. Physiol,* 2003, 285, H418-H423.

[63] Strecker, T; Dieterle, A; Reeh, PW; Weyand, M; Messlinger, K. Stimulated release of calcitonin gene-related peptide from the human right atrium in patients with and without diabetes mellitus. *Peptides,* 2006, in press.

[64] Nemeth, J; Szilvassy, Z; Thanb, M; Oroszi, G; Sari, R; Szolcsanyi; J. Decreased sensory neuropeptide release from trachea of rats with streptozotocin-induced diabetes. *Eur. J. Pharmacol,* 1999, 369, 221-224.

[65] Ralevic, V; Belai, A; Burnstock, G. Impaired sensory-motor nerve function in the isolated mesenteric arterial bed of streptozotocin-diabetic and ganglioside-treated streptozotocin-diabetic rats. *Br. J. Pharmacol,* 1993, 110, 1105-1111.

[66] Lu, R; Hu, CP; Peng, J; Deng, HW; Li, YJ. Role of calcitonin gene-related peptide in ischaemic preconditioning in diabetic rat hearts. *Clin. Exp. Pharmacol. Physiol,* 2001, 28, 392-396.

[67] Hopkins, DA; Armour, JA. Ganglionic distribution of afferent neurons innervating the canine heart and cardiopulmonary nerves. *J. Auton. Nerv. Syst,* 1989, 26, 213-222.

[68] Rittenhouse, PA; Marchand, JE; Chen, J; Kream, RM; Leeman, SE. Streptozotocin-induced diabetes is associated with altered expression of peptide-encoding mRNAs in rat sensory neurons. *Peptides,* 1996, 17,1017-1022.

[69] Murakawa, Y; Zhang, W; Pierson, CR; Brismar, T; Ostenson, CG; Efendic, S; Sima, AA. Impaired glucose tolerance and insulinopenia in the GK-rat causes peripheral neuropathy. *Diabetes Metab. Res. Rev,* 2002, 18, 473-83.

[70] Van Rossum, D; Hanisch, UK; Quirion, R. Neuroanatomical localization, pharmacological characterization and functions of CGRP, related peptides and their receptors. *Neurosci. Biobehav. Rev,* 1997, 21, 649-678.

[71] Yang, Y; Ozawa, H; Lu, H; Yuri, K; Hayashi, S; Nihonyanagi, K; Kawata, M. Immunocytochemical analysis of sex differences in calcitonin gene-related peptide in the rat dorsal root ganglion, with special reference to estrogen and its receptor. *Brain Res,* 1998, 791, 35-42.

[72] Gangula, PRR; Zhao, H; Wimalawansa, SJ; Supowit, SC; DiPette, DJ; Yallampalli, C. Pregnancy and steroid hormones enhance the systemic and regional hemodynamic effects of calcitonin gene-related peptide (CGRP) in rats. *Biol. Reprod,* 2001, 64, 1776-1783.

[73] Spinetti, A; Margutti, A; Bertolini, S; Bernardi, F; BiFulco G; Degli Uberti, EC; Petraglia, F; Genazzani, AR. Hormonal replacement therapy affects calcitonin gene-related peptide and atrial natriuretic peptide secretion in postmenopausal women. *Eur. J. Endocrinol,* 1997, 137, 664-669.

[74] Valdemarsson, S; Edvinsson, L; Hedner, P. Hormonal influence on calcitonin gene-related peptide in man: effects of sex difference and contraceptive pills. *Scand. J. Clin. Lab. Incest,* 1990, 50, 385-388.

[75] Wells, CC; Riazi, S; Mankhey, RW; Bhatti, F; Ecelbarger, C; Maric, C. Diabetic nephropathy is associated with decreased circulating estradiol levels and imbalance in the expression of renal estrogen receptors. *Gend. Med,* 2005, 2, 227-237.

[76] Atkins, BZ; Silvestry, SC; Samy, RN; Shah, AS; Sabiston, DC; Glower, DD. Calcitonin gene-related peptide enhances the recovery of contractile function in stunned myocardium. *J. Thorac. Cardiovasc. Surg,* 2000, 119, 1246-1254.

[77] Li, YJ; Peng, J. The cardioprotetion of calcitonin gene-related peptide-mediated preconditioning. *Eur. J. Pharmacol,* 2002, 442, 173-177.

[78] Lu, R; Hu, CP; Deng, HW; Li, YJ. Calcitonin gene-related peptide-mediated ischemic preconditioning in the rat heart: influence of age. *Regul. Pept,* 2001, 99, 183-189.

[79] Pradhan, AD; Skerrett, PJ; Manson, JE. Obesity, diabetes, and coronary risk in women. *J. Cardiovasc. Risk,* 2002, 9, 323-330.

[80] Juutilainen, A; Kortelainen, S; Lehto, S; Rönnemaa, T; Pyörälä, K; Laakso, M. Gender difference in the impact of type 2 diabetes on coronary heart disease risk. *Diabetes Care*, 2004, 27, 2898–2904.

In: Neuropeptide Research Trends
Editor: Bernice A. Levine, pp. 169-188
ISBN: 978-1-60021-640-4

Chapter VI

The Role of Dynorphins in Homeostasis: Implications for Seizures and Postictal Phenomena

Marylou V. Solbrig[*1,2,3] ***and Marco Bortolato***[4]
[1] Department of Neurology, [2] Medicine, [3] Pharmacology, University of California-Irvine, Irvine, CA, USA and [4] Department of Neuroscience "B.B.Brodie", University of Cagliari, Cagliari, Italy

Abstract

Epilepsy remains a major medical problem. Despite widespread use of both classic and newer pharmacologic agents that target ion channels, amino acid transmission or receptors, there are many examples of therapeutic failures. The study of the neuropharmacologic actions of opioids in CNS has yielded new insights into epileptogenesis and epilepsy treatment. In animals, the dynorphin system in the hippocampus is known to regulate excitability and epilepsy vulnerability has been associated with a dynorphin promoter region polymorphism, or low dynorphin expression genotype, in man. Encephalitic temporal lobe viruses, such as herpes and Borna Disease virus in rats, cause seizures by reproducing this neurochemical marker of epilepsy, namely low dynorphin tone, and restoration of dynorphin tone with a kappa opioid receptor agonist reverses their seizures. This work has renewed interest in the role of dynorphin dysregulation in the pathogenesis of epilepsy and kappa opioids in anticonvulsant protection, but leaves open to question a full understanding of kappa opioid mechanisms of action and side effect profiles. In particular, recent human and animal studies indicate that activation of kappa opioid receptors also elicits dysphoric and hallucinatory responses, the same responses reported to accompany cessation of seizures in temporal lobe epilepsy. Altogether, the evidence is consistent with dynorphin system and kappa opioid receptors being key factors in the regulation of excitability in

* Correspondence to: Marylou V. Solbrig, Department of Neurology, 3226 Gillespie Neuroscience Research Building, University of California-Irvine, Irvine, CA 92697-4292, USA; Tel 949 824-6270 FAX 949 824-3135 email msolbrig@uci.edu

hippocampus, and in shaping the clinical presentation of the psychiatric disorders in postictal phenomena.

Introduction

Opiates were first derived from extracted juice of the opium poppy (*Papaver somniferum*), and "opiates" were preparations or semisynthetic derivatives of opium. "Opioid" signified synthetic drugs with morphine-like actions, and the term was applied to all drugs, natural and synthetic, with morphine-like actions. Opioids also include the endogenous "morphine-like" substances (such as endorphins) that bind the same brain site (receptors) as opioids, and antagonists of opioid drugs. Major historical uses of opioids were treatment of pain by the ancient Greeks and medieval physicians, and treatment of diarrhea during dysentery epidemics, recorded in medical literature in the 1600's [Koob and LeMoal 2006].

The opioid system includes several families of related neuropeptides: β-endorphins, enkephalins, dynorphins; and mu, delta, kappa opioid receptors. The presence of Tyr-Gly-Gly-Phe-Leu (Leu-enkephalin) in the first five residues of dynorphin A, for example, categorizes the peptide as an endogenous ligand for opioid receptors, and kappa opioid peptide receptor activation accounts for many of the biological activities of dynorphin [Chavkin 1982]. Dynorphin is involved in a range of biological functions, nociception, mood, motor activity, homeostatic response to injury, hypothermia, inflammation, neuroprotection, spatial memory and seizures.

Electrophysiology and animal models have been critical for understanding neuropharmacological mechanisms of opioids in seizures and seizure control. For example, electrical and chemical convulsant experiments showing release of dynorphin in hippocampus led to a hypothesized role for dynorphin and kappa opioids in seizure control. These observations would have implications for understanding neuroadaptive responses of the brain to overstimulation, for developing anticonvulsant treatments, for epilepsy risk assessment, and for understanding ictal and postictal behaviors associated with temporal lobe epilepsy.

Electrophysiology

The pioneering cellular electrophysiological work on opioid peptides used the hippocampus in slice preparations as a model system. Opioids produced an excitation of neurons in CA1 and CA3 regions of hippocampus, with these effects mediated by indirect inhibitory action on interneurons and/or presynaptic terminals. These observations gave rise to a disinhibition theory of mu and delta opioid peptide action at the cellular level [Zieglgansberger et al. 1979, Siggins and Zielgansberger, 1981]. Yet, *in vitro* cellular recordings from rat hippocampus showed that at least 10% of CA3 neurons exhibit inhibitory responses to electrophoretically applied dynorphin peptides [Henriksen et al. 1982a, Henriksen et al. 1982b]. Subsequent studies showed that dynorphin has a strong inhibitory effect in the hippocampus [Wagner et al. 1993, Weisskopf et al. 1993] with inhibition mediated by via presynaptic inhibition of glutamate release or postsynaptic current stabilization.

Dynorphin in the hippocampus is known to regulate excitability by several mechanisms. Activation of the KOR by dynorphin causes activation of the Shaker-type potassium channel (a dendrotoxin-sensitive Kv1 type delayed rectifying potassium channel) on mossy fibers in the hippocampus, thereby limiting calcium influx and presynaptic neurotransmitter release [Simmons and Chavkin, 1996]. On pyramidal neurons, dynorphin acting on kappa opioid receptors was reported to augment voltage-dependent K M currents (I_M) [Moore et al. 1994, Madamba et al. 1999]. However, these results await confirmation with cellular recordings from cloned Kv7 (KCNQ) potassium channels in heterologous expression systems. Because I_M opposes membrane depolarization, compounds that increase I_M diminish neuronal excitability.

These electrophysiology observations have taken on added significance when it was realized the study of dynorphin could elucidate cellular and pharmacologic mechanisms of temporal lobe epilepsy.

Endogenous Dynorphins in Brain

Dynorphin, as other neuropeptides, occurs in the medial domain or core regions of brain. The core includes greater limbic system, neostriatum, hippocampus, protions of neocortex, and medial brainstem [Nieuwenhuys 2000]. The core region is involved in the organization of specific, motivated or goal-oriented behaviors directly aimed at the maintenance of homeostasis and survival. For example, opioids in hippocampus are involved in consolidation of memory. Dynorphin disrupts and enkephalin is important in acquisition and retrieval of spatial memory [McDaniel et al. 1990, Sandin et al. 1998, Meilandt et al. 2004, Nguyen et al. 2005]. Opioids are in areas of hippocampus that receive, gate or process strong excitatory input. Dynorphin is present in granule cells, enkephalin in granule cells, and enkephalin in CA1 interneurons that receive Schaffer collaterals.

Prodynorphin-derived peptides are synthesized in granule cells of the dentate gyrus in the hippocampus of rodents and humans and are transported to axon terminals of the CA3 region. Granule cells are the only neural cell type in the hippocampus to express and release dynorphin [McGinty et al. 1983]. Their axons are the mossy fibers that release glutamate and other peptides (enkephalin or cholecystokinin) and terminate in the *stratum lucidum* proximal to the CA3 pyramidal cell bodies [McGinty et al. 1983, Chavkin et al. 1985, Gall et al. 1986, McLean et al. 1987]. Studies of *in vitro* and animal models of epilepsy show that hippocampal dynorphin is transported and released within the mossy fiber system during seizures, redistributing to CA3 region axons [Gall et al. 1990, Weisskopf et al. 1993, Simmons and Chavkin, 1996], recurrent mossy fiber collaterals [Terman et al. 2000], as well as mossy fiber dendrites [Drake et al. 1994] in a manner consistent with it acting to counter or attenuate increases in excitatory neurotransmission.

Dynorphin-mediated inhibition occurs at several different dentate granule cell synapses or loci. Granule cell excitation releases dynorphin from granule cell axon terminals in the CA3 region [Weisskopf et al. 1993, Simmons and Chavkin, 1996], binding to presynaptic kappa receptors of mossy fibers, providing feedback inhibition of both excitatory transmission and long term potentiation [Weisskopf et al. 1993]. Here the targets of released dynorphin are the mossy fiber terminals themselves. At a different site, in the dentate molecular layer, dendritic stores of dynorphin are released by granule cell excitation and pass retrogradely. Dendritic dynorphin release is by L-type calcium channel stimulation [Simmons et al. 1995]. Release of dynorphin peptides from dense core vesicles in granule cell dendrites activate kappa opioid receptors present on perforant path terminals and inhibit excitatory amino acid release, thereby inhibiting further excitatory neurotransmission and long-term potentiation induction at this synapse [Drake et al. 1994]. Mossy fiber collateral excitatory hilar input to granule cells is distinct from perforant path inputs. In this, the third case, hilar dynorphin from mossy fiber collaterals inhibits hilar path granule cell excitation. Dynorphin released from axonal stores in both the hilus and CA3 is independent of L-type calcium channels, relying instead, at least in CA3, on N-type calcium channels [Simmons et al. 1995].

Dynorphin (Kappa Opioid) Receptors in Brain

In the normal brain, dynorphin and kappa opioid receptors (KORs) appear strategically placed to protect against overstimulation, preferentially distributed to hippocampal regions with strong excitatory synaptic input [Solbrig and Koob 2004]. In hippocampus, KORs are in dentate gyrus, CA3 and CA1 subfields [Tempel and Zukin, 1987, Zukin et al. 1988, Mansour et al. 1994, Arvidsson et al. 1995, Drake et al. 1996], enabling dynorphin or any other KOR ligands to function as endogenous anticonvulsants as channel and transmitter neuromodulators.

Kappa opioid receptors in dentate gyrus are located presynaptically on perforant pathway afferents (in guinea pig [Wagner et al. 1993, Drake et al. 1994, Drake et al. 1996]) on granule cell (mossy fiber) terminals (in rat [Tempel and Zukin, 1987, Zukin et al. 1988] and guinea pig [Weisskopf et al. 1983, Drake et al. 1994, Drake et al. 1996, Castillo et al. 1996]). In CA3 and CA1 subfields, kappa opioid receptors are found in pyramidal cell layers of rat [Tempel and Zukin, 1987, Zukin et al. 1988] and at postsynaptic sites on CA3 and CA1 pyramidal neurons in rat [Mansour et al. 1994] and guinea pig [Arvidsson et al. 1995]. In addition, kappa opioid receptors are thought to exist in the hilar mossy fiber collateral-granule cell network (in guinea pig [Terman et al. 2000]), on supramammillary afferents that terminate on granule cell dendritic shafts and perikarya (in guinea pigs [Drake et al. 1987]), and on dendrites and soma of GABA, somatostatin and neuropeptide Y (GABA/SOM/NPY)-containing interneurons of the dentate hilus and the *stratum oriens* layer of CA1 (in rat) [Halasy et al. 2000, Racz and Halasy, 2002].

Dynorphin released from granule cells binding KOR on perforant pathway terminals in the molecular layer, on mossy fiber terminals in hilus, or on granule cell axon terminals would counter hippocampal excitation. Site-specific release and actions of dynorphin were discussed in the previous section.

Dynorphin action on postsynaptic KOR is not as well established. Both CA3 and CA1 pyramidal neurons possess mRNA that encodes KOR [Mansour et al. 1994]. It has been reported that KOR agonists mediate increases in M-type potassium currents to stabilize membrane potentials of CA3 pyramidal neurons [Moore et al. 1994]. Since most CA1 neurons are remote from the release of dynorphin from mossy fibers which terminate proximal to CA3 pyramidal cells, either the enkephalins of CA1 interneurons [Gall et al. 1984, McLean et al. 1987] or dynorphin volume conduction might be involved in activating KOR of CA1 pyramidal neurons.

KOR are also reported to be present on a class of GABA-containing interneurons of CA1. KOR stimulation of selective GABA-containing interneurons could desynchronize GABA inputs to their postsynaptic network. Desynchronization of signals from GABA-containing interneurons is one mechanism of enhanced inhibition of principal cells [Aradi et al. 2002]. How GABA-containing interneurons actually respond, sensing residual dynorphin by volume (non-synaptic) transmission, is possible but not yet proven.

Besides synaptic release, non-synaptic interneural communication is thought to play an important role in the organization and regulation of behavior by the neuropeptides of core brain structures [Nieuwenhuys 2000]. Released dynorphin can act synaptically or at a distance, but most study has been devoted to actions of dynorphin at receptor sites close to sites of release. That dynorphin is the most basic of neurotransmitters [Hauser et al., 2005] raises the possibility that extracellular dynorphin could also function for pathological conditions during which severe acidosis occurs, to maintain neural excitation at moderate levels for example, through inhibition of acid-sensing ion channels (ASIC) [Yermolaieva et al., 2004]. Acidosis, as well as excitatory stimulation, has been implicated in seizures [Xiong et al., 2004].

Dynorphin as Endogenous Anticonvulsant

Together, anatomic, electrophysiologic, pharmacologic, and molecular studies have backed up a role for dynorphin as a regulatory of excitability [Kanamatsu et al., 1986, Tortella et al. 1986, Tortella 1988, Gall et al. 1990, Houser et al. 1990, Douglass et al. 1991, Hong et al. 1992, Przewlocka et al. 1994, Romualdi et al. 1995, Simonato and Romualdi, 1996, deLanerolle et al. 1997, Bausch et al. 1998, Mazarati et al. 1999, Pierce et al., 1999]. The basic relation between seizures and hippocampal dynorphin is release of dynorphin in first minutes after seizure, followed by activation of prodynorphin gene expression, new peptide synthesis with overproduction, then rebound decrease in prodynorphin mRNA, although different seizure models may produce varying quantitative and temporal relationships [Kanamatsu et al. 1986, Morris et al. 1987, Xie et al. 1989, Douglass et al. 1991, Harrison et al. 1995, Romualdi et al. 1995, Simonato and Romualdi 1996, Rocha and Maidement 2003]. Actions of dynorphin on its receptor protects against seizures, with KOR agonists having anticonvulsant efficacy in a broad range of experimental seizure models. KOR agonists are effective against maximal electroshock- [Tortella et al. 1986, Manocha et al. 2003] pilocarpine- [Przelocka et al. 1994, Bausch et al. 1998] bicuculline- [Yajima et al. 2000] and excitatory amino acid-induced convulsions [Von Voigtlander et al. 1987] and limit electrical kindling [Albertson et al. 1984] or kindling of seizures by the GABA A antagonist pentylenetetrazol [Becker et al. 1999]. Our own work has demonstrated a role for KOR agonist stimulation and dynorphin in controlling seizures associated with CNS viral injury and encephalitis in rodent models [Solbrig et al. 2006a, Solbrig et al. 2006b].

Evidence for the involvement of dynorphin in temporal lobe epilepsy has also been obtained from human studies. Aberrant dynorphin immunoreactivity in mossy fiber terminals in the molecular layer in *post-mortem* specimens is characteristic of temporal lobe epilepsy in man [deLanerolle et al. 1997]. In this distribution, dynorphin could function as an endogenous anticonvulsant and dampen perforant path excitatory input. The surgical specimen of a patient who had 2-3 complex partial seizures in the 48 hours before surgery had temporal lobe dynorphin depletion [Houser et al. 1990], to suggest either dynorphin reduction was the proximate cause of seizures, or seizures lead to further dynorphin depletion and an accelerating cycle of seizures. Recently, an allelic variant in man with a dynorphin promoter region polymorphism, presumed to confer a low dynorphin expression genotype, was identified. Furthermore, its association with temporal lobe epilepsy in familial-risk patients (Stogmann et al., 2002) supports a role for dynorphin tone in seizure control.

Genetics

The "candidate gene approach" which is based on the knowledge of biological mechanisms or pathways of disease, is a way to search for specific genetic variations that contribute to complex disorders [Lander and Schork, 1994]. Individual variations in genes encoding proteins involved in regulation of excitability or excitatory transmission have been sought to explain some of the genetic bases for individual differences in vulnerability to epilepsy. Molecular variation in many genes, including the prodynorphin gene, have been examined.

A 68-base pair repeat polymorphism within the core promoter region of the human prodynorphin gene has been reported to occur in alleles containing one, two, three, or four copies. This repeat contains a putative AP-1 transcription factor binding site, and reporter gene constructs with three or four, but not one or two copies of the tandem repeats were shown to be associated with approximately 50% increases in transcriptional activation in *in vitro* cellular assays [Zimprich et al. 2000].

Two studies have shown that allelic variation at the promoter region of the prodynorphin gene may contribute to individual variation in vulnerability to seizures or epilepsy. L-alleles, or low expression alleles, conferred an increased risk for temporal lobe epilepsy in patients with a family history of seizures and L-homozygotes had a higher risk for secondary generalized seizures and *status epilepticus*, irrespective of family background [Stogmann et al. 2002]. L-homozygotes were more likely to have suffered from febrile convulsions [Stogmann et al. 2002], and L-homozygotes were twice as prevalent in the seizure group that had a family history [Gambardella et al. 2003]. Thus, a risk factor for disease may be encoded in regulatory regions. Gene promoter variability, i.e. weakness of promoter, may lead to sporadic forms of disease, While gene x environment interactions may lead to premature disease expression or progression.

The identification of genes or genetic polymorphisms that either cause or increase risk of seizures will be important for improving methods for diagnosis and treatment. The identification of genes or genetic polymorphisms that either cause or increase risk of seizures will also be important for risk assessment and for developing hypotheses of epilepsy susceptibility.

Opioid Mediated Viral Risk Factors

An estimated half of encephalitis patients experience seizures and about 4% develop *status epilepticus*. Possibly viruses can be environmental triggers for the development of seizures in genetically vulnerable individuals, with risk associated with variant opioid expression. In our lab, the specific hypothesis that reduced dynorphin expression in dentate gyrus of hippocampus due to virus exposure leads to epileptic responses has been tested using rodent models of encephalitis. This was explored with 2 different neurotropic viruses, Borna Disease virus (BDV) and herpes simplex virus type-1 (HSV-1).

BDV is an encephalitic veterinary pathogen, possibly a human pathogen, with wide natural and experimental host range. BDV causes hippocampal injury in many mammalian species [Narayan et al. 1983, Gosztonyi and Ludwig 1995] and has been reported in hippocampal sclerosis in humans [Czygan et al. 1999]. BDV in rats causes a persistent infection with time-based neurological sequelae, one of which has an epileptiform phenotype [Solbrig et al. 2006a].

BDV infection of rats decreases hippocampal dynorphin and causes limbic seizures. The chief damaging elements were viral-induced dentate granule cell loss, failed neurogenesis, poor survival of new neurons, depletion of dynorphin with seizures, and inability to restore dynorphin to meet demand. Treatment with a selective kappa opioid agonist prevented behavioral and electrographic seizures while administration of a kappa antagonist duplicated the convulsive syndrome [Solbrig et al. 2006a]. Thus, viral infection reproduced a neurochemical marker of epilepsy, namely, low dynorphin tone.

The possibility of other encephalitic viruses causing low dynorphin and seizures was explored using another virus with tropism for the hippocampus, HSV-1. HSV-1 is the most common cause of sporadic encephalitis in the US, and up to 89% of patients with HSV-1 encephalitis can have seizures [Panagariya et al. 2001]. In rats, HSV-1 infection caused loss of dynorphin A-like immunoreactivity in hippocampus and seizures. As in Borna rats, a kappa opioid receptor agonist effectively blocked ictal activity, linking absence of dynorphin to propensity for epileptic activity. However, in contrast to Borna rats, the effect of HSV-1 encephalitis on granule cell dynorphin was independent of direct viral interference and cell loss. Dentate hilar cells and not granule cells were infected, and granule cell layers appeared intact [Solbrig et al. 2006b]. The presence of NeuN positive but Dynorphin A negative granule cells would be consistent with dynorphin depletion with seizures and inability to restore dynorphin to meet demand, similar to the suggestion for the BD rats. The findings show a vulnerability of hippocampal dynorphin during infection in both acute encephalitis due to HSV-1 and chronic encephalitis due to Borna Disease Virus, suggesting a neurochemical bases for seizures that may be generalizable to still other encephalitic viruses.

Opioid Withdrawal Seizures

The encephalitic rats present an interesting parallel to opiate withdrawal seizures. Systemic injection of the competitive opioid antagonist naloxone precipitates seizures in rats with chronic encephalitis due to Borna Disease (BD) virus [Solbrig et al., 1996, 2006a] and in rats with acute herpes simplex virus type-1 (HSV-1) encephalitis [Solbrig et al., 2006b].

Seizures often accompany alcohol or barbiturate withdrawal, but are not typically part of narcotic withdrawal or abstinence syndromes. Although the adaptive changes in excitatory synaptic transmission during acute opiate withdrawal do not usually produce seizures, there are exceptions. Seizures occur in up to 8% of neonates exposed to heroin or methadone *in utero* [Osborn et al., 2005] and one-third of critically ill adults after sudden withdrawal of narcotic agents [Wijdicks and Sharbrough, 1993]. In man, the recommendations for treating neonatal abstinence seizures with paregoric, a mixture of opium and alcohol [Osborn et al. 2005] supports a role for hypofunction or imbalance of endogenous opioid systems in seizure genesis.

Recently, *in utero* opioid exposure in rats was reported to produce long-lasting changes in neurons associated with hippocampal opioid circuits and sensitized to naloxone seizures [Schindler et al., 2004]. BDV infected rats, through similar changes in hippoccampal dynorphin, were sensitized to naloxone seizures (Solbrig et al., 2006a) and HSV-1 infection in rats reduced hippocampal dynorphin to produce naloxone seizures [Solbrig et al., 2006b]. Taken together, the experiments are consistent with a role of hippocampal opioids in seizure development, and support the possibility of common seizure mechanisms in opioid withdrawal seizures and naloxone seizures in viral models.

Imaging

Imaging studies are revealing neurochemical and functional changes in the brains of epileptic subjects that provide insights into mechanisms and treatments. Just as human genetic studies have suggested roles for genes encoding opioid (dynorphin) elements involved in hippocampal circuits in temporal lobe epilepsy (TLE) vulnerability, imaging studies have also pointed to participation of hippocampal opioids in TLE.

Benzodiazepine and opioid receptors have been the most studied in PET research of epilepsy [Duncan, 1997]. [^{11}C]diprenorphine is a weak opiate receptor partial agonist with similar *in vivo* affinities for the mu, delta, and kappa receptor subtypes. [^{11}C]carfentanil selectively binds mu receptors. An increase in the binding of [^{11}C]carfentanil to mu-receptors in the lateral temporal cortex, and no changes in [^{11}C]diprenorphine binding have been found in patients with TLE [Frost et al. 1988, Mayberg et al. 1991]. There are several possibilities for the differential binding of mu and non-mu opioid receptors in patients with temporal lobe epilepsy. One is upregulation of mu receptors and occupation of kappa opioid receptors by an endogenous kappa opioid ligand such as dynorphin, a consequence of dynorphin release during recent seizures. Another possibility is reduction in number or affinity of kappa receptors, due to attrition of dynorphin system. Studies of ligand-receptor binding before and immediately after seizures may distinguish between these possibilities, and confirm whether temporal changes in kappa receptor binding are consistent with dynorphin release during seizures.

Dynorphin and Psychosis

Recent studies have set the stage for consideration of a role of dynorphin in seizure-associated neuropsychiatric disorders, with preliminary evidence that dynorphin might be related to the pathophysiology of psychotic and affective phenomena. Human studies have shown that dynorphin concentration is markedly elevated in CSF of schizophrenic patients, as compared to controls [Heikkila et al. 1990, Lindstrom, 1996]. Moreover, longitudinal analyses have revealed that high levels of dynorphin are a predictor of poor outcome, in association with negative symptoms [Lindstrom, 1996]. Dynorphin might indeed play a role in shaping negative symptoms of psychotic disorders, since it is postulated to reduce striatal levels of dopamine and to elicit anhedonia [Todtenkopf et al. 2004]. Accordingly, systemic administration of phencyclidine, the main pharmacological model of negative psychotic symptoms in rats, increases hippocampal levels of dynorphins [Zhou et al. 1986]. Furthermore, dynorphin A reduces spontaneous locomotion, rearing and grooming behaviors in mice [Ukai et al. 1984], while inactivation of dynorphin B increases exploratory behavior in DBA/2 mice [Van Daal et al. 1987], a strain regarded as a model for the study of the pathophysiologic substrates of psychosis [Olivier et al. 2001, Zhang et al. 2006].

While investigations on the direct impact of dynorphins on animal models are hindered by the presence of an enzymatic system aimed at the rapid degradation of opioid peptides, an interesting alternative is offered by the studies on agents activating kappa-opioid receptors.

Kappa opioid receptor agonists elicit psychosis-like reactions in man. The KOR agonist enadoline, for example, induces negative symptoms and hallucinosis [Walsh et al. 2001] and psychotomimetic properties of KOR agonists has been corroborated by the discovery that the main psychoactive ingredient of the "magic mint", the hallucinogenic plant *Salvia divinorum,* activates KOR [Roth et al. 2002]. Salvinorin is reported to produce potent and intense hallucinations, with very rapid onset and duration [Siebert, 1994, Valdes, 1994]. The subjective experience produced by salvinorin is generally described as qualitatively different from other hallucinogens, being rich in hallucinations and accompanied by depersonalization, derealization and dissociation [Yan and Roth, 2004].

In man and experimental animals, the best validated behavior paradigm to evaluate and measure sensorimotor gating is the prepulse inhibition (PPI) of the acoustic startle reflex, consisting of the reduction of the acoustic startle reflex that occurs when the eliciting stimulus is immediately preceded by a nonstartling prestimulus [Graham 1975]. PPI is impaired in neuropsychiatric disorders characterized by attentional or somatosensory deficits such as schizophrenia spectrum disorders [Braff et al. 2001]. Probes of PPI in rats with KOR agonists have shown that U50488, a selective KOR agonist, dose-dependently impairs gating functions and informational filtering in the rat model of PPI of the acoustic startle. The phenomenon is prevented by the specific KOR antagonist NOR-binaltorphimine and by the atypical antipsychotic clozapine, but not by haloperidol [Bortolato et al, 2005]. This result is particularly interesting if dynorphin actually contributes to induce negative symptoms of schizophrenia. Unlike haloperidol and other traditional neuroleptics, clozapine and other atypical antipsychotics are efficacious in the treatment of negative symptoms.

Dynorphin and Depression

As in psychosis, the hypothesis of a role of dynorphin in depression and dysphoria in humans is mainly based on evidence from studies on KOR agonists. In humans, pharmacological activation of KOR induces dysphoria [Pfeiffer et al. 1986, Dortch-Carnes and Potter, 2005]. In rodents, KOR agonists elicit depressive-like, dysphoric or aversive behaviors [Shippenberg and Herz, 1987, Mague et al. 2003, Todtenkopf et al. 2004]. Similarly, salvinorin increases immobility in the rat forced swim test [Cryan et al. 2002] and increases intracranial self-stimulation thresholds [Carlezon et al. 2006]. The mechanism underlying the dysphoric effect of KOR may be based on the reduction of dopaminergic mesolimbic signaling [Di Chiara and Imperato, 1988, Donzanti et al. 1992], although other neurotransmitters such as glutamate may also be involved [Todtenkopf et al. 2006].

Dynorphin and Postictal Psychiatric Disorders

Epilepsy is comorbid with a number of heterogeneous psychiatric disorders. A recent metanalysis has established that 6% of the epileptic patients are affected by a psychiatric disturbance, and this percentage rises to 10-20% in the population suffering from temporal lobe epilepsy [Gaitatzis et al. 2004]. The clinical association between epilepsy and behavioral abnormalities can be related to factors that include psychological disadvantages associated with the neurological condition or side effects of anticonvulsant drugs. Nevertheless, recent studies provides evidence for the existence of psychiatric conditions disorders represented only in epilepsy, particularly in the clusters of psychotic and depressive disorders.

Psychotic symptoms are a common complicating factor in epilepsy with an estimated 0.5-9% incidence [Tebartz van Elst et al. 2002]. In the neurology literature, psychotic disturbances in epilepsy were originally described under the broad rubric of "epilepsy psychoses", but a more detailed study of temporal relation between ictal phenomena and behavior has recently led to the identification of at least two entities, postictal psychosis (PIP) and interictal psychotic disorder (IPD) [Logsdail and Toone, 1988, Devinsky 2004, Kanner 2004]. PIP emerges shortly after the seizure extinction and is characterized by antecedent seizures, lucid intervals, short episodes, and mild confusion with subsequent amnesia [Prendergast et al. 1999]. In contrast, IPD occurs between epileptic episodes, is long-lasting and is generally classified as a schizophrenia-like syndrome [Roy et al. 2003]. PIP, however, can evolve to IPD, suggesting that the two disorders may be related or share common neural substrates.

Depression and mood disorders affect almost 50% of patients with complicated epilepsy [Schmitz, 2005]. Depressive symptoms can have major impact on the severity of epileptic symptoms [Cramer et al. 2003] and are strong predictors of quality of life in epileptic patients [Perrine et al. 1995, Lehrner et al. 1999]. Of note, the nosographic description of depression in epilepsy does not completely overlap with the standard definition of depression, but includes an array of atypical symptoms, such as irritability and mood lability [Blumer and Altshuler, 1998]. In particular, Blumer has described an intermittent and pleomorphic disorder, featuring anger, irritability and frank dysphoric responses, typically associated to the interictal phases of temporal lobe epilepsy [Blumer, 2002, Blumer et al. 2004].

The frequency of psychotic and depressive symptoms in patients with temporal lobe epilepsy raises the possibility that the temporal lobe is involved in these neuropsychological phenomena. Increased and ectopic expression of dynorphins in hippocampus is characteristic of temporal lobe epilepsy in man [Houser et al. 1990, de Lanerolle et al. 1997], and previous work in animal models of seizures has suggested that overproduction of endogenous opioids, particularly overstimulation of hippocampal kappa opioid receptors, might induce rebound post-ictal and interictal dysphoric effects [Hong et al. 1988, Engel and Rocha, 1992]. Indeed, if the comorbidity between temporal lobe epilepsy and psychotic disorders signifies aspects of neurochemical destabilization due to dynorphin dysregulation in hippocampus, the development of postictal psychotic states may be the result of natural mechanisms of endogenous seizure control through excess dynorphin release and kappa opioid receptor hyperactivation.

Should further preclinical and clinical studies provide evidence for seizure-induced dynorphin release in psychotic and depressive postictal disorders, therapeutic strategies to control seizure disorders and treat comorbid psychiatric disturbances might include KOR partial agonists or alternative interventions to restore a more balanced mu/delta/kappa profile. Other therapeutic strategies might also include enhancers of dynorphin degradation. The possibility that KOR blockade might elicit antidepressive effects is suggested by numerous studies on animal models of depression, such as the forced swim test [Pliakas et al. 2001, Mague et al. 2003] and the learned helplessness paradigm [Newton et al. 2002, Shirayama et al. 2004]. Moreover, antipsychotic effects of KOR antagonists are supported by experiments in mice showing a KOR antagonist prevents some of the behavioral effects mediated by the NMDA receptor antagonist dizocilpine [Qi et al. 2006].

Conclusion

Most genetic epilepsies have a complex mode of inheritance, and genes identified so far account only for a small number of families and sporadic cases. The strength of the association of dynorphin genotypes and epilepsies may not translate to prodynorphin constituting a susceptibility gene for temporal lobe epilepsy at the population level as yet. But the range of experimental evidence supporting dynorphin as an important inhibitory/modulatory transmitter makes the relation plausible. Human dynorphin promoter polymorphisms and low expression alleles exist. In models studied, dynorphin acts in a counter-regulatory fashion to attenuate excitability, and reduced dynorphin expression in the hippocampus due to virus exposure leads to epileptic responses. Under pathophysiologic conditions such as infection, the brain is awash in soluble factors, including cytokines, neurotrophins. and many biologically active neuropeptides. Understanding that extracellular spaces can bring to conventional or wired synapses dynorphin as well as a wide range of factors capable of regulating synaptic function may even elucidate seizure control strategies based on volume transmitters in the future. Extrasynaptic actions of dynorphin, other opioids, or other neuropeptides can complement existing drugs that target fast synaptic transmission. Finally, as epilepsy is further explored with functional neuroimaging, use of PET with opioid receptor ligands has the potential to link endogenous opioid system not only to seizures but also to comorbid and postictal behavioral phenomena. Such studies will add a new level of specificity to identification of mechanisms of disease and targets for treatment.

Acknowledgement

This work was supported by National Institutes of Health grant NS042307 from the National Institute of Neurological Disorders and Stroke to MVS.

References

Albertson, T.E., Joy, R.M., Stark, L.G. (1984) Modification of kindled amygdaloid seizures by opiate agonists and antagonists. *J. Pharmacol. Exp. Ther. 228*(3), 620-627.

Aradi, I., Soltesz, I. (2002). Postsynaptic effects of GABAergic synaptic diversity: regulation of neuronal excitability by changes in IPSC variance. *Neuropharmacology. 43*(4), 511-522.

Arvidsson, U., Riedl, M., Chakrabarti, S., Vulchanova, L., Lee, J.H., Nakano, A.H., Lin, X., Loh, H.H., Law, P.Y., Wessendorf, M.W., et al. (1995). The kappa-opioid receptor is primarily postsynaptic: combined immunohistochemical localization of the receptor and endogenous opioids. *Proc. Natl. Acad. Sci. USA. 91*(11), 5062-5066.

Bausch, S.B., Esteb, T.M., Terman, G.W., Chavkin, C. (1998). Administered and endogenously released kappa opioids decrease pilocarpine-induced seizures and seizure-induced histopathology. *J. Pharmacol. Exp. Ther. 284*(3), 1147-1155.

Becker, A., Braun, H., Schroder, H., Grecksch, G., Hollt, V. (1999). Effects of enadolline on the development of pentylenetetrazol kindling, learning performance, and hippocampal morphology. *Brain Res. 823*(1-2), 191-197.

Blumer, D. (2002) Psychiatric aspects of intractable epilepsy. *Adv. Exp. Med. Biol.* 497,133-147.

Blumer, D., Altshuler, L.L. Affective disorders. In: Engel, J., Pedley, T.A., eds. *Epilepsy: a comprehensive textbook;* Philadelphia: Lippincott Raven, 1988: pp. 2083-2099.

Blumer, D., Wakhlu, S., Davies, K., Hermann, B. (1998). Psychiatric outcome of temporal lobectomy for epilepsy: incidence and treatment of psychiatric complications. *Epilepsia.* 39(5),478-486.

Bortolato, M., Aru, G.N., Frau, R., Orru, M., Fa, M., Manunta, M., Puddu, M., Mereu, G., Gessa, G.L. (2005). Kappa opioid receptor activation disrupts prepulse inhibition ofo othe acoustic startle in rats. *Biol. Psychiatry. 57*(12), 1550-1558.

Braff, D.L., Geyer, M.A., Light, G.A., Sprock, J., Cadenhead, K.S., Swerdlow, N.R. (2001). Impact of prepulse characteristics on the detection of sensorimotor gating deficits in schizophrenia. *Schizophr. Res. 49*(1-2), 171-178.

Carlezon, W.A. Jr., Beguin, C., DiNieri, J.A., Baumann, M.H., Richards, M.R., Todtenkopf, M.S., Rothman, R.B., Ma, Z., Lee, D.Y., Cohen, B.M. (2005) Depressive-like effects of the kappa-opioid receptor agonist salvinorin A on behavior and neurochemistry in rats. *J. Pharmacol. Exp. Ther.* 316(1),440-447.

Castillo, P.E., Salin, P.A., Weisskopf, M.G., Nicoll, R.A. (1996). Characterizing the site and mode of action of dynorphin at hippocampal mossy fiber synapses in the guinea pig. *J. Neurosci. 16*(19), 5942-5950.

Chavkin, C., James, I. F., Goldstein, A. (1982) Dynorphin is a specific endogenous ligand of the k opioid receptor. *Science. 215*(4531), 413-415.

Chavkin, C., Shoemaker, W.J., McGinty, J.F., Bayon, A., Bloom, F.E. (1985). Characterization of the prodyorphin and proenkephalin neuropeptide systems in rat hippocampus. *J. Neurosci. 5*(3), 8-8-816.

Cramer, J.A., Blum, D., Reed, M., Fanning, K. (2003). The influence of comorbid depression on seizure severity. *Epilepsia.* 44(12),1578-1584.

Cryan, J.F., Page, M.E., Lucki, I. (2002) Noradrenergic lesions differentially alter the antidepressant-like effects of reboxetine in a modified forced swim test. *Eur. J. Pharmacol.* 436(3),197-205.

Czygan, M., Hallensleben, W., Hofer, M., Pollak, S., Sauder, C., Bilzer, T., Blumcke, I., Riederer, P., Bogerts, B., Falkai, P., Schwarz, M.J., Masliah, E., Staeheli, P., Hufert, F.T., Lieb, K. (1999). Borna disease virus in human brains with a rare form of hippocampal degeneration but not in brains of patients with common neuropsychiatric disorders. *J. Infect. Dis. 180*(5), 1695-1699.

De Lanerolle, N.C., Williamson, A., Meredith, C., Kim, J.H., Tabuteau, H., Spencer, D.D., Brines, M.L. (1997). Dynorphin and the kappa 1 ligand [3H]U69,593 binding in the human epileptogenic hippocampus. *Epilepsy Res. 28*(3), 189-205.

Devinsky, O. (2004). Therapy for neurobehavioral disorders in epilepsy. *Epilepsia. 45*(Suppl. 2), 34-40.

Di Chiara, G., Imperato, A. (1988) Drugs abused by humans preferentially increase synaptic dopamine concentrations in the mesolimbic system of freely moving rats. *Proc. Natl. Acad. Sci. U S A.* 85(14),5274-5278.

Donzanti, B.A., Althaus, J.S., Payson, M.M., Von Voigtlander, P.F. Kappa agonist-induced reduction in dopamine release: site of action and tolerance. *Res. Commun. Chem. Pathol. Pharmacol.* 78(2),193-210.

Dortch-Carnes, J., Potter DE. (2005) Bremazocine: a kappa-opioid agonist with potent analgesic and other pharmacologic properties. *CNS Drug Rev.* 11(2), 195-212.

Douglass, J., Grimes, L., Shook. J., Lee, P.H., Hong, J.S. (1991). Systemic administration of kainic acid differentially regulates the levels of prodynorphin and proenkephalin mRNA and peptides in the rat hippocampus. *Brain Res. Mol. Brain Res. 9*(1-2), 79-86.

Drake, C.T., Terman, G.W., Simmons, M.L., Milner, T.A., Kunkel, D.D., Schwartzkroin, P.A., Chavkin, C. (1994). Dynorphin opioids present in dentate granule cells may function as retrograde inhibitory neurotransmitters. *J. Neurosci. 14*(6), 3736-3750.

Drake, C.T., Patterson, T.A., Simmons, M.L., Chavkin, C., Milner, T.A. (1996). Kappa opioid receptor-like immunoreactivity in guinea pig brain: ultrastructural localization in presynaptic terminals in hippocampal formation. *J. Comp. Neurol.* 370(3), 377-395.

Drake, C.T., Chavkin, C., Milner, T.A. (1997). Kappa opioid receptor-like immunoreactivity is present in substance P-containing subcortical afferents in guinea pig dentate gyrus. *Hippocampus.* 7(1), 36-47.

Duncan, J.S. (1997) Imaging and epilepsy. *Brain, 120*(Pt 2), 339-377.

Engel, J. Jr., Rocha, L.L. (1992). Interictal behavioral disturbances : a search for molecular substrates. *Epilepsy Res. Suppl. 9*, 341-349.

Frost, J.J., Mayberg, H.S., Fisher, R.S., Douglass, K.H., Dannals, R.F., Links, J.M., Wilson, A.A., Ravert, H.T., Rosenbaum, A.E., Snyderm S.H., et al. (1988) Mu-opiate receptors measured by positron emission tomography are increased in temporal lobe epilepsy. *Ann. Neurol. 23*(3), 231-237.

Gaitatzis, A., Trimble, M.R., Sander, J.W. (2004) The psychiatric comorbidity of epilepsy. *Acta Neurol. Scand.* 110(4),207-220.

Gall, C., Brecha, N., Chang, K.J., Karten, H.J. (1984). Ontogeny of enkephalin-like immunoreactivity in the rat hippocampus. *J. Comp. Neurol. 225*(2), 212-227.

Gall, C., Berry, L.M., Hodgson, L.A. (1986). Cholecystokinin in the mouse hippocampus : localization in the mossy fiber and dentate commissural systems. *Exp. Brain Res. 62*(2), 431-437.

Gall, C., Lauterborn, J., Isackson, P., White, J. (1990). Seizures, neuropeptide regulation, and mRNA expression in the hippocampus. *Prog. Brain Res. 83*, 371-390.

Gambardella, A., Manna, I., Labate, A., Chifari, R., Serra, P., La Russa, A., LePiane, E., Cittadella, R., Andreoli, V., Sasanelli, F., Zappia, M., Aguglia, U., Quattrone, A., (2003). Prodynorphin gene promoter polymorphism and temporal lobe epilepsy. *Epilepsia. 44*(9), 1255-1256.

Graham, F.K. (1975). Presidential address, 1974. The more or less startling effects of a weak prestimulation. *Psychophysiology. 12*(3), 238-248.

Gosztonyi, G., Ludwig, H. (1995). Borna disease-neuropathology and pathogenesis. *Curr. Top. Microbiol. Immunol. 190*, 39-73.

Halasy, K., Racz, B., Maderspach, K. (2000). Kappa opioid receptors are expressed by interneurons in the CA1 area of the rat hippocampus : a correlated light and electron microscopic immunocytochemical study. *J. Chem. Neuroanat. 19*(4), 233-241.

Harrison, M.B., Shumate, M.D., Lothman, E.W. (1995). Opioid peptide expression in models of chronic temporal lobe epilepsy. *Neuroscience. 65*(3), 785-795.

Hauser, K.F., Aldrich, J.V., Anderson, K.J., Bakalkin, G., Christie, M.J., Hall, E.D., Knapp, P.E., Scheff, S.W., Singh, I.N., Vissel, B., Woods, A.S., Yakovleva, T., Shippenberg, T.S. (2005). Pathobiology of dynorphins in trauma and disease. *Front. Biosci. 10,* 216-235.

Heikkila, L., Rimon, R., Terenius, L. (1990). Dynorphin A and substance P in the cerebrospinal fluid of schizophrenic patients. *Psychiatry Res.* 34(3),229-236.

Henriksen, S. J., Chouvet, G., Bloom, F. E. (1982). In vivo cellular responses to electrophoretically applied dynorphin in the rat hippocampus. *Life Sci. 31*(16-17), 1785-1788.

Henriksen, S. J., Chouvet, G., McGinty, J., Bloom, F. E. (1982). Opioid peptides in the hippocampus: anatomical and physiological considerations. *Ann. NY Acad. Sci. 398*, 207-220.

Hong, J.S., McGinty, J.F., Grimes, L., Kanamatsu, T., Obie, J., Mitchell, C.L. (1988) Seizure-induced alterations in the metabolism of hippocampal opioid peptides suggest opioid modulation of seizure-related behaviors. *NIDA Res. Monogr. 82*,48-66

Hong, J.S. (1992). Hippocampal opioid peptides and seizures. *Epilepsy Res. Suppl. 7*, 187-195.

Houser, C.R., Miyashiro, J.E., Swartz, B.E., Walsh, G.O., Rich, J.R., Delgado-Escueta, A.V. (1990). Altered patterns of dynorphin immunoreactivity suggest mossy fiber reorganization in human hippocampal epilepsy. *J. Neurosci. 10*(1), 267-282.

Kanamatsu, T., McGinty, J.F., Mitchell, C.L., Hong, J.S. (1986). Kainic acid alters the metabolism of Met5-enkephalin and the level of dynorphin A in the rat hippocampus. *J. Neurosci. 6*(10), 3094-3102.

Kanner, A.M. (2004). Recognition of the various expressions of anxiety, psychosis, and aggression in epilepsy. *Epilepsia. 45*(Suppl. 2), 22-27.

Koob, G.F.; LeMoal, M. *Neurobiology of Addiction*; Amsterdam: Elsevier, 2006; pp. 121-122.

Lander, E.S., Schork, N.J. (1994). Genetic dissection of complex traits. *Science, 265*(5181), 2037-2048.

Lehrner, J., Kalchmayr, R., Serles, W., Olbrich, A., Pataraia, E., Aull, S., Bacher, J., Leutmezer, F., Groppel, G., Deecke, L., Baumgartner, C. (1999) Health-related quality of life (HRQOL), activity of daily living (ADL) and depressive mood disorder in temporal lobe epilepsy patients. *Seizure.* 8(2),88-92.

Lindstrom L.H., (1996). Clinical and biological markers for outcome in schizophrenia: a review of a longitudinal follow-up study in Uppsala schizophrenia research project. *Neuropsychopharmacology.* 14(3 Suppl):23S-26S.

Logsdail, S.J., Toone, B.K. (1988). Post-ictal psychoses. A clinical and phenomenological description. *Br. J.Psychiatry. 152*, 246-252.

Mayberg, H.S., Sadzot, B., Meltzer, C.C., Fisher, R.S., Lesser, R.P., Dannals, R.F., Lever, J.R., Wilson, A.A., Ravert, H.T., Wagner, H.N., Bryan, N., Cromwell, C.C., Frost, J.J. (1991). Quantification of mu and non-mu opiate receptors in temporal lobe epilepsy using Positron Emission Tomography. *Ann. Neurol. 30*(1), 3—11.

Madamba, S.G., Schweitzer, P., Siggins, G.R. (1999). Dynorphin selectively augments the M-current in hippocampal CA1 neurons by an opiate receptor mechanism. *J. Neurophysiol. 82*(4), 1768-1775.

Mague, S.D., Pliakas, A.M., Todtenkopf, M.S., Tomasiewicz, H.C., Zhang, Y., Stevens, W.C. Jr., Jones, R.M., Portoghese, P.S. Carlezon, W.A. Jr. (2003) Antidepressant-like effects of kappa-opioid receptor antagonists in the forced swim test in rats. *J Pharmacol Exp Ther.* 305(1),323-330

Manocha, A., Mediratta, P.K., Sharma, K.K. (2003). Studies on the anticonvulsant effect of U50488H on maximal electroshock seiziure in mice. *Pharmacol. Biochem. Behav. 76*(1), 111-117.

Mansour, A., Fox, C.A., Meng, F., Akil, H., Watson, S.J. (1994). K1 receptor mRNA distribution in the rat CNS: comparison to k receptor binding and prodynorphin mRNA. *Mol. Cell Neurosci.* *5*(2), 124-144.

Mazarati, A., Liu, H., Wasterlain, C. (1999). Opioid peptide pharmacology and immunocytochemistry in an animal model of self-sustaining status epilepticus. *Neuroscience, 89*(1), 167-173.

McDaniel, K.L., Mundy, W.R., Tilson, H.A. (1990). Microinjection of dynorphin into the hippocampus impairs spatial learning in rats. *Pharmacol. Biochem. Behav. 35*(2), 429-435.

McGinty, J.F., Henriksen, S.J., Goldstein, A., Terenius, L., Bloom, F.E. (1983). Dynorphin is contained within hippocampal mossy fibers: immunochemical alterations after kainic acid administration and colchicines-induced neurotoxicity. *Proc. Natl. Acad. Sci. U.S.A., 80*(2), 589-593.

McLean, S., Rothman, R.B., Jacobson, A.E., Rice, K.C., Herkenham, M. (1987). Distribution of opiate receptor subtypes and enkephalin and dynorphin immunoreactivity in the hippocampus of squirrel, guinea pig, rat, and hamster. *J. Comp. Neurol. 255*(4), 497-510.

Meilandt, W.J., Barea-Rodriguez, E., Harvey, S.A.K., Martinez, J.L. Jr. (2004). Role of hippocampal CA3 mu-opioid receptors in spatial learning and memory. *J. Neurosci. 24*(12), 2953-2962.

Moore, S.D., Madamba, S.G., Schweitzer, P., Siggins, G.R. (1994). Voltage-dependent effects of opioid peptides on hippocampal CA3 pyramidal neurons in vitro. *J. Neurosci. 14(*2), 809-820.

Morris, B.J., Moneta, M.E., ten Bruggencate, G., Hollt, V. (1987). Levels of prodynorphin mRNA in rat dentate gyrus are decreased during hippocampal kindling. *Neurosci. Lett. 80*(3), 298-302.

Narayan, O., Herzog, S., Frese, K., Scheefers, H., Rott, R. (1983). Behavioral disease in rats caused by immunopathological responses to persistent borna virus in the brain. *Science. 220*(4604), 1401-1403.

Newton, S.S., Thome, J., Wallace, T.L., Shirayama, Y., Schlesinger, L., Sakai, N., Chen, J., Neve, R., Nestler, E.J., Duman, R.S. (2002) Inhibition of cAMP response element-binding protein or dynorphin in the nucleus
accumbens produces an antidepressant-like effect. *J. Neurosci.* 22(24):10883-10890.

Nguyen, X.V., Masse, J., Vijitruth, R., Kulik, C., Liu, M., Choi, D.Y., Foster, T.C., Usynin, I., Balkalkin, G., Bing, G. (2005). Prodynorphin knockout mice demonstrate diminished age-associated impairment in spatial water maze performance. *Behav. Brain Res.* 161(2), 254-262.

Nieuwenhuys, R. Comparative aspects of volume transmission , with sidelight on other forms of intercellular communication. In: Agnati LF, Fuxe K, Nicholson C, Sykova E editors. *Volume transmission revisited.* Progress in Brain Research vol. 125. Amsterdam: Elsevier Science BV; 2000; 49-126.

Olivier, B., Leahy, C., Mullen, T., Paylor, R., Groppi, V.E., Sarnyai, Z., Brunner, D. (2001) The DBA/2J strain and prepulse inhibition of startle: a model system to test antipsychotics? *Psychopharmacology.* (Berl). 156(2-3):284-290.

Osborn, D.A., Jeffery, H.E., Cole, M. (2005). Opiate treatment for opiate withdrawal in newborn infants. *Cochrane Database Syst. Rev. 20*(3), CD002059.

Panagariya, A., Jain, R.S., Gupta, S., Garg, A., Sureka, R.K., Mathur, V. (2001). Herpes simplex encephalitis in North West India. *Neurol. India 49*(4), 360-365.

Perrine, K., Hermann, B.P., Meador, K.J., Vickrey, B.G., Cramer, J.A., Hays, R.D., Devinsky, O. (1995) The relationship of neuropsychological functioning to quality of life in epilepsy. *Arch. Neurol.* 52(10),997-1003.

Pfeiffer, A., Brantl, V., Herz, A., Emrich, H.M. (1986) Psychotomimesis mediated by kappa opiate receptors. *Science.* 233(4765),774-776.

Pierce, J.P., Kurucz, o.S., Milner, T.A. (1999). Morphometry of a peptidergic transmiiter system: dynorphin B-like immunoreactivity in the rat hippocampal mossy fiber pathway before and after seizures. *Hippocampus. 9*(3), 255-276.

Pliakas, A.M., Carlson, R.R., Neve, R.L., Konradi, C., Nestler, E.J., Carlezon, W.A. Jr. (2001). Altered responsiveness to cocaine and increased immobility in the forced swim test associated with elevated cAMP response element-binding protein expression in nucleus accumbens. *J. Neurosci.* 21(18),7397-7403.

Prendergast, J., Spira, P., Schnieden, V. (1999). Post-ictal psychosis: an unusual clinical entity. *Aust. N. Z. J.Psychiatry 33*(3), 433-435.

Przewlocka, B., Machelska, H., Lason, W. (1994). Kappa opioid receptor agonists inhibit the pilocarpine-induced seizures and toxicity in the mouse. *Eur. Neuropsychopharmacol., 4*(4), 527-533.

Qi, C.T., Zou, H., Zhang, C.H., Xie, Q.L., Jin, M.L., Yu, L. (2006). Effect of GNTI, a kappa opioid receptor antagonist, on MK-801-induced hyperlocomotion and stereotypy in mice. *Acta Pharmacol. Sin. 27*(11), 1401-1408.

Racz, B., Halasy, K. (2002). Kappa opioid receptor is expressed by somatostatin- and neuropeptide Y-containing interneurons in the rat hippocampus. *Brain Res. 931*(1), 50-55.

Rocha, L., Maidment,m N.T. (2003). Opioid peptide release in the rat hippocampus after kainic acid-inducing status epilepticus. *Hippocampus. 13*(4), 472-480.

Romualdi, P., Donatini, A., Bregola, G., Bianchi, C., Beani, L., Ferri, S., Simonato, M. (1995). Early changes in prodynorphin mRNA and ir-dynorphin A levels after kindled seizures in the rat. *Eur. J. Neurosci.* 7(9), 1850-1856.

Roth, B.L., Baner, K., Westkaemper, R., Siebert, D., Rice, K.C., Steinberg, S., Ernsberger, P., Rothman, R.B. (2002). Salvinorin A: A potent naturally occurring nonnitrogenous kappa opioid selective agonist. *Proc. Natl. Acad. Sci. USA 99*(18), 11934-11939.

Roy A.K., Rajesh, S.V., Iby, N., Jose, J.M., Sarma, G.R. (2003). A study of epilepsy-related psychosis. *Neurol. India 51*(3), 359-360.

Sandin, J., Nylander, I., Georgieva, J., Schott, P.A., Ogren, S.O., Terenius, L. (1998). Hippocampal dynorphin B injections impair spatial learning in rats: a kappa-opioid receptor-mediated effect. *Neuroscience. 85*(2), 375-382.

Schindler, C.J., Slamberova, R., Rimanoczy, A., Hnactzuk, O.C., Riley, M.A., Vathy, I. (2004). Field-specific changes in hippocampal opioid mRNS, peptides, and receptors due to prenatal morphine exposure in adult male rats. *Neuroscience. 126*(2), 355-364.

Schmitz B. (2005) Depression and mania in patients with epilepsy. *Epilepsia.* 46 Suppl 4,45-49.

Shippenberg, T.S., Herz, A. (1987). Place preference conditioning reveals the involvement of D1-dopamine receptors in the motivational properties of mu- and kappa-opioid agonists. *Brain Res.* 436(1),169-172.

Shirayama, Y., Ishida, H., Iwata, M., Hazama, G.I., Kawahara, R., Duman, R.S. (2004) Stress increases dynorphin immunoreactivity in limbic brain regions and dynorphin antagonism produces antidepressant-like effects. *J. Neurochem.* 90(5),1258-1268.

Siebert, D.J. (1994). Salvia divinorum and salvinorin A: new pharmacologic findings. *J. Ethnopharmacol. 43*(1), 53-56.

Siggins, G. R. and Zieglagansberger, W. (1981) Morphine and opioid peptides reduce inhibitory synaptic potentials in hippocampal pyramidal cells in vitro without alteration og membrane potential. *Proc. Natl. Acad. Sci. U.S.A. 78*(8), 5235-5239.

Simmons, M.L., Terman, G.W., Gibbs, S.M., Chavkin, C. (1995). L-type calcium channels mediate dynorphin neuropeptide release from dendrites but not axons of hippocampal granule cells. *Neuron 14*(6), 1265-1272.

Simmons, M.L., Chavkin, C. (1996). K-opioid receptor activation of a dendrotoxin-sensitive potassium channel mediates presynaptic inhibition of mossy fiber neurotransmitter release. *Mol. Pharmacol. 50*(1), 80-85.

Simonato, M., Romualdi, P. (1996). Dynorphin and epilepsy. *Prog. Neurobiol. 50*(5-6), 557-583.

Solbrig, M.V., Koob, G.F., Lipkin, W.I. (1996). Naloxone-induced seizures in rats infected with Borna disease virus. *Neurology. 46*(4), 332-338.

Solbrig, M.V., Adrian, R., Baratta, J., Lauterborn, J.C., Koob, G.F. (2006a). Kappa opioid control of seizures produced by a virus in an animal model. *Brain. 129*(Pt 3), 642-654.

Solbrig, M.V., Adrian, R., Chang, D.Y., Perng, G.C. (2006b). Viral risk factor for seizures: pathobiology of dynorphin in herpes simplex viral (HSV-1) seizures in an animal model. *Neurobiol. Dis. 23*(3), 612-620.

Stogmann, E., Zimprich, A., Baumgartner, C., Aull-Watschinger, S., Hollt, V., Zimprich, F. (2002). A functional polymorphism in the prodynorphin gene promoter is associated with temporal lobe epilepsy. *Ann. Neurol., 51*(2), 260-263.

Tebartz van Elst, L.H., Baeumer, D., Lemieux, L., Woermann, F.G., Koepp, M., Krishnamoorthy, S., Thompson, P.J., Ebert, D., Trimble, M.R. (2002). Amygdala pathology in psychosis of epilepsy: A magnetic resonance imaging study in patients with temporal lobe epilepsy. *Brain. 125* (Pt 1), 140-149

Tempel, A., Zukin, R.S. (1987). Neuroanatomical patterns of the mu, delta, and kappa opioid receptors of rat brain as determined by quantitative in vitro autoradiography. *Proc. Natl. Acad. Sci. USA. 84*(12), 4308-4312.

Terman, G.W., Drake, C.T., Simmons, M.L., Milner, T.A., Chavkin, C. (2000). Opioid modulation of recurrent excitation in the hippocampal dentate gyrus. *J. Neurosci. 20*(12), 4379-4386.

Todtenkopf, M.S, Marcus, J.F., Portoghese, P.S., Carlezon, W.A. Jr. (2004) Effects of kappa-opioid receptor ligands on intracranial self-stimulation in rats. *Psychopharmacology.* 172(4), 463-470.

Todtenkopf, M.S., Parsegian, A., Naydenov, A., Neve, R.L., Konradi, C., Carlezon, W.A. Jr. Brain reward regulated by AMPA receptor subunits in nucleus accumbens shell. (2006) *J. Neurosci.* 26(45)11665-11669.

Tortella, F.C., Robles, L., Holaday, J.W. (1986). U50,488, a highly selective kappa opioid.: anticonvulsant profile in rats. *J. Pharmacol. Exp. Ther. 237*(1), 49-53.

Tortella, F.C. (1988). Endogenous opioid peptides and epilepsy: quieting the seizing brain? *Trends Pharmacol. Sci. 9*(10), 366-372.

Ukai, M., Kameyama, T. (1984). The antagonistic effects of naloxone on hypermotility in mice induced by dynorphin-(1-13) using a multi-dimensional behavioral analysis. *Neuropharmacology. 23* (2A),165-168.

Valdex, L.J.3rd (1994). Salvia divinorum and the unique diterpene hallucinogen, (Salvinorin (divinorin) A. *J. Psychoactive Drugs. 26*(3), 277-283.

Van Daal, J.H., De Kok, Y.J., Jenks, B.G., Wendelaar Bonga, S.E., Van Abeelen, J.H. (1987). A genotype-dependent hippocampal dynorphinergic mechanism controls mouse exploration. *Pharmacol. Biochem. Behav.* 28(4),465-468.

VonVoigtlander, P.F., Hall, E.D., Ochoa, M.C., Lewis, R.A., Triezenberg, H.J. (1987). U-54494A: a unique anticonvulsant related to kappa opioid agonists. *J. Pharmacol. Exp. Ther. 243*(2), 542-547.

Wagner, J.J., Terman, G.W., Chavkin, C. (1993). Endogenous dynorphins inhibit excitatory neurotransmission and block LTP induction in the hippocampus. *Nature. 363*(6328), 451-454.

Walsh, S.L., Strain, E.C., Abreu, M.E., Bigelow, G.E. (2001). Enadoline, a selective kappa opioid agonisit: comparison with butorphanol and hydromorphone in humans. *Psychopharmacology. (Berl) 157*(2), 151-162.

Weisskopf, M.G., Zalutsky, R.A., Nicoll, R.A. (1993). The opioid peptide dynorphin mediates heterosynaptic depression of hippocampal mossy fiber synapses and modulates long-term potentiation. *Nature. 362*(6419), 423-427.

Wijdicks, E.F, Sharbrough, F.W. (1993). New-onset seizures in critically ill patients. *Neurology. 43*(5), 1042-1044.

Xie, C.W., Lee, P.H., Takeuchi, K., Owyang, V., Li, S.J., Douglass, J., Hong, J.S. (1989). Single or repeated electroconvulsive shocks alter the levels of prodynorphin and proenkephalin mRNAs in rat brain. *Brain Res. Mol. Brain Res. 6*(1), 11-19.

Xiong, Z.G., Zhu, X.M., Chu, X.P., Minami, M., Hey, J., Wei, W.L., MacDonald, J.F., Wemmie, J.A., Price, M.P., Welsh, M.J., Simon, R.P. (2004). Neuroprotection in ischemia: blocking calcium-permeable acid-sensing ion channels. *Cell. 118*(6), 687-698.

Yajima, Y., Narita, M., Takahashi-Nakano, Y., Misawa, M., Nagase, H., Mizoguchi, H., Tseng, L.F., Suzuki, T. (2000). Effects of differential modulation of mu-, delta- and kappa opioid systems on bicuculline-induced convulsions in the mouse. *Brain Res. 862*(1-2), 120-126.

Yan, F., Roth, B.L. (2004). Salvinorin A: a novel and highly selective kappa-opioid receptor agonist. *Life Sci. 75*(22), 2615-2619.

Yermolaieva, O., Leonard, A.S., Schnizler, M.K., Abboud, F.M., Welsh, M.J. (2004). Extracellular acidosis increases neuronal cell calcium by activating acid-sensing ion channel 1a. *Proc. Natl. Acad. Sci. U.S.A., 101*(17), 6752-6757.

Zieglgansberger, W., French, E. D., Siggins, G. R., Bloom, F. E. (1979) Opioid peptides may excite hippocampal pyramidal neurons by inhibiting adjacent inhibitory interneurons. *Science. 205*(4404), 415-417.

Zimprich, A., Kraus, J., Woltje, M., Mayer, P., Rauch, E., Hollt, V. (2000). An allelic variation in the human prodynorphin gene promoter alters stimulus-induced expression. *J. Neurochem., 74*(2), 472-477.

Zhang, M., Ballare, M.E., Kohlhaas, K.L., Browman, K.E., Jongen-Relo, A.L., Unger, L.V., Fox, G.B., Gross, G., Decker, M.W., Drescher, K.U., Rueter, L.E. (2006). Effect of dopamine D3 antagonists on PPI in DBA/2J mice or PPI deficit induced by neonatal ventral hippocampal lesions in rats. *Neuropsychopharmacology*. 31(7), 1382-1392.

Zhou, G.Z., Jiang, C.H., Yin, S.H., Sun, Q.F. (1986) Effects of phencyclidine on levels of dynorphin 1-17 and dynorphin 1-8 immunoreactivities in various brain regions. *Acta Pharmacol. Sin.,* 7(6),491-495.

Zukin, R.S., Eghbali, M., Olive, D., Unterwald, E.M., Tempel, A. (1988). Characterization and visualization of rat and guinea pig brain kappa opioid receptors: evidence for kappa 1 and kappa 2 opioid receptors. *Proc. Natl. Acad. Sci. USA. 85*(11), 4061-4065.

In: Neuropeptide Research Trends
Editor: Bernice A. Levine, pp. 189-212

ISBN: 978-1-60021-640-4

Chapter VII

The Role of Vasopressin in Chronic Stress-Induced Hypothalamo-Pituitary-Adrenal Axis Hyperactivity: Studies on Brattleboro Rats with Repeated Restraint

Dóra Zelena*[*], *Ágnes Domokos, Zsuzsa Mergl, István Barna, Anna Földes, Krisztina J. Kovács and Gábor B. Makara
Institute of Experimental Medicine, Hungarian Academy of Sciences, Budapest, Hungary

Abstract

Hypothalamo-pituitary-adrenal axis (HPA) is a crucial component of the homeostasis. During chronic stress the regulation of the HPA seems to be relying on a pronounced role of vasopressin. Our hypothesis is that during prolong stress regiment in the absence of vasopressin the chronic stress symptoms will be diminished. Therefore we studied stress-related variables after 14 and 28 days of daily restraint in littermate heterozygous and vasopressin-deficient (di/di) Brattleboro rats.

Stressed heterozygote rats showed retarded growth, decreased thymus and spleen weights, and hypertrophied adrenals. After 2 weeks these changes were equal in di/di and di/+ rats with some reduction in di/di after prolongation of the restraint to 4 week. In chronically stressed animals the POMC mRNA in the anterior lobe of the pituitary and the slightly, but not significantly elevated resting plasma corticosterone level were equal in di/+ and di/di rats after 2 weeks restraint. Prolongation of the stress regime for 4 weeks leads to significantly smaller elevation of CRH and vasopressin mRNA in the nucleus paraventricularis hypothalami, diminished enhancement of POMC mRNA in the anterior lobe of the pituitary and reduced rise in basal plasma corticosterone levels in di/di rats. In addition, the acute response to the first, 15th and 29th restraint was also smaller (but not

[*] Contact information: Dóra Zelena; Institute of Experimental Medicine; H-1083 Budapest; Szigony 43; Hungary; Phone: +36-1-210-9950; Fax.:+36-1-210-9951; e-mail: zelena@koki.hu

abolished) in vasopressin-deficient as compared with heterozygous rats. The difference between ACTH and corticosterone responses (small decrease in ACTH but no habituation in corticosterone) hypothesised that the sensitivity of the adrenal gland to ACTH was changed but from superfusion experiments it seems that after 4 weeks the sensitivity was first decreased rather than increased with no significant effect of vasopressin deficiency.
The results from 14 days repeated restraint suggested that vasopressin is not indispensable for activating the HPA system by chronic stress although the prolongation of the stress lead to different conclusion. These later results supported the view that during long term repeated restraint the role of vasopressin becomes more important (but not exclusive) in the maintenance of the resting HPA axis activity. However, its effects become evident after prolonged stress exposure only.

Keywords: Brattleboro, vasopressin, ACTH, corticosterone, chronic stress.

Introduction

The activation of the hypothalamo-pituitary-adrenal axis (HPA) is crucial for the maintenance of homeostasis. Under the effect of acute and chronic stressful stimuli the parvocellular neurons of the hypothalamic paraventricular nucleus (PVN) release both corticotropin-releasing hormone (CRH) and arginine vasopressin to the portal blood (Antoni, 1993; Harbuz and Lightman, 1989). In response to these neuropeptides plasma adrenocorticotropin (ACTH, cleaved from proopiomelanocortin (POMC) in the anterior lobe of the pituitary) and glucocorticoid (in rodents mainly corticosterone from adrenal cortex) levels rapidly increase.

Vasopressin exerts its effect through plasma membrane receptors of which two major subtypes have been identified. V_{1a} receptors are widely distributed on blood vessels, while V_2 receptors are predominantly located in the principal cells of the renal collecting system. Vasopressin-induced ACTH release occurs via the V_{1b} (V_3) receptor subtype.

Vasopressin is released primarily from the magnocellular system which is activated in response to osmotic stressors. Vasopressin produced by the parvocellular neurones of the PVN are co-localised with CRH-containing neurons. In control, non-stressed rats, within the pool of CRH neurosecretory cells, 50% co-express vasopressin. These neurosecretory granules undergo remarkable changes in size and peptide composition under experimental manipulations of the HPA. After adrenalectomy the number of CRH+/vasopressin+ cells in PVN increases (Swanson et al., 1983; Whitnall et al., 1987). Similar shift towards vasopressin production in parvocellular neurones appears after various types of chronic stress and also in depressed patients (Raadsheer et al., 1994).

The relative contributions of CRH and vasopressin to ACTH release during stress seem to be stressor-specific (Scott and Dinan, 1998; Antoni, 1993; Kovács et al., 2000). The sensitivity of CRH and vasopressin transcription to glucocorticoid feedback is different. Acute stress responses are restrained by negative feedback effects of glucocorticoids, which reduce the secretory activity of the endocrinomotor neurons of the hypothalamus and the POMC producing cells of the anterior pituitary (Kjaer et al., 1993). This negative feedback is manifested on CRH system, as CRH mRNA level in PVN and CRH1 receptor mRNA level in the pituitary are reduced by elevated glucocorticoids (Kretz et al., 1999). In chronic stress states there is a facilitatory mechanism that helps maintaining the chronic hyperactivity of the HPA axis despite the inhibitory feedback of glucocorticoids (Aguilera, 1994; Aguilera and Rabadan-Diehl, 2000; Dallman, 1993). One of the key components of this facilitatory mechanism could be the vasopressinergic system since vasopressin-stimulated ACTH release is relatively resistant to feedback inhibition and pituitary V_{1b} receptor mRNA levels are stimulated by glucocorticoids. Thus vasopressinergic regulation of the HPA axis is critical for sustaining corticotrope responsiveness in the presence of relatively high circulating glucocorticoid levels during chronic stress (Aguilera, 1994; Dallman, 1993; Lightman et al., 2002; Scott and Dinan, 1998). This hypothesis is well supported by circumstantial evidence, such as preferential increase of vasopressin in the median eminence in adjuvant-induced arthritis (Chowdrey et al., 1995), repeated hypoglycemia and immobilization (De Goeij et al., 1991; 1992), increased vasopressin binding and V_{1b} receptor mRNA levels in the anterior pituitary (Aguilera and Rabadan-Diehl, 2000) and increased vasopressin mRNA and hnRNA in the parvocellular PVN of the chronically stressed rats (Bartanusz et al., 1993; Ma et al., 1997; 1999; Makino et al., 1995; Pinnock and Herbert, 2001). Vasopressin of magnocellular origin (from the magnocellular PVN and the supraoptic nucleus) may also contribute to the relative increase of the vasopressin:CRH ratio, as vasopressin released from these cells can reach the portal circulation (Holmes et al., 1986). The involvement of magnocellular vasopressin in certain acute stress conditions was demonstrated (Engelmann et al., 2004), but its role in chronic stress responses is questionable (Aguilera, 1994). Despite the increasing number of indirect evidences direct supporting data with blocking the role of vasopressin (e.g. immunoneutralization with vasopressin antibody, V_{1b} receptor antagonist or in V_{1b} receptor knockout mice) are missing.

To characterize the role of vasopressin in stress activation the vasopressin-deficient Brattleboro strain provides a good tool (Sokol and Valtin, 1982). The Brattleboro strain is a natural vasopressin knockout derived from Long Evans rats in the 1960s. Vasopressin deficiency is a result of a single nucleotide deletion in the neurophysin coding region of the vasopressin gene. Because of this mutation homozygous diabetes insipidus (di/di) Brattleboro rats synthesize an altered vasopressin precursor, which is unable to enter the secretory pathway (Bohus and de Wied, 1998; Evans et al., 2000). A proper control for di/di animals is difficult to obtain because of possible genetic drift as Long Evans strain was separated more than 40 years and Wistar rats are even more far from Brattleboros. In our previous experiments we used di/+ animals as controls, while di/+ rats have enough vasopressin for normal water balance and genetically are close to their di/di littermates. To exclude the influence of maternal genotype littermates from di/+ mother and di/di father were compared (Zelena et al., 2003b).

The aim of the present experiments was to assess chronic stress symptoms in vasopressin-deficient Brattleboro rats. Our working hypothesis was that the presence and upregulation of vasopressin is necessary for maintaining HPA activity during chronic stress. We used repeated restraint as a combined psychological-physical stress stimulus. If vasopressin signalling is instrumental for the maintenance of adrenocortical responsiveness during chronic stress then the vasopressin deficient animals are expected to show markedly impaired HPA responses in a chronic stress situation.

In the first series we investigated the role of vasopressin in chronic stress responses by exposing vasopressin deficient Brattleboro rats to daily physical restraints for 14 days (Zelena et al., 2004). In these experiments, vasopressin deficiency did not affect the chronic stress response, which casts doubt on the role of vasopressin in chronic stress responses. Despite the fact that even shorter stress periods (e.g. 4-5 days) were considered "chronic" in some reports (Akana et al., 1996; Schwartz et al., 1997), here we assumed that the role of vasopressin remained hidden in these earlier experiments, because 14 days of daily restraint might be too short for revealing it. We hypothesized that longer stress regimen (e.g. 28 days of daily restraint) would be necessary to uncover the role of vasopressin in chronic stress responses. Therefore the present study was focused on the role of vasopressin during prolonged (4 weeks) restraint procedure. We tested also the hypothesis that adrenal hypertrophy may be accompanied by increased adrenocortical sensitivity to ACTH.

Materials and Methods

Animals

3-4 month old Brattleboro rats were bred in our Institute from a colony originating from Harlan (Indianapolis, USA). We used male rats throughout the experiments because of the possible confounding variations during the estrous cycle. Rats were maintained in a controlled environment (temperature 23±1°C, humidity 50-70 %, day:night schedule of 12:12h, with lights on at 07:00h), were fed by a commercial rat chow (Charles River, Hungary), and had free access to tap water. We compared the vasopressin deficient homozygous (di/di) rats with congenital diabetes insipidus to heterozygous (di/+) control rats from the same litters selected upon their water consumption in 5-6 weeks of age (Bohus and de Wied, 1998; Zelena et al., 2003b). Due to high level of urination in di/di rats wood shaving bedding was changed daily. The animals were individually housed from the beginning of the experiment (cage size: 42.5 x 26.6 x 18.5cm, 1291 Eurostandard Type III/h). All studies were carried out in accordance with the European Communities Council Directive of 24 November 1986 (86/609/EEC) and were reviewed and approved by the Animal Welfare Committee of the Institute of Experimental Medicine.

Stress Procedure

Restraint stress was applied as described earlier using transparent plastic tubes (5–6 cm ID) fitted to the size of the animals (Zelena et al., 2003a). The tubes had a 4-cm-long conical head part ending with a large breathing hole (2 cm ID). The animals were restrained by loosely packing with paper towels the rear end of the tube behind the body. This restraint procedure minimized the space around the animal, provided a rather strong stressful stimulus without being harmful in the long run. Restraint sessions lasted for 60 min, were applied between 0900-1200h, and were repeated daily for 14 or 28 days. Controls were nonstressed, individually housed rats. Animals were sacrificed by decapitation 24h after the last session (i.e. on day 15 (14R) or 29 (28R)) or immediately after the first (1R) or an additional acute restraint performed on day 15 (15R) or 29 (29R). Rats were weighted on day 14 or 28 to avoid additional stress by the time of decapitation. Groups and treatments are presented in table 1.

Table 1. Groups and treatments studied in the experiment

Group	Vasopressin status	chronic stress (14 or 28 restraints)	acute stress
di/+C	Heterozygote (di/+; normal vasopressin levels)	-	-
di/+14R or di/+28R		+	-
di/+1R		-	+
di/+15R or di/+29R		+	+
di/diC	Vasopressin-deficient (di/di; homozygous for the mutant gene)	-	-
di/di14R or di/di28R		+	-
di/di1R		-	+
di/di15R or di/di29R		+	+

Detection of mRNA Level

Northern Hybridization

Pituitaries from the first series (14R) were quickly removed and anterior lobes were immediately frozen on dry ice. Samples were stored at -70°C until RNA isolation. The Northern blot analysis was the same as previously described (Zelena et al., 2003a). The tissue samples were homogenized and lyzed in RNA-Clean solution (AGS GmbH, Heidelberg, Germany) and total RNA was extracted with chloroform according to the manufacturer's instructions.

For preparation of Northern blots, 20 μg of total RNA samples were size-fractionated on a 1% agarose, 8% formaldehyde denaturing gel and transferred onto Hybond N membranes (Amersham, UK) using the capillary transfer method. The POMC 48-mer synthetic oligonucleotide probe (kindly provided by G. Aguilera, NICHD, NIH) was labelled by terminal transferase (Boehringer Mannheim) and the β-actin riboprobe was labelled by the random primer method (HexaLabel DNA Labeling Kit, MBI Fermentas) using ^{32}P-dCTP (Izinta, Hungary). Prehybridizations (42°C, 4 h) and hybridizations (42°C, overnight) were carried out in a solution containing 50% (vol./vol.) formamide, 6x SSC, 5x Denhardt's solution, 0.5% SDS, 50 mM Na-phosphate buffer, 100μg/ml tRNA, 10μg/ml polyU-homopolymer, 7.5μg/ml denatured salmon-sperm DNA. The labeled probes were added to the hybridization solution at 1x 10^6 cpm/ml. Filters were washed at high stringency conditions, (RT in 2x SSC/0.1 % SDS for 5 min, 68°C 2x SSC/0.1 % SDS for 30 min, 68°C in 0.2x SSC/0.1 % SDS for 30 min). Blots were exposed to X-ray films (Sigma, Kodak XAR) for one or two days at -70°C using intensifying screens. Between hybridizations, filters were washed in a solution containing 5mM Na-phosphat/0.1% SDS at 100°C for 30 min to remove the labeled probe. Quantification of the hybridization signals was achieved by using an image analysis system (ScionImage, Scion Corporation) to obtain plot profiles of autoradiograms.

All comparisons were made from RNA samples hybridized on the same filter and normalized to the content of β-actin RNA detected in each individual sample.

In Situ Hybridization

Tissue Preparation

The brains were collected on dry ice-cooled n-hexan, frozen to −80°C and stored until cryostat sectioning. Every sixth 16 μm section from the anterior commissure to the end of the amygdala was mounted on silanized slides. Pituitaries were quickly dissected from decapitated animals and the separated anterior lobes (AL) were frozen to −80°C in embedding medium. These specimens were mounted on a cryostate and cut in 16 μm sections. Six sections per animals were collected.

Hybridization

The hybridization technique was derived from that described by Simmons et al. (1989) and was already detailed elsewhere (Zelena et al., 2006). Tissue sections were fixed with 4% paraformaldehyde, then digested with proteinase K (1μg/ml in 50 mM Tris, pH 8, 5 mM EDTA), acetylated (0.25% acetic anhydride in 0.1 M triethanolamine, at pH=8) and dehydrated using ascending ethanol series (50–100%). The hybridization mixture (50% formamide, 0.3 M NaCl, 10mM Tris, pH 8, 2 mM EDTA, 1×Denhardt's, 10% dextran sulfate, 0.5 mg yeast tRNA, 10^7 dpm/ml specific probe) was pipetted onto the slides and left overnight at 58 °C. On the next day, the slides were rinsed four times with SSC (1×SSC: 0.15 M, 15mM trisodium citrate, pH 7), digested with ribonuclease A (10μg/ml in 5 mM Tris, 1 mM EDTA, 0.5 M NaCl, pH 8), gradually desalted, and washed in 0.1×SSC at 65 °C for 30 min. After a final wash in 0.1×SSC at room temperature for 1 min slides were air-dried.

CRH mRNA levels were quantified by means of ^{35}S-UTP containing riboprobes complementary to the exonic sequences of the CRH gene.

The cRNA probe for the matured vasopressin mRNA was generated from the vasopressin-specific 3' end (exon C) of the rat vasopressin gene. The 230 bp PstI/DraI cDNA fragment was cloned into pSP 65 vector (cloning site: SmaI/PstI). The probe was obtained from Dr. D Richter (University of Hamburg, Germany). The ^{35}S-UTP labeled antisense riboprobe was transcribed from the linearized (Hind III) plasmid by SP6 polimerase.

POMC mRNA levels in the second experimental series (28R) were quantified by in situ hybridization using ^{35}S-UTP -labeled riboprobes corresponding to a 1.2 kb exonic sequence within the POMC gene (POMC plasmid was a generous gift of Dr. J. Eberwine, University of Pennsylvania).

Analysis

After hybridization, slides were exposed to imaging plates (Fujifilm, BAS-IP, MS 2340) for 72 h (except for POMC where exposure lasted 16h), and the plates were scanned by a fluorescent image analyzer (FLA 3000, Fujifilm, scanning resolution 50 μm). Radiograms were evaluated by the ImageJ program (written by Wayne Rasband at the US National

Institutes of Health and available from the Internet by anonymous FTP from zippy.nimh.nih.gov).

CRH mRNA. The integrated densities for the PVN and the central amygdala (CeA) were calculated. The boundary of the examined region was outlined, and the average grayness value was corrected by the background taken from the neighbouring hypothalamic tissue. mRNA expression was evaluated by summing up the grayness values measured over the whole extent of the structure (integrated density) (Barna et al., 2003; Zelena et al., 2003a). In addition, the spatial extension of the signal was also evaluated by multiplying the number of sections showing a distinguishable CRH mRNA signal by 6 (as every sixth section was collected) and 16 (the thickness of the sections). E.g., the spatial extension of the signal was 960 μm if 10 sections showed a distinguishable CRH mRNA signal.

Vasopressin mRNA. The hybridized slides were dipped into Kodak NTB3 nuclear emulsion and exposed for three days. The intensity of the signal above parvocellular PVN was measured through a microscope by the ImageJ program. The results were obtained as pixels occupied by silver grain above the examined cell (pixel/cell). These measurements were done on two sections of the left and right CeA. The values of these four CeA sections were averaged (Palkovits et al., 1998).

POMC mRNA expression in the anterior pituitary in second series (28R) was evaluated as described above for CRH mRNA hybridization.

Hormone Analysis

Blood samples were collected once a week from the tail vein (in case of 28R) and, at the end of the stress regimen by decapitation (trunk blood in both experimental series). Blood was sampled on ice-cold K_2-EDTA containing tubes and centrifuged at 3000 rpm/min for 20 min. The plasma was stored at -20°C until hormone measurements.

Plasma ACTH was measured by radioimmunoassay (RIA) in 50μl unextracted plasma as described earlier (Zelena et al., 1999). The ACTH antibody (no. 8514) directed against the midportion of the h-$ACTH_{1-39}$ molecule, was raised in rabbit in the Institute of Experimental Medicine, Hungarian Academy of Sciences (Budapest, Hungary). It is highly specific showing 0.2% cross-reaction with α-MSH and no significant cross-reaction with γ-MSH, CLIP, $ACTH_{11-24}$, $ACTH_{25-39}$, $ACTH_{1-14}$, and $ACTH_{1-19}$. The intra- and interassay coefficients of variation were 4.7 and 7%, respectively.

Plasma corticosterone was measured from 10μl unextracted plasma by a RIA using specific antiserum developed in our Institute as described earlier (Zelena et al., 2003a). The corticosterone antiserum was raised in rabbits against corticosterone-carboximethyloxime bovine serum albumin. ^{125}I-labelled corticosterone-carboxymethyloxime-tyrosinemethyl ester was used as tracer. The interference with plasma transcortin was eliminated by inactivating transcortin at low pH. Assay sensitivity was 1 pmol.. The intra- and interassay coefficient of variation was 12.3 and 15.33%, respectively.

Prolactin was assessed by RIA using materials kindly provided by the Pituitary Program (National Institute of Diabetes, Digestive and Kidney diseases, NIH, Bethesda, MD). The intra- and inter-assay coefficients of variation were 15.8 and 24.5%, respectively.

The vasopressin content of the hypophysis was also assessed by RIA. The neurointermediate lobe of the pituitary was homogenized in 100µl 0.1N HCl and, after centrifugation, the supernatant was kept at -20°C until measurement. The anti-vasopressin antiserum was produced in rabbit and obtained from dr. M. Vecsernyés (Szent-Györgyi Med. Univ., Szeged, Hungary). The intra- and inter-assay coefficients of variation were 5.42 and 19%, respectively.

The Assessment of Corticosterone Secretion by the Adrenal Gland

We used a modified version of the method of Ács et al. (1987) as described earlier (Barna et al., 2004). In brief, the adrenal gland was immediately transferred into Dulbecco's modified Eagle's medium (Sigma-Aldrich, Budapest, Hungary) containing 2.5g bovine serum albumin/l (Fraction V, Calbiochem, La Jolla) (DMEM). The adrenal gland was subsequently minced with a razor blade into approximately 0.5 mm segments. The segments were transferred on perifusion columns.

Preparatory perifusion. The segments were perifused with DMEM for 1h at approximately 100µl/min flow rate. During the next hour, segments were perifused at 200µl/min (1ml/5min) to allow the basal corticosterone release to stabilize.

Basal secretion of corticosterone. After the second hour of the preparatory perifusion corticosterone release was measured under basal conditions for 15 min.

ACTH-induced increases in corticosterone secretion. After measuring basal secretion, the adrenal glands were stimulated with $2*10^{-10}$mol/ml ACTH for 5 min. After 60 min washout, glands were again stimulated with 10^{-9}mol/ml ACTH.

Column effluent was collected every 5 min on ice. Transit time from the medium reservoir to the fraction collector was 150 s. Media and columns were maintained at 37 °C. Column effluent was stored at -20 °C until assayed. Eight columns filled up with adrenal glands from different groups (di/+C, di/+28R, di/diC, di/di28R) were used in parallel and the experiment was repeated until each group had data from 5-6 columns (n=5-6).

Histological Examination of the Adrenal Cortex

15µm sections of paraffin-embedded adrenal glands were mounted on gelatin coated slides and were stained by the hematoxylin-eosin. The depth of the adrenal cortex was calculated by using the Image J analyzer. We subtracted the area of the medulla from the area of the whole adrenal gland at its highest diameter.

Statistical Analysis

Values were presented as mean±S.E.M. Hormonal data underwent logarithmic transformation. Data were analyzed by analysis of variance (ANOVA) using the ANOVA/MANOVA modul of the STATISTICA 6.0 software package (Tulsa, OK, USA).

Pairwise comparisons were made by the Newman-Keuls method and indicated on figures and tables.

Results

Two Weeks Repeated Restraint

Body and Organ Weights

The initial body weight at age of 8-10 weeks of the di/di animals was smaller (di/+: 337±7g, di/di: 277±5g; n=70; $p<0.01$) and gain less weight during 14 days ($p=0.047$) than their di/+ littermates (figure 1A). Two weeks repeated restraint induced a reduction in the body weight with similar effect in di/+ and di/di rats (effect of 14R $p<0.01$ without interaction).

Repetition of restraint for 14 days resulted in a decrease in thymus weight and in adrenal gland hypertrophy to the similar extent in both genotype (figure 1B,C; effect of 14R $p<0.01$). Unstressed di/di rats showed lower thymus and adrenal weight probably due to their smaller body weight (effect of genotype $p<0.01$).

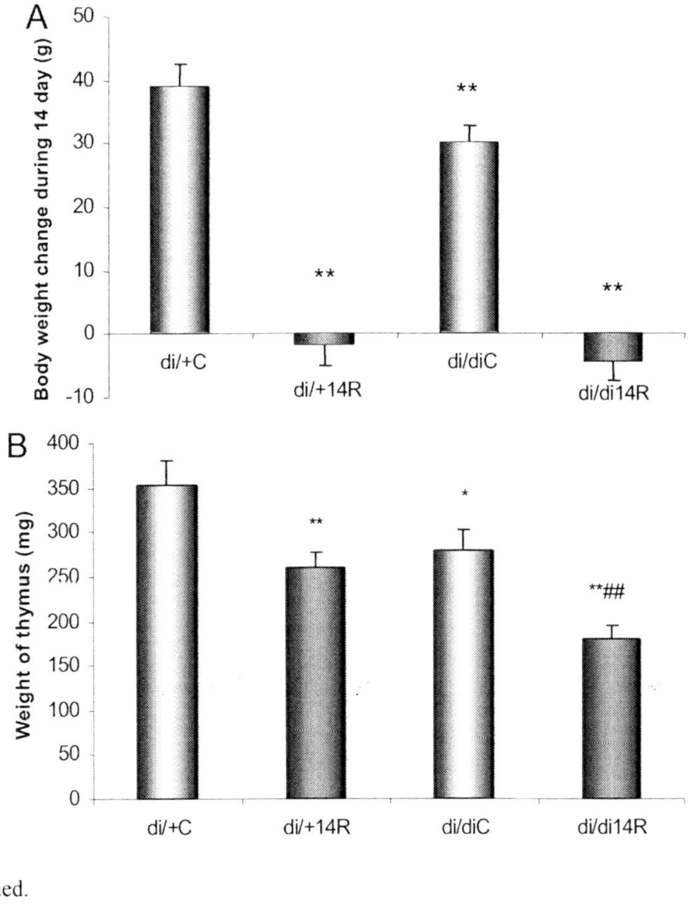

Figure 1. Continued.

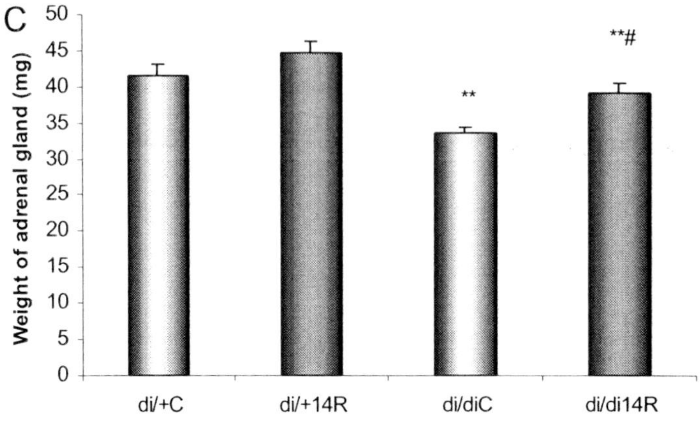

Figure 1. Somatic changes after 14 day repeated restraint (14R) in vasopressin deficient Brattleboro rats. A: Body weight changes (n=32-39). The effect of genotype was close to significance (p=0.065) while 14R induce significant changes (p<0.01) without interaction with genotype. B. Weight of thymus (n=19-22) Both the effect of genotype and 14R was significant (p<0.01) without interaction. C. Weight of adrenal gland (n=19-22). Both the effect of genotype and 14R was significant (p<0.01) without interaction. *p<0.05; **p<0.01 compared to control; #p<0.05; ##p<0.01 compared to di/+.

Changes in HPA Axis

In the anterior lobe of the pituitary the POMC mRNA levels were diminished after acute -, and were elevated after chronic stimuli (table 2, effect of acute stress p=0.02 and effect of chronic stress p<0.01) without any effect or influence of genotype.

Plasma ACTH levels were dramatically elevated by acute restraint (p<0.01). After 14R, under resting conditions the ACTH levels was not changed, while previous chronic stress diminished the ACTH rise induced by the last stress (acute*chronic stress interaction p<0.01). The lack of vasopressin did not influence the basal ACTH levels or the resting levels after 14R, however the acute stress-induced rise was smaller in chronically stressed vasopressin deficient animals.

The plasma corticosterone levels were approximately 7-times elevated by 60 min acute restraint (effect of acute stress p<0.01). The repetition of the stress was unable to induce a significant rise in resting levels (effect of chronic stress p=0.067) and did not influence the acute stress induced elevation (no interaction between acute and chronic stress). There was no effect of vasopressin deficiency on corticosterone levels independently from acute and chronic stress.

Prolactin levels in plasma were significantly elevated by acute challenge (p<0.01) without any influence of chronic stress (14R) or the lack of vasopressin.

Table 2. HPA axis changes during 14 days repeated restraint

Genotype	di/+				di/di			
Chronic stress	C		14R		C		14R	
Acute stress	+C	+1R	+C	+1R	+C	+1R	+C	+1R
POMC	105±4	69±11	131±16	112±10	90±9	71±12	127±9	121±20$^{\#}$
ACTH	11±3	335±93**	16±4	246±73**	17±3	350±82**	19±3	139±35$^{*\#\#}$
Cort.	193±38	1385±83**	244±35	1222±84**	149±24	1188±107**	228±39	1193±104**
Prolactin	6.7±1	41.2±16**	9.4±2	35.9±6**	4.0±1	46.4±8**	6.1±1	32.6±7**

POMC mRNA levels in the anterior lobe of the pituitary (% of control) measured by Northern hybridization revealed a significant acute ($p=0.02$) and chronic stress ($p<0.01$) effect without any genotype effect or interaction (n=4-8).

Plasma ACTH levels (fmol/ml) were significantly elevated in an acute stress situation ($p<0.01$) while the effect of chronic stress (14R) were also significant ($p=0.014$) without significant in pairwise comparisons. The adaptation was visible on acute*chronic stress interaction ($p<0.01$). The vasopressin deficient rats had reduced reactivity after chronic challenge (1R compared to 29R, $p<0.01$) (n=11-22).

Corticosterone plasma levels (Cort. pmol/ml) was elevated by acute challenge ($p<0.01$) while the chronic stress was able to induce a marginally elevation ($p=0.067$). Genotype had no effect at all (n=11-22).

Plasma prolactin levels (ng/ml) were influenced just by acute restraint (effect of acute stress $p<0.01$) (n=11-22).

*$p<0.05$; **$p<0.01$ compared to respective control (C); #$p<0.05$; ##$p<0.01$ compared to respective 1R.

Four Weeks Repeated Restraint

Body and Organ Weights

The initial body weight of the di/di animals was smaller than that of their di/+ littermates (di/+: 337±13g, di/di: 296±8g; n=39; $p<0.01$). In controls, body weight increased linearly during the experiment (figure 2A). Repeatedly restrained di/+ rats did not show weight gain during the first two weeks, but their weight started to increase from the third week onwards. The weight of stressed heterozygotes was still decreased at the end of the experiment. In stressed di/di rats the initial reduction in body weight was smaller and the difference from their homozygote controls (di/diC) disappeared during the following two weeks (genotype*28R interaction $p<0.01$).

Changes seen in thymus and spleen weights were similar in the two genotypes (effect of 28R $p<0.05$), despite the fact that unstressed di/di controls showed lower thymus and spleen weights than unstressed di/+ rats (figure 2.B,C, effect of genotype $p<0.01$).

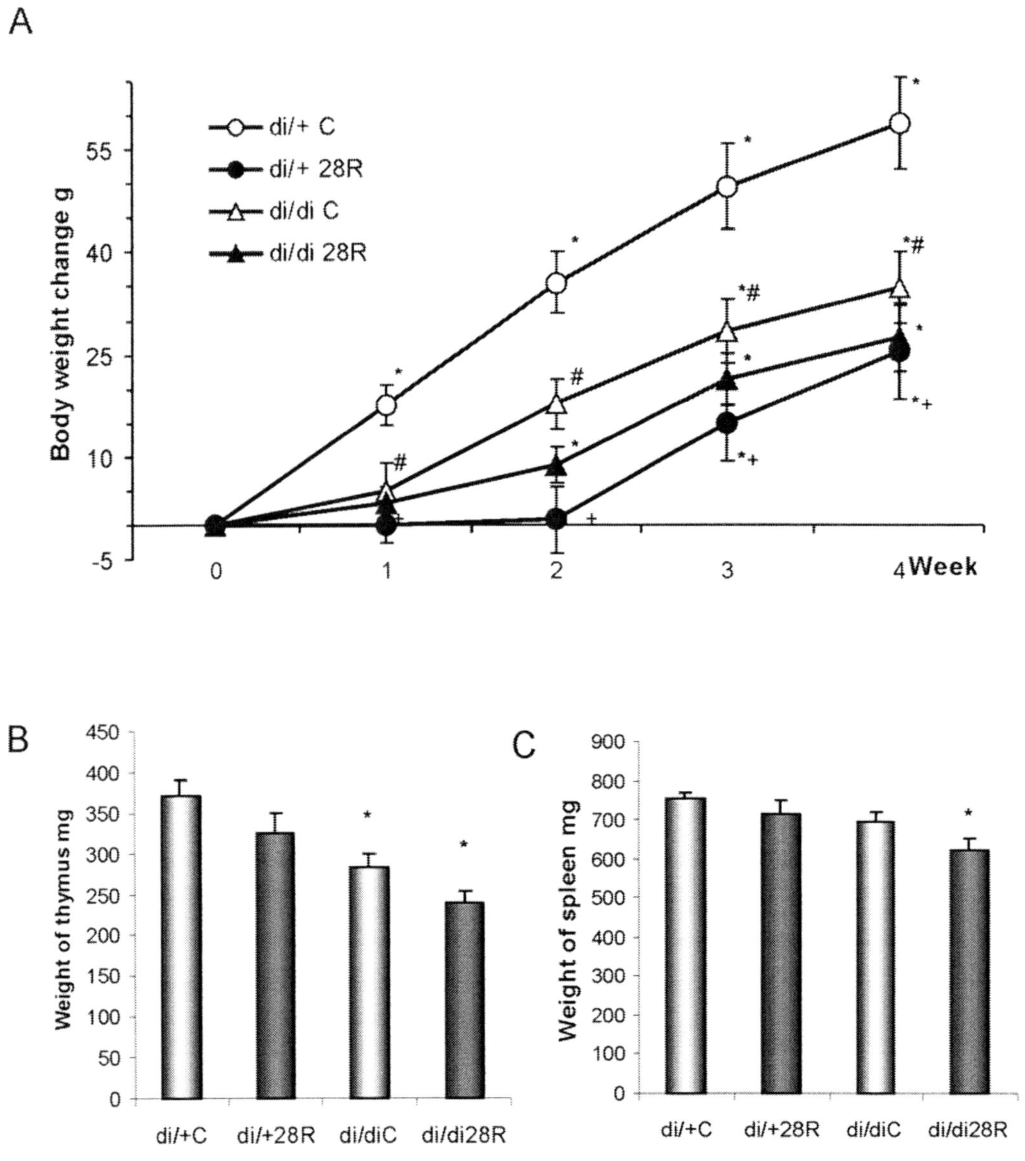

Figure 2. Somatic changes after 28 days repeated restraint (28R) in vasopressin deficient Brattleboro rats. A: Body weight changes (n=18-21). The effect of genotype was not significant while the effects of 28R; time; time*28R and genotype*28R were ($p<0.01$). B: Weight of thymus (n=18-21). There was a significant effect of genotype ($p<0.01$) and 28R ($p=0.017$). C: Weight of spleen (n=14-17). There was a significant effect of genotype ($p<0.01$) and 28R ($p=0.042$). *$p<0.01$ compared to initial body weight or control (thymus, spleen); #$p<0.05$ compared to control, non-stressed group or di/+ rats (thymus, spleen); +$p<0.01$ compared to di/+ group.

Changes in Adrenal Glands

Repeated restraint increased the weight of the adrenals in heterozygotes (figure 3A). Although di/di rats had smaller adrenals, the effect of 28R was still visible ($p<0.01$). In addition to changes in adrenal weight, repetition of restraint for 28 days also increased the cross-sectional area of the adrenal cortex (figure 3B). This effect was smaller in di/di rats.

The basal *in vitro* secretion of corticosterone decreased after 28R (figure 3C; $p<0.01$). Both the first and second ACTH stimulus induced smaller corticosterone secretion from adrenals of previously stressed rats (1. stimulus: effect of 28R $p<0.01$; 28R*time interaction $p=0.038$; 2. stimulus: effect of 28R $p=0.057$; 28R*time interaction $p=0.018$). There was a slight, not significant elevation of corticosterone secretion in di/di rats to both ACTH stimuli ($p=0.15$) with significantly different time course to the second stimulus (genotype*time interaction $p=0.02$).

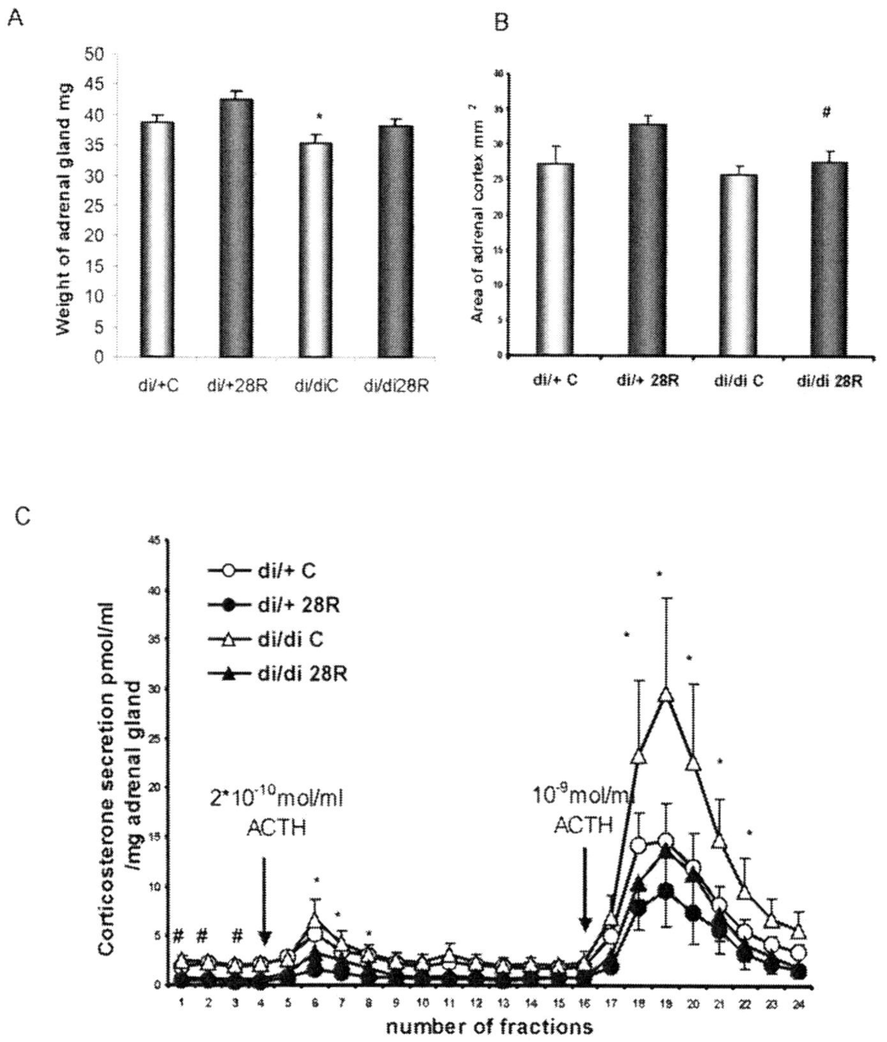

Figure 3. Changes in adrenal gland after 28R. A: Weight of adrenal glands. n=18-21 By two way ANOVA the effect of genotype ($p<0.01$) and chronic stress ($p=0.023$) was significant without interaction. B: The adrenal cortical area measured at the largest diameter in percentage of the whole adrenal gland. n=5-9 The effect of chronic stress was significant ($p=0.042$). C: The *in vitro* adrenal gland sensitivity to ACTH measured by superfusion method. n=5-6 Repeated doses of ACTH were able to induce corticosterone rises. Although the increment seemed to be higher in di/di than di/+ rats just the reducing effect of chronic stress was significant both on basal as well as on stimulated secretion. *$p<0.05$ v.s. control (Part A) or basal secretion (Part C). #$p<0.05$ v.s. di/+28R (Part B) or effect of chronic stress (Part C).

Changes in mRNA Levels

CRH mRNA density in the PVN was similar in di/+ and di/di controls (figure 4A). Chronic stress increased CRH mRNA levels in the PVN, but the increase was smaller in vasopressin deficient rats (genotype*chronic stress interaction $p<0.01$). Neither the intensity nor the extension of amygdalar CRH mRNA was affected by genotype or chronic stress (data not shown).

Although the mRNAs of di/+ and di/di animals may have slightly different composition (due to the mutation that prevents peptide synthesis in di/di rats), in our study there was no significant difference in vasopressin mRNA levels in the parvocellular part of the PVN between di/+ and di/di control animals (figure 4B). 28 days of restraint significantly elevated the vasopressin mRNA levels in individual paraventricular parvocellular cells in the hypothalamus of the di/+ rats. The elevation was significantly lower in vasopressin-deficient animals ($p<0.01$). The magnocellular vasopressin mRNA levels in the PVN -measured on imaging plates - were very high both in control and repeatedly stressed rats. No effect of genotype or stress were detected (data not shown).

The POMC mRNA levels of the anterior pituitary was higher in di/di, as compared with di/+ controls (figure 4C). After chronic stress, POMC mRNA levels were increased in di/+ rats, whereas they remained unchanged in vasopressin-deficient rats (genotype*chronic stress interaction $p<0.01$). Although 1h restraint is not optimal to see changes in mRNA levels but we could reveal that the lack of vasopressin lead to reduced POMC mRNA activation even during 1h restraint (table 4.; genotype*acute stress interaction $p=0.028$).

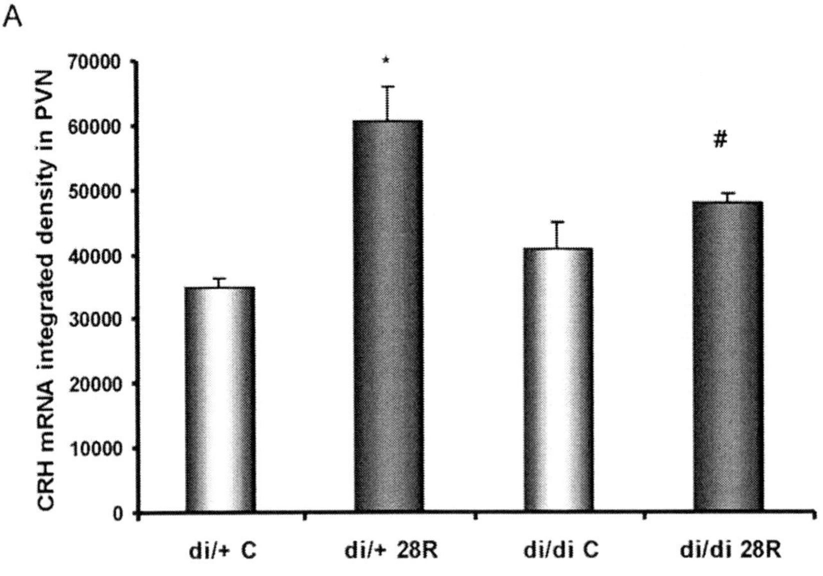

Figure 4 Continued.

B

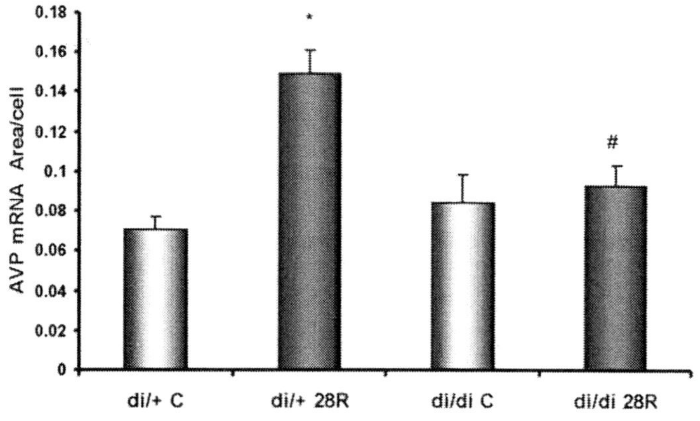

C

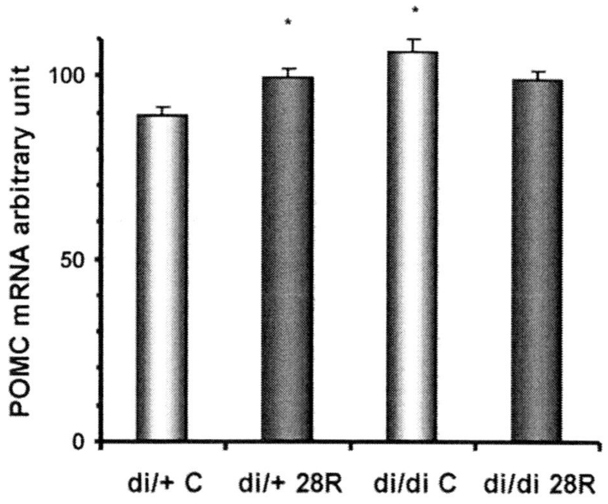

Figure 4. mRNA levels measured by in situ hybridization in in the PVN and the anterior pituitary gland. A: CRH mRNA integrated density in PVN, n=7-9. The effect of chronic stress and chronic stress*genotype interaction was significant ($p<0.01$). B: Vasopressin mRNA levels, area/cell. n=8-10 rat/group. The effect of chronic stress and chronic stress*genotype interaction was significant ($p<0.01$). C: POMC mRNA levels in AL. n=7-10 POMC mRNA level was significantly increased in di/+ stressed group ($p<0.05$) and the genotype*chronic stress interaction was also significant ($p<0.01$). *$p<0.01$ v.s. control; #$p<0.01$ v.s. di/+28R.

Changes in Hormone Levels

Resting ACTH levels of control animals were higher in di/di than in di/+ animals (figure 5A). Single restraint increased ACTH levels in both genotypes (acute stress $p<0.01$), but the increase was smaller in vasopressin-deficient rats than in heterozygotes (acute stress*genotype interaction $p<0.01$). 28 days of restraint *per se* had no significant effect on resting ACTH plasma levels in either genotype, although the differences between the genotypes were maintained. There was a small habituation to the 29th restraint (acute stress*chronic stress interaction $p<0.01$). An additional acute restraint had a larger effect on ACTH plasma levels in di/+ as compared with di/di rats.

Basal corticosterone levels were higher in vasopressin deficient rats throughout the experiment (table 3). Basal levels were increased by repeated restraint in di/+ rats from week three onwards. No similar changes were noticed in di/di rats. A single acute restraint increased plasma corticosterone levels in both genotypes ($p<0.01$) but the increase was smaller in vasopressin-deficient rats as compared with heterozygotes (figure 5B, acute stress*genotype interaction $p<0.01$). Pre-exposure of rats to repeated restraint did not diminish the response to an additional acute restraint.

Table 3. Corticosterone plasma levels (pmol/ml) changes during 4 week repeated restraint measured weekly via tail cut

	di/+		di/di	
	C	28R	C	28R
1st week	89.5 ± 24.6	69.2 ± 31.6	122.7 ± 15.0 *	131.65 ± 29.9 *
2nd week	123.8 ± 15.0	211.3 ± 50.7	289.8 ± 30.2	198.9 ± 53.5
3rd week	57.0 ± 27.6	148.5 ± 79.9 #	137.3 ± 25.5 #	266.4 ± 47.9
4th week	89.8 ± 32.4	203.7 ± 58.2 #	196.9 ± 34.2 #	195.0 ± 33.5

n=5-10; *$p<0.05$ effect of genotype; #$p<0.05$ compared to di/+ control.

Table 4. Plasma prolactin levels and neurointermediate lobe vasopressin content at the end of 28R with or without acute challenge

Chronic stress	C		28R	
Acute stress	C	R	C	R
	POMC mRNA in anterior pituitary (arbitrary unit) n=7-10			
di/+	89.2±2.1	99.6±2.7	99.4±2.2	103.4±3.6
di/di	106.6±3.3	101.2±4.7	98.7±2.7	99.4±2.0
Prolactin in plasma (ng/ml) n=9-11				
di/+	0.81± 0.1	18.92± 2.7*	1.31± 0.5	13.88± 3.2*
di/di	1.92± 0.7	16.44± 1.9*	1.03±0.3	15.67± 2.2*
	Vasopressin in neurointermediate lobe of pituitary (ng/ml) n=6-11			
di/+	6909±517	6116±340	5823±602	5260±544
di/di	45±11^{+}	46±6^{+}	28±9^{+}	104±20$^{*\#+}$

The lack of vasopressin diminished the acute and chronic stress induced POMC mRNA elevation in the anterior lobe of the pituitary (genotype*acute stress $p=0.02$; genotype*chronic stress $p<0.01$) without any other significant effect.

The plasma prolactin levels were elevated by the acute stress ($p<0.01$). Neither the genotype nor the chronic stress had an influence on these changes. The vasopressin content of the neurointermediate lobe was very low in di/di rats (effect of genotype $p<0.01$). Moreover the effect of acute stress and acute stress*genotype interaction was also significant ($p<0.01$). In di/+ rats the effect of chronic stress become significant ($p=0.04$).

*$p<0.01$ compared to control (C) #$p<0.01$ compared to acute stress ® $^{+}p<0.01$ compared to di/+.

Plasma prolactin levels (table 4) were increased by acute restraint irrespective of pre-exposure to repeated restraint or genotype ($p<0.01$).

The vasopressin content of the neurointermediate lobe of di/di rats was close to our detection limit (table 4, effect of genotype $p<0.01$). Both acute and repeated restraint decreased the vasopressin content of the neurointermediate lobe in di/+ animals (effect of chronic stress $p<0.01$). Due to low basal expression, no similar changes were noticed in di/di rats.

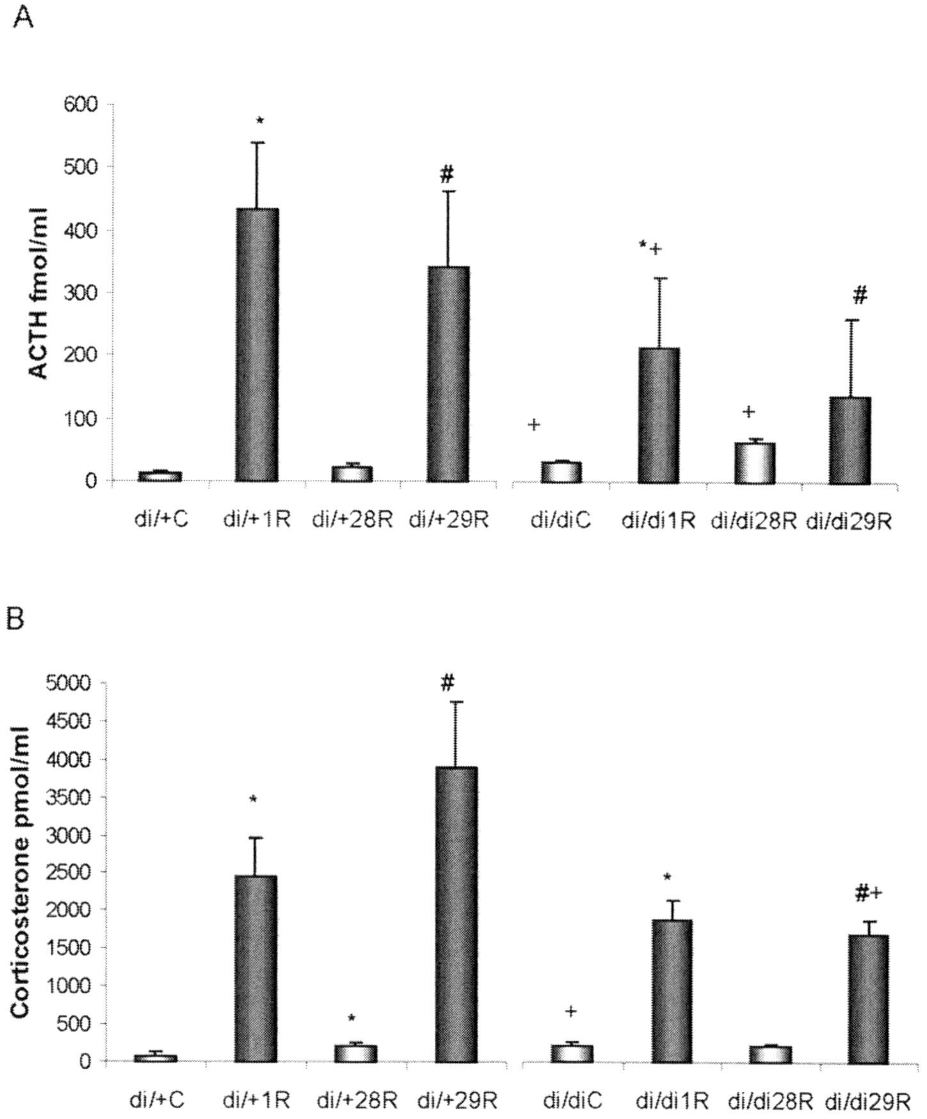

Figure 5. Plasma ACTH (Part A) and corticosterone (Part B) levels. n=9-11. A: The effect of acute stress, acute stress*genotype and acute stress*chronic stress interactions were significant ($p<0.01$). B: The effect of chronic stress ($p=0.0585$) approached significance, and acute stress and acute stress*genotype interaction ($p<0.01$) was significant. Symbols: C: control; 1R: single, acute restraint; 28R: chronic restraint stress; 29R: acute stress after previously chronic stress. *$p<0.01$ v.s. control; #$p<0.01$ v.s. 28R; $p<0.01$ v.s. di/+.

Conclusion

Already two week of daily restraint resulted in a chronic stress state characterized by typical somatic changes (body weight reduction, thymus involution, adrenal gland hypertrophy). Both the somatic changes and the POMC mRNA elevation in the anterior lobe of the pituitary as well as the slight, non significant resting corticosterone elevation visible after 14R was not influenced by the absence of vasopressin. From this first series we can conclude that the role of vasopressin in maintenance the chronic hyperactivity of the HPA is less than predicted from earlier literature. In the second series, the prolongation of the stress regime up to 4 weeks lead to some habituation as the body weight gain started to increase and the thymus involution was also diminished at the end of 4 week. This later state (28R) was characterized by reduced weight gain, thymus and spleen involution, adrenal and adrenocortical hypertrophy, increased CRH and vasopressin mRNA levels in the PVN, enhanced POMC mRNA levels in the anterior lobe of the hypophysis and elevated basal corticosterone levels. Weight gain, adrenocortical (but not adrenal) hypertrophy, CRH-, vasopressin- and POMC mRNA expression and basal corticosterone levels were less markedly changed in vasopressin-deficient rats suggesting that the role of vasopressin in maintenance the resting activity of the HPA on chronic stress background enhanced with time.

In unstressed resting rats, vasopressin-deficiency resulted in interesting (and somewhat surprising) changes in stress-related variables with a marked difference between the two series. CRH mRNA levels were normal in the parvocellular part of the PVN, but in the second series (28R) POMC mRNA levels in the anterior lobe of the pituitary, as well as ACTH and corticosterone plasma levels were higher in the mutant di/di genotype as compared with di/+ heterozygotes. It was repeatedly shown that vasopressin alone is able to stimulate ACTH secretion (see e.g. Buckingham, 1987). If this effect of vasopressin played a role in maintaining basal glucocorticoid secretion, than di/di rats should have shown lower rather than higher ACTH and glucocorticoid plasma levels. This finding together with the work of Laulin et al (Laulin et al., 1985) casts doubt on the role of vasopressin in maintaining basal stress hormone production. Contrasting results were obtained earlier, although findings obtained in control (unstressed) rats were in many cases similar with ours (Bohus and de Wied, 1998). A large part of the inconsistency in the reports may be explained by the choice of control animals from a different strain (Bohus and de Wied, 1998). Another possible explanation may be related to maternal care. As shown previously, the maternal genotype has a large impact on the stress responses of offspring (Casolini et al., 1993; Zelena et al., 2003b). To minimize confounding factors here we used littermate di/+ and di/di rats, i.e. we standardized maternal care and compared di/di rats with littermate di/+ animals. It is also possible that the lack of vasopressin shifts the circadian rhythm and sampling at different time point may lead to different conclusions. It is well known that the "endogen clock", the suprachiasmatic nucleus contains a large number of vasopressinerg cells (Reghunandanan et al., 1991) and the lack of vasopressin in Brattleboro rats might interfere with rhythmicity of e.g. melatonin secretion (Schroder et al., 1988). On the other hand the ACTH rhythm of di/di rats seems to be similar to other strains with greater fluctuation amplitude (Tankosic et al., 1982).

During chronic stress the enhanced release of vasopressin has been proposed to explain how chronic stressful stimulation results in "normal" HPA axis regulation despite the elevated circulating glucocorticoid levels (Aguilera, 1994; Dallman, 1993). Two week repetition of daily 60 min restraint induced typical changes also occuring in other chronic stress situations (somatic changes (Tache et al., 1976), resting POMC mRNA elevation in the anterior lobe of the pituitary, basal plasma corticosterone elevation). However, in contrast to our hypothesis, the absence of vasopressin did not influence these changes. The examination of two weeks repeated restraint (Zelena et al., 2004) as well as other data from streptozotocin induced diabetes mellitus for two weeks (Zelena et al., 2006) let us concluded that the role of vasopressin in maintaining the chronic hyperactivity of the HPA is smaller than it was supposed. Prolongation of the restraint procedure for 28 days also resulted in an increase of the HPA-axis activity demonstrated at all three levels of the axis: in the PVN (elevated CRH and vasopressin mRNA expression), in the hypophysis (elevated POMC mRNA in the anterior lobe of the pituitary and increased plasma ACTH) and in the adrenal cortex (adrenocortical hypertrophy and increased corticosterone plasma levels). Although Ma et al. (1999) did not find CRH mRNA elevation after 14 days repeated restraint, Makino et al. (1995) did observe increased CRH mRNA under similar conditions. The effect was not large, but it was suggested that even a moderate increase in CRH output is able to effectively stimulate POMC production (Aguilera, 1994). Although the possible role of CRH was acknowledged, Aguilera (1994) suggested in this paper that chronic stress responses are maintained predominantly by the vasopressin-containing parvocellular neurons of the PVN. In line with this assumption, we noticed a higher increase in vasopressin mRNA than in CRH mRNA after 28 daily restraints ($112\pm17\%$ vs. $74\pm16\%$ increase, respectively). Beyond this difference, vasopressin-deficiency in Brattleboro rats clearly reduced the above mentioned parameters of prolonged chronic stress (28R), suggesting that vasopressin indeed plays an important role especially in the later phases of the development of chronic stress state. In contrast, the effects of 4 weeks repeated restraint on body and organ weights were similar in di/di and di/+ rats in most cases, except for the diminished body weight reduction and smaller increase in adrenocortical area in di/di rats. We may conclude that the typical somatic changes occurring during chronic stress (body weight loss, thymus and spleen involution, adrenal gland hypertrophy) are under complex regulation. Although these changes are fundamentally influenced by circulating level of glucocorticoids (Reul et al., 1998) many other factors should be taken into consideration and/or an even more prolonged time is needed to see an effect as the role of vasopressin became important just up to 2 week.

Acute restraint increased the plasma level of all three studied hormones, ACTH, corticosterone and prolactin in both experiments. The lack of vasopressin resulted in a mild reduction of ACTH and corticosterone elevation although the difference reaches the level of significance only in a portion of experiments. It was shown earlier that the repeated application of a stressor often leads to tolerance towards the endocrine effects of the same (i.e. homotypic) stressor (Aguilera, 1994). Pre-exposure to repeated restraint only slightly, not significantly intensified the effect of vasopressin deficiency except after two week repeated restraint when the role of vasopressin seems to be specially important in the maintenance the acute stress reactivity upon chronic stress base. Our acute stress study was focused on the plasma hormone levels as 1h restraint is optimal for those parameters while mRNA levels were found to change just after 4h in many studies (e.g. Harbuz and Lightman, 1989). Therefore the extrapolation from our results is limited but we have to mentioned that the lack of vasopressin interacted with the acute stress induced changes in the ACTH precursor POMC mRNA in the anterior lobe of the pituitary, although this was found just in the second experimental series (28R). Taken together, vasopressin plays a role in acute ACTH and corticosterone stress responses in rats however its effect is strongly influenced by various factors. Noteworthy, the effect of restraint on prolactin secretion was not affected by vasopressin background, suggesting that the effects of vasopressin-deficiency were specific to the glucocorticoid stress response. This is in line with earlier findings on the participation of both CRH and vasopressin in the induction of acute stress responses (Makara et al., 2004) however the contribution of vasopressin seems to be smaller than it was predicted in earlier literature (e.g Tilders et al., 1985).

The difference between ACTH and corticosterone responses (small decrease in ACTH but no habituation in corticosterone) suggests that the sensitivity of adrenal gland to ACTH and/or to other neural or humoral factors can be changed. From our superfusion experiments it seems that after 4 week chronic stress the sensitivity of the gland to the main secretagogue (ACTH) was decreased rather than increased, thus further experiments may be needed to elucidate why the ACTH stress response habituates while the corticosterone response does not. Moreover our work suggested that the lack of vasopressin may lead to a slightly increased sensitivity of the adrenals to ACTH. It is in contrast to the work of Wiley et al. (1974) who found decreased adrenal sensitivity in Brattleboro rats. The discrepancies could be due to methodological differences or to some changes in the strain during the past 30 years.

The vasopressin prohormone is processed into vasopressin, neurophysin and a glycoprotein. This processing is disrupted in Brattleboro rats which express a mutant vasopressin precursor due to a single nucleotide deletion in the neurophysin region of the vasopressin gene resulting autosomal vasopressin deficiency. As a consequence, these animals (homozygous di/di) suffer from diabetes insipidus (Evans et al., 2000). The sequence of our vasopressin riboprobe was able to catch the mutated gene. It turned out that the level of the mutated vasopressin mRNA in the parvocellular PVN is normal although not regulated in di/di animals. We could not confirm the role of magnocellular vasopressin although according to literature repeated restraint stress could induce structural plasticity in magnocellular cells (Miyata et al., 1994). The reduction of the neurointermediate lobe vasopressin content to an acute stimulus was expected. On the other hand, the decreased vasopressin content in the neurointermediate lobe of chronically stressed rats is in contrast to the unchanged magnocellular vasopressin mRNA content. A possible explanation is that the secretion of the hormone is enhanced but synthesis cannot keep up with the increased release.

In summary, according to our data vasopressin plays a minor (if any) role in maintaining basal ACTH and corticosterone secretion, but the present findings support the hypothesis that vasopressin plays an important role in the development of chronic stress state, although its effects become prominent only after a prolonged exposure to daily stressful events. Two weeks of repeated restraint were not sufficient to reveal the role of vasopressin, suggesting that some critical changes occurred during the third and fourth week of stress exposure. One can still hypothesize that the decreased stress responses seen in di/di rats was due to exhaustion, as these rats may be less able to maintain elevated CRH, vasopressin, POMC mRNA expression and increased corticosterone secretion on the long run. However, the decrease in weight gain was less dramatic with this genotype and their acute stress responsiveness was not markedly impaired by pre-exposure to chronic stress. These findings make the "exhaustion" hypothesis less likely. It seems more likely that the role of vasopressin in maintaining chronic stress responses develops over time, a process that requires more than two weeks.

Acknowledgements

We thank Ms. Hajni Beko for excellent assistance with the animals and the Hormone Laboratory of the Institute of Experimental Medicine for the hormone analyses. F.A. László (University of Szeged, Hungary) provided valuable help with establishing the Brattleboro breeding colony. These studies were supported by OTKA grants T 025845 and T043161 to GBM

References

Ács, Zs; Szabo, B; Kapocs, G; Makara, GB. Gamma-Aminobutyric acid stimulates pituitary growth hormone secretion in the neonatal rat. A superfusion study. *Endocrinology.* 1987, 120:1790–1798.

Aguilera, G. Regulation of pituitary ACTH secretion during chronic stress. *Front. Neuroendocrinol.* 1994, 15: 321-350.

Aguilera, G; Rabadan-Diehl, C. Vasopressinergic regulation of the hypothalamic-pituitary-adrenal axis: implications for stress adaptation. *Regul. Pept.* 2000, 96:23-29.

Akana, SF; Hanson, ES; Horsley, CJ; Strack, AM; Bhatnagar, S; Bradbury, MJ; Milligan, ED; Dallman, MF. Clamped Corticosterone (B) Reveals the Effect of Endogenous B on Both Facilitated Responsivity to Acute Restraint and Metabolic Responses to Chronic Stress. *Stress.* 1996, 1: 33-49.

Antoni, FA. Vasopressinergic control of pituitary adrenocorticotropin secretion comes of age. *Front. Neuroendocrinol.* 1993, 14: 76-122.

Barna, I; Balint, E; Baranyi, J; Bakos, N; Makara, GB; Haller, J. Gender-specific effect of maternal deprivation on anxiety and corticotropin-releasing hormone mRNA expression in rats. *Brain Res. Bull.* 2003, 62:85-91.

Barna, I; Zelena, D; Arszovszki, AC; Ledent, C. The role of endogenous cannabinoids in the hypothalamo-pituitary-adrenal axis regulation: in vivo and in vitro studies in CB1 receptor knockout mice. *Life Sci.* 2004, 24:2959-2970.

Bartanusz, V; Jezova, D; Bertini, LT; Tilders, FJ; Aubry, JM; Kiss, JZ. Stress-induced increase in vasopressin and corticotropin-releasing factor expression in hypophysiotrophic paraventricular neurons. *Endocrinology.* 1993, 132: 895-902.

Bohus, B; de Wied, D. The vasopressin deficient Brattleboro rats: a natural knockout model used in the search for CNS effects of vasopressin. *Prog. Brain Res.* 1998, 119: 555-573.

Buckingham, JC. Vasopressin receptors influencing the secretion of ACTH by the rat adenohypophysis. *J. Endocrinol.* 1987, 113: 389-396.

Casolini, P; Patacchioli, FR; Domenici, MR; Angelucci, L. Brain adrenocorticoid receptor binding capacity in the diabetes insipidus Brattleboro rat is dependent on maternal genotype. *Neuroendocrinology.* 1993, 57: 658-662.

Chowdrey, HS; Larsen, PJ; Harbuz, MS; Jessop, DS; Aguilera, G; Eckland, DJ; Lightman, SL. Evidence for arginine vasopressin as the primary activator of the HPA axis during adjuvant-induced arthritis. *Br. J. Pharmacol.* 1995, 116: 2417-2424.

Dallman, MF. Stress update. Adaptation of the hypothalamic-pituitary-adrenal axic to chronic stress. *Trends Endocrinol. Metab.* 1993, 4:62-69.

De Goeij, DCE; Kvetnansky, R; Whitnall, MH; Jezova, D; Berkenbosch, F; Tilders, FJH. Repeated stress-induced activation of corticotropin-releasing factor neurons enhances vasopressin stores and colocalization with corticotropin-releasing factor in the median eminence of rats. *Neuroendocrinology.* 1991, 53: 150-159.

De Goeij, DCE; Binnekade, R; Tilders, FJH. Chronic stress enhances vasopressin but not corticotropin-releasing factor secretion during hypoglycemia. *Am. J. Physiol.* 1992, 263: E394-E399.

Engelmann, M; Landgraf, R; Wotjak, CT. The hypothalamic-neurohypophysial system regulates the hypothalamic-pituitary-adrenal axis under stress: an old concept revisited. *Front. Neuroendocrinol.* 2004, 3-4: 132-49.

Evans, DAP; De Bree, FM; Nijenhuis, M; Van der Kleij, AAM; Zalm, R; Korteweg, N; Van Leeuwent, FW; Burbach, JPH. Processing of frameshifted vasopressin precursors *J. Neuroendocrinology.* 2000, 12: 685-663.

Harbuz, MS; Lightman, SL. Responses of hypothalamic and pituitary mRNA to physical and psychological stress in the rat. *J. Endocrinol.* 1989, 122: 705-711.

Holmes, MC; Antoni, FA; Aguilera, G; Catt, KJ. Magnocellular axons in passage through the median eminence release vasopressin. *Nature.* 1986, 319: 326–329.

Kjaer, A; Knigge, U; Bach, FW; Warberg, J. Impaired histamine- and stress-induced secretion of ACTH and beta-endorphin in vasopressin-deficient Brattleboro rats. *Neuroendocrinology.* 1993, 57: 1035-1041.

Kovács, KJ; Földes, A; Sawchenko, PE. Glucocorticoid negative feedback selectively targets vasopressin transcription in parvocellular neurosecretory neurons. *J. Neurosci.* 2000, 20: 3843-3852.

Kretz, O; Reichardt, HM; Schutz, G; Bock, R. Corticotropin-releasing hormone expression is the major target for glucocorticoid feedback-control at the hypothalamic level. *Brain Res.* 1999, 818: 488-491.

Laulin, JP; Ezzarani, EA; Brudieux, R. Production of corticosterone in male homozygous Brattleboro rats. *Horm. Metab. Res.* 1985, 17 :342-345.

Ma, XM; Levy, A; Lightman, SL. Emergence of an isolated arginine vasopressin (AVP) response to stress after repeated restraint: a study of both AVP and corticotropin-releasing hormone messenger ribonucleic acid (RNA) and heteronuclear RNA. *Endocrinology.* 1997, 138: 4351-4357.

Ma, XM; Lightman, SL; Aguilera, G. Vasopressin and corticotropin-releasing hormone gene responses to novel stress in rats adapted to repeated restraint. *Endocrinology*. 1999, 140: 3623-3632.

Makara, GB; Mergl, Zs; Zelena, D. The role of vasopressin in hypothalamo-pituitary-adrenal axis activation during stress: an assessment of the evidence. *Ann. N Y Acad. Sci.* 2004, 1018: 151-161.

Makino, S; Smith, MA; Gold, PW. Increased expression of corticotropin-releasing hormone and vasopressin messenger ribonucleic acid (mRNA) in the hypothalamic paraventricular nucleus during repeated stress: association with reduction in glucocorticoid receptor mRNA levels. *Endocrinology*. 1995, 136: 3299-3309.

Miyata, S; Itoh, T; Matsushima, O; Nakashima, T; Kiyohara, T. Not only osmotic stress but also repeated restraint stress causes structural plasticity in the supraoptic nucleus of the rat hypothalamus. *Brain Res. Bull.* 1994, 33: 669-675.

Palkovits, M; Young, WS 3rd; Kovacs, K; Toth, Z; Makara, GB. Alterations in corticotropin-releasing hormone gene expression of central amygdaloid neurons following long-term paraventricular lesions and adrenalectomy. *Neuroscience*. 1998, 85:135-147.

Pinnock, SB; Herbert, J. Corticosterone differentially modulates expression of corticotropin releasing factor and arginine vasopressin mRNA in the hypothalamic paraventricular nucleus following either acute or repeated restraint stress. *Eur. J. Neurosci.* 2001, 13: 576-584.

Raadsheer, FC; Hoogendijk, WJ; Stam, FC; Tilders, FJ; Swaab, DF. Increased numbers of corticotropin-releasing hormone expressing neurons in the hypothalamic paraventricular nucleus of depressed patients. *Neuroendocrinology*. 1994, 60: 436-444.

Reghunandanan, V; Reghunandanan, R; Marya, RK. Vasopressin: its possible role in circadian time keeping. *Chronobiologia*. 1991, 18: 39-47.

Reul, JM; Labeur, MS; Wiegers, GJ; Linthorst, AC. Altered neuroimmunoendocrine communication during a condition of chronically increased brain corticotropin-releasing hormone drive. *Ann. N Y Acad. Sci.* 1998, 840: 444-455.

Schroder, H; Stehle, J; Henschel, M. Twenty-four-hour pineal melatonin synthesis in the vasopressin-deficient Brattleboro rat. *Brain Res.* 1988, 459: 328-332.

Schwartz, MW; Strack, AM; Dallman, MF. Evidence that elevated plasma corticosterone levels are the cause of reduced hypothalamic corticotrophin-releasing hormone gene expression in diabetes. *Regul. Pep.* 1997, 72: 105-112.

Scott, LV; Dinan, TG. Vasopressin and the regulation of hypothalamic-pituitary-adrenal axis function: implications for the pathophysiology of depression. *Life Sci.* 1998, 62: 1985-1998.

Simmons, A. Complete protocol for in situ hybridization of messenger RNAs in brain and other tissues with radiolabeled single-stranded RNA probes. *J. Histotechnol.* 1989, 12: 169–181.

Sokol, HW; Valtin, H. The Brattleboro rat. Ann N Y Acad Sci, New York, 1982

Swanson, LW; Sawchenko, PE; River, J; Vale, WW. Organization of ovine corticotropin releasing immunoreactive cells and fibers in the rat brain: An immunohistochemical study. *Neuroendocrinology*. 1983, 36:165-186.

Tache, Y; Du Ruisseau, P; Tache, J; Selye, H; Collu, R. Shift in adenohypophyseal activity during chronic intermittent immobilization of rats. *Neuroendocrinology*. 1976, 22: 325-336.

Tankosic, P; Burlet, A; Jegou, S; Chateau, M; Vaudry, H; Burlet, C; Boulange, M. Fetal and postnatal maturation of corticotrope function in the vasopressin-deficient rat (Brattleboro strain): a radioimmunological, immunocytochemical, and morphometric study. *Ann. N Y Acad. Sci.* 1982, 394: 560-573.

Tilders, FJ; Berkenbosch, F; Vermes, I; Linton, EA; Smelik, PG. Role of epinephrine and vasopressin in the control of the pituitary-adrenal response to stress. *Fed. Proc.* 1985, 44: 155-160.

Wiley, MK; Pearlmutter, AF; Miller, RE. Decreased adrenal sensitivity to ACTH in the vasopressin-deficient (Brattleboro) rat. *Neuroendocrinology*. 1974, 14: 257-270.

Whitnall, MH; Key, S; Gainer, H. Vasopressin-containing and vasopressin-deficient subpopulations of corticotropin-releasing factor axons are differentially affected by adrenalectomy. *Endocrinology*. 1987, 120: 2180-2182.

Zelena, D; Makara, GB; Jezova, D. Simultaneous blockade of two glutamate receptor subtypes (NMDA and AMPA) results in stressor-specific inhibition of prolactin and corticotropin release. *Neuroendocrinology*. 1999, 69: 316-323.

Zelena, D; Mergl, Zs; Makara, GB. Maternal genotype influences stress reactivity of vasopressin-deficient brattleboro rats. *J. Neuroendocrinol.* 2003b, 15: 1105-1110.

Zelena, D; Mergl, Zs; Foldes, A; Kovacs, KJ; Toth, Zs; Makara, GB. Role of hypothalamic inputs in maintaining pituitary-adrenal responsiveness in repeated restraint. *Am. J. Physiol. Endocrinol. Metab.* 2003a, 285: E1110-1117.

Zelena, D; Földes, A; Mergl, Zs; Barna, I; Kovács, KJ; Makara, GB. Effects of repeated restraint stress on hypothalamo-pituitary-adrenocortical function in vasopressin deficient Brattleboro rats. *Brain Res. Bull.* 2004, 63: 521-530.

Zelena, D; Filaretova, L; Mergl, Z; Barna, I; Toth, ZE; Makara, GB. Nucleus paraventricularis hypothalami but not vasopressin participates in chronic hyperactivity of the hypothalamo-pituitary-adrenal axis in diabetic rats. *Am. J. Physiol. Endocrinol. Metab.* 2006, 290: E243-250.

In: Neuropeptide Research Trends
Editor: Bernice A. Levine, pp. 213-227
ISBN: 978-1-60021-640-4

Chapter VIII

Disturbed Cross-Talk between Hypothalamic Neuropeptides, Nitric Oxide and other Factors May Significantly Contribute to the Hyperactivity of the HPA Axis in Depression and Schizophrenia

Hans-Gert Bernstein [*]***, Henrik Dobrowolny and Bernhard Bogerts***
Department of Psychiatry, Medical School, University of Magdeburg,
Leipziger Str.44, D-39120 Magdeburg, Germany

Abstract

Affective disorder and schizophrenia are associated with dysregulation of the hypothalamus-pituitary-adrenal (HPA) axis, which leads to repeated episodes of hypercortisolemia. Hypothalamic paraventricular neurons are believed to trigger these processes by aberrant generation and/or release of corticotrophin releasing hormone (CRH), vasopressin and oxytocin. Classically, this persisting overactivation of the HPA axis in neuropsychiatric disorders is thought to be maintained by a disturbed glucocorticoid-mediated feedback action. However, the discovery of strongly reduced numbers of nitric oxide synthase containing paraventricular (PVN) and suprachiasmatic (SCN) neurons as well as of fewer beta-endorphin generating arcuate nucleus neurons in post mortem brains of depressive and schizophrenic patients point to other possible mechanisms by which neuropeptide-synthesizing hypothalamic neurons may become disinhibited in affective disorder and schizophrenia, namely by disturbed cross-talk with local hypothalamic factors. We propose a model which suggests that at least four different factors may contribute to enhanced expression of paraventricular neuropeptides: reduced generation of nitric oxide (NO), reduced generation and intrahypothalamic

[*] To whom correspondence should be addressed. E-mail. Hans-Gert.Bernstein@med.uni-magdeburg.de

release of beta-endorphin, disturbed neuronal control of the PVN by SCN neurons and, last but not least, reduced GABAergic inhibition of PVN neurons.

Keywords: affective disorder, schizophrenia, hypothalamus, HPA-axis, nitric oxide, vasopressin, oxytocin, beta-endorphin, glutamic acid decarboxylase.

1. Affective Disorder and HPA Axis – The Established View

HPA axis activity is governed by the secretion of the corticotrophin-releasing hormone (CRH) and vasopressin (AVP) from the hypothalamus, which synergistically bring about the release of adrenocorticotrophic hormone (ACTH) from the pituitary. ACTH in its turn evokes the secretion of the glucocorticoids from the adrenal cortex. Glucocorticoids (with cortisol being the most important hormone in humans) influence a plethora of physiological events by interacting with their receptors in multiple neural and extraneural target tissues. This includes proper functioning of the HPA axis, where they are responsible for feedback inhibition of the secretion of ACTH from the pituitary and CRH from the hypothalamus. Besides, direct catecholaminergic, serotonergic and dopaminergic innervation of the CRh neurons in the hypothalamus has been described, and these and other neurotransmitters have been shown to significantly influence CRH and vasopressin release (for overview see Lopez et al. 1998).

Hyperactivity of the HPA axis is one of the key biological abnormalities described in major depressive disorder, occurring in 30-50% of depressed subjects (Scott and Dinan 2002). An increased activity of the HPA axis has also been observed in patients with bipolar disorder and schizophrenia (Walker and Diforio 1997; Ryan et al. 2004), albeit less frequently than in patients with major depression. Based on elevated cellular expression of CRH and vasopressin (but also oxytocin) in post mortem hypothalamic PVN nuclei of depressed patients (Raadsheer et al. 1995; Purba et al. 1996; Swaab et al. 2005) and increased cerebrospinal fluid levels of CRH it is postulated that excess CRH and vasopressin form the basis of neuroendocrine dysregulation in depression. The increased levels of CRH (and, in addition, vasopressin) in the hypothalamus and bodily fluids are commonly thought to be related to altered feedback inhibition by endogenous glucocorticoids (as excellently reviewed by Swaab et al. 2005). Data supporting the notion that glucocorticoid-mediated feedback inhibition is impaired in major depression comes from several studies demonstrating, for example, no suppression of cortisol secretion after the dexamethasone test (Raison and Miller 2003 and many others). Since the highest concentrations of glucocorticoid receptors are found in the hippocampus and other limbic structures, a disturbed interplay between limbic structures and hypothalamus is assumed to form the basis of this dysregulation. This glucocorticoid overproduction and the resulting cellular changes "down stream" to neuronal glucocorticoid binding (often referred as to the "glucocorticoid cascade hypothesis") is believed to significantly contribute to many behavioral and pathophysiological sequelae associated not only with major depression, but with chronic stress in general (Holsboer 2000; Raison and Miller 2003).

Indeed, most (but not all) studies show that hippocampal glucocorticoid receptor density is altered in major depression and other neuropsychiatric disorders – with consequences not only for HPA axis functioning but also for memory storage of the patients (Lopez et al. 1998; Raison and Miller 2003; Bremner et al. 2004). And finally, it is known that long-term antidepressant treatment "normalizes" the HPA axis by positively influencing the cerebral glucocorticoid level, the immune system (Holsboer 2000; Gold et al. 2002) and, possibly, even the proposed impaired hippocampal neurogenesis (for recent considerations, see Norman 2006). Of note there is evidence for the expression of glucocorticoid receptors at hypothalamic vasopressin producing neurons which may be part of the feedback loop (Swaab et al. 2005). In rat hypothalamus glucocorticoids increase vasopressin mRNA (Swanson and Simmons 1989)., but it remains to be clarified, if these receptors are altered in human affective disorder and/or schizophrenia.

Thus, disturbances in the cerebral glucocorticoid system seeminingly play a predominant role in the dyregulation of the HPA axis in stress and depression. However, this does not exclude the possibility that other factors also contribute to the pathophysiology of the overactivated HPA axis in affective disorder and other forms of mental illness. We will try to show here that the gas nitric oxide is a candidate molecule to do so.

2. Nitric Oxide Has Multiple Functions in the Brain

The biologically active molecule nitric oxide (NO) is a membrane-permeable gas with unique chemistry. It is formed by the conversion of one molecule of L-arginine to one molecule of L-citrulline, with the release of one NO molecule (Bredt and Snyder, 1992). The enzymatic oxidation of L-arginine to L-citrulline takes place in the presence of oxygen and NADPH using flavin adenine dinucleotide (FAD), flavin mononucleotide (FMN), heme, thiol and tetrahydrobiopterin as cofactors (Akyol et al. 2004). The enzyme responsible for the generation of NO is nitric oxide synthase (E.C. 1.7.99.7; NOS). Three NOS isoforms have been described and shown to be encoded on three distinct genes: Neuronal NOS (nNOS, NOS type I), inducible NOS (NOS type II) and endothelial NOS (eNOS, NOS type III). Two of them are constitutively expressed and are dependent on the presence of calcium ions and calmodulin to function, while iNOS is considered non-constitutive and calcium-independent (Bredt and Snyder, 1990). Neuronal NOS accounts for over 90% of the cerebral NO production (Hara et al. 1996). The density of nNOS expressing neurons is region dependent. In the cerebral cortex, nNOs is located in about 2% of the neurons (Snyder and Bredt, 1991), whereas in the human hypothalamic paraventricular nucleus up to 20% of the neurons express the enzyme (Bernstein et al. 1998a). Functionally, NO is an "universal soldier" that may function as a hormone, neurotransmitter, paracrine messenger, mediator, cytoprotective and cytotoxic molecule. NO has multiple molecular targets in the immune, cardiovascular and nervous systems of the body, through which it influences the activity of transcription factors, modulates upstream signaling cascades, mRNA stability and translation, as well as processing of the primary gene products (Bogdan 2001; Abbott and Nahm 2004). As to the brain, many physiological and pathophysiological processes are linked to NO. NO plays important roles in different signal transduction cascades of the CNS. It activates soluble guanylate cyclase by binding to it. The stimulation of this enzyme leads to increased synthesis of the second messenger cGMP, which in its turn activates cGMP-dependent kinases in target cells (Hobbs and Ignarro, 1996). NO exerts strong influence on glutamatergic neurotransmission by directly interacting with the N-methyl-D-aspartate (NMDA) receptor. Neuronal NOS is connected to NMDA receptors and increases NO production after the activation of this receptor. Thus, the level of endogenously produced NO around NMDA-receptor–nNOS-containing synapses reflects the activity of glutamate-mediated neurotransmission (Akyol et al. 2004). Besides its influence on glutamate, NO is known to have considerable effects on the storage, uptake and/or release of most other classical neurotransmitters in the CNS (acetylcholine, dopamine, noradrenaline, GABA, and taurine and glycine, for review Boehning and Snyder, 2003) as well as of certain neuroactive peptide hormones (Bernstein et al. 1998a). Moreover, since NO is a diffusible molecule it may reach extrasynaptic receptors at target cell membranes at some distance from the place of NO synthesis (Kiss and Vizi 2001). Thus NO is capable of mediating both synaptic and non-synaptic communication processes. Another remarkable facet of NO with considerable biological relevance is its appearance as a free radical. NO may exert toxic effects at higher concentrations (Bredt and Snyder 1994).

3. NO Metabolism Is Altered in Schizophrenia and Affective Disorder

3. 1. Genetic Aspects of NO Metabolism in Schizophrenia

There exists an allelic association between a polymorphism in exon 29 of the gene coding for nNOS and schizophrenia. A single nucleotide polymorphism on chromosome 12q24 (C→T) transition located 276 base pairs downstream from the translation termination site) was shown to occur more frequently in schizophrenics. The authors conclude that nNOS gene polymorphism may confer increased susceptibility to schizophrenia (Shinkai et al., 2002). However, this finding was not confirmed by others (Liou et al. 2003). Another nNOS haplotype associated with schizophrenia was recently identified by Reif and colleagues (Reif et al. 2006). Furthermore, it has been reported that the development of tardive dikinesia in some of drug-treated patients with schizophrenia is related to nNOS gene polymorphism (Wang et al. 2004). Another candidate gene for a genetic linkage with schizophrenia encodes the carboxyl-terminal PDZ-ligand of nNOS, CAPON, which is located on chromosome 1q22 (Brzustowicz et al., 2004; Zheng et al., 2005).

3.2. NO Metabolism in Schizophrenia

Numerous papers describe qualitative and quantitative alterations in the cellular expression of brain-associated NOS in schizophrenia. Remarkably, both decreases and increases in NOS activity, NOS protein and/or mRNA content were found in schizophrenia.

Besides an abnormal distribution of NOS expressing cortical neurons was seen in schizophrenia (Akbarian et al. 18996).

Elevated levels of nNOS and mRNA were measured in postmortem cerebellum, brain stem, the cerebral cortex and the caudate nucleus, while decreased levels were reported for the hypothalamus (see below) and the cerebral cortex (!). For a comprehensive review on this topic the reader is referred to Bernstein et al. (2005a). Conflicting results exist with regard to NO metabolism in bodily fluids. A majority of papers demonstrates increased NOS acticvity in the blood of patients (Das et al. 1996 and others).

3.3. Genetic Aspects of NO Metabolism in Affective Disorder

Genetic studies have shown that polymorphisms in the nNOS gene are neither linked to major depression (Yu et al., 2003) nor to bipolar disorder (Buttenschon et al., 2004).

3.4. NO Metabolism in Affective Disorder

In patients with major depression, reduced nNOS activity and protein content have been found in various brain regions such as neocortex, hypothalamus (see above) and locus coeruleus, while no alterations were seen in the cerebellum (reviewed in Akyol 2004). Measuring NO in blood samples of patients with unipolar depression Selley (2004) and other groups found decreased NO concentrations.

4. Human Hypothalamic nNOS is Partly Co-Expressed with Vasopressin, Oxytocin and CRH, and is Altered in Affective Disorder and Schizophrenia

The hypothalamus is a very complex brain region encompassing nuclei with different functions in the regulation of endocrine, autonomic and behavioural activities. Although important phylogenetic differences exist, the general organization of the hypothalamus is amazingly uniform in most mammalian species including man. The hypothalamus is known to be extraordinarily rich in various regulatory neuropeptides, which partly appear co-localized with classical neurotransmitters in the same neurons thus forming a complex neuroanatomical puzzle Neuronal NOS is widely distributed in the hypothalamus, being co-localized with various neuropeptides (somatostatin, substance P, cholecystokinin, galanin, enkephalin, CRH, oxytocin and vasopressin and *N*-methyl-D -aspartate (NMDA) receptors (for review see Bernstein et al. 1998a). There is good histochemical evidence for a wide distribution of nNOS in human hypothalamic neurons (Bernstein et al. 1998a, 2000 and 2005). In the PVN, more than one fifth of all neurons express the enzyme (figure 1). When immunohistochemically co-localizing nNOS with either CRH or neurophysins (the carrier proteins of vasopressin and oxytocin) one can detect a considerable portion of cells which co-express the neuropeptides and NOS (Bernstein et al. 1998a). In the human supraoptic nucleus (SON), where NOS is less frequently found than in rodent SON, enzyme immunoreactivity is mostly attributable to oxytocinergic neurons (Bernstein et al. 2000). Finally, a remarkable overlap between NOS and vasopressin/neurophysin expression was seen in neurons of the human SCN (Bernstein et al. 2005b). The intimate anatomical relation between NOS expressing neurons and nerve cells generating hypothalamic peptides known to be altered in affective disorder, as well as the multiple roles played by NO in the brain, encouraged us to look at NOS expressing hypothalamic neurons in neuropsychiatric disorders. When morphometrically analyzing nNOS-containing PVN neurons of depressives, schizophrenics and controls we have seen a significant reduction in cell numbers, both in depressives and schizophrenics (Bernstein et al. 1998a). A similar reduction in NOS immunoreactive neurons was recently found in another hypothalamic area, the SCN (figure 2), whereas nNOS expression was normal in neurons of the SON (Bernstein et al. 2000). The SCN was included because (i) it is the zeitgeber of our brain triggering circadian and saisonal rhythms and (ii) a considerable portion of its neurons project to the PVN to inhibit PVN neurons (reviewed in detail in Zhou et al. 2001). We conclude from the reduced expression of NOS and the altered expression of neurophysin/vasopressin (Zhou et al. 2001; Bernstein et al. 2002a, 2005) that the inhibitory control of the SCN on PVN is lowered in depression and, possibly, in schizophrenia. The idea suggests that there may be a common pathophysiological background behind the similar changes in the cellular expression of nNOS in neuropsychiatric disorders. Since NO is known to inhibit the generation and the release of CRH, oxytocin and vasopressin (at least in rodents), one might speculate that diminished hypothalamic NO production might contribute to the altered neuropeptide expression and release in affective disorder and schizophrenia. Interestingly, evidence is accumulating that the successful treatment of major depression and other psychiatric disorders by electroconvulsive therapy is in part due to an increase of the endocrine effects of hypothalamic NO in patients (Rosen et al. 2003; Bernstein et al. 2004).

To learn more about the interplay between hypothalamic NO and disease-relevant neuropeptides we decided to study these peptides in nNOS knockout mice.

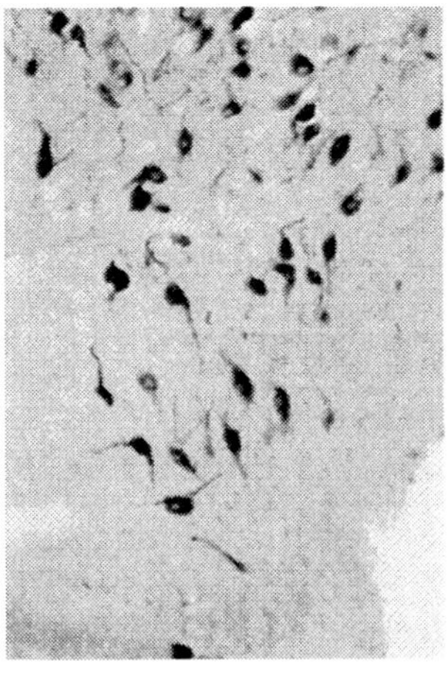

Figure 1. Multiple PVN neurons express nNOS protein.

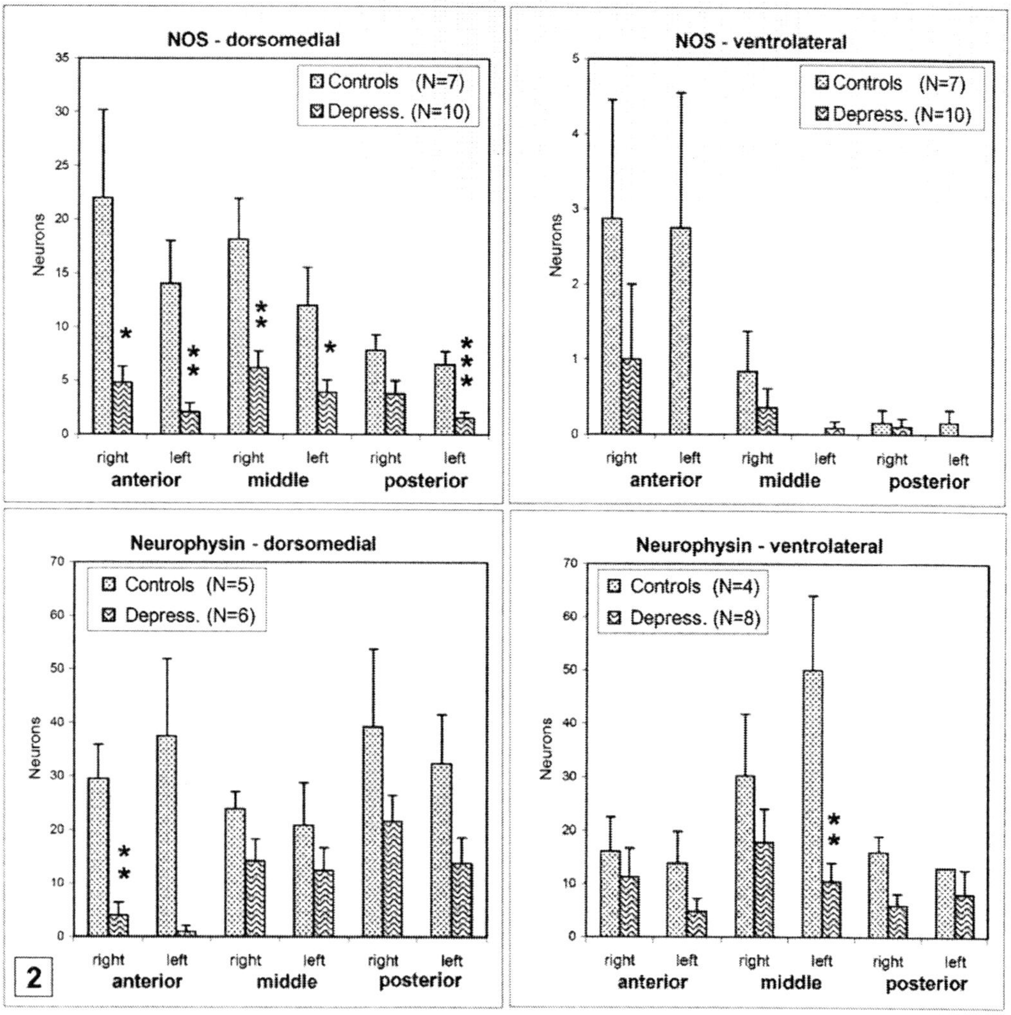

Figure 2. The number of nNOS immunoreactive and neurophysin-immunopositive neurons is reduced in the SCN of depressives in comparison to controls. The same can be found in cases with schizophrenia (not shown).

5. Of Mice and Men: Lessons Learned from nNOS Null Mice

Mice knocked out for the nNOS gene are doing remarkably well. They are resistant to cerebral ischemia, while having only mild impairments in the limbic and neuroendocrine systems (Elibol et al. 2001; Bernstein et al. 1998b; Keilhoff et al. 2001a, b). However, recent findings with nNOS deficient mice point to seriously impaired cognitive performance in these animals (Kirchner et al. 2004), and, more importantly for the current topic, show depression-like behavior in an experimental stress situation (forced swimming, Orlando et al. 2006) and schizophrenia-like behavior with regard to the effect of phencyclidine on acoustic startle and prepulse inhibition (Klamer et al. 2005). But is this aberrant, "psychotic" behavior of nNOS null mice accompanied by any changes in the expression of hypothalamic peptides relevant for HPA axis overactivation? We have investigated this particular problem in some detail and have got disappointing results in that neither CRH, nor neurophysins nor pro-opiomelanocortin (POMC)-derived peptides were altered at the level of the hypothalamus. This is in line with more recent findings of Nomura et al. (2005) and Orlando et al. (2006) showing that disruption of the nNOS gene has only little if any influence of the expression and the release of vasopressin and oxytocin. However, what we saw was a drastic decline of beta-endorphin (β-E) expression in the arcuate nucleus in nNOS null mice (Bernstein et al. 1998b, figure 3A-C, figure 4). In addition to that, the number of β-E-immunoreactive nerve terminals on PVN neurons was reduced in knock out mice (Bernstein et al. 1998b). This is an important finding in the light of recent findings showing that POMC peptides play a central role in the regulation of the HPA axis (Smart et al. 2006). Moreover, when looking for POMC expression in the pituitaries of these mice, a strong reduction in comparison to wild-type animals was quite evident (Keilhoff et al. 2001a). These unexpected results prompted us to look for the opioid system in schizophrenics and depressives.

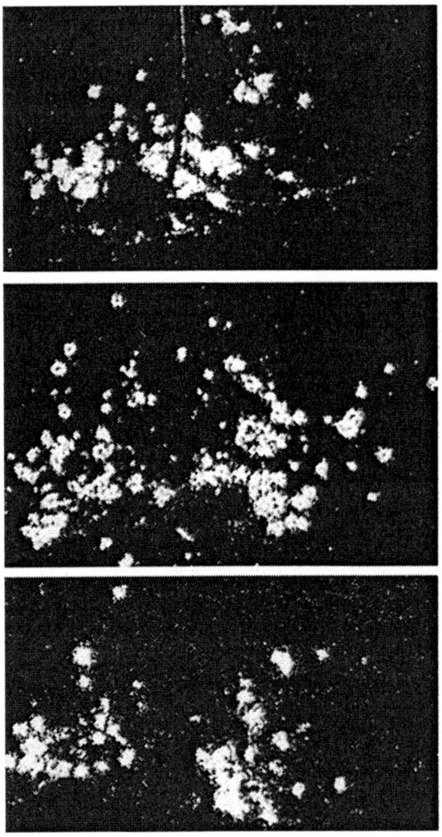

Figure 3. Dark-field microphotographs of in situ hybridization histochemistry of POMC mRNA in neurons of the arcuate nuclei of mice. A. POMC mRNA in the anterior part of the nucleus, normal mouse, B. POMC mRNA in the intermediate part of the structure, normal mouse, C. POMC mRNA in arcuate neurons of a nNOS minus mouse. No statistically significant differences were found between the density of silver grains in wild-type and knockout mice. Reproduced from Neuroendocrinology 1998b, with permission of Karger.

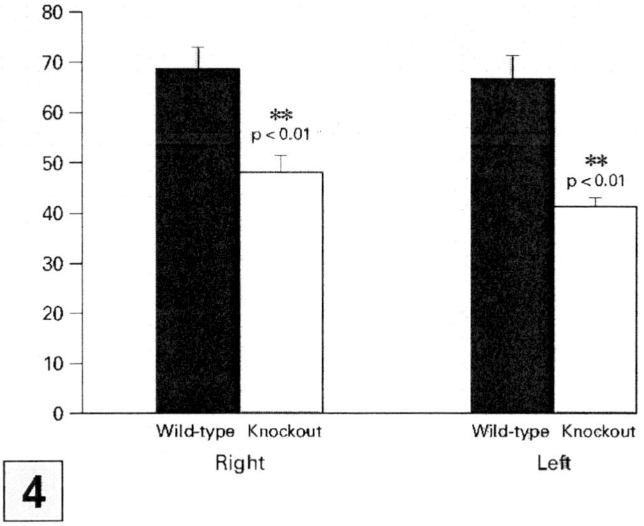

Figure 4. Amount of β-E-immunoreactive arcuate nucleus neurons per section in wild-type and knockout mice. Reproduced from Neuroendocrinology 1998b, with permission of Karger.

6. Neuropsychiatric Disorders and Endogenous Opioides – A Contribution to an Old Discussion

Beta-endorphin (β-E) is an endogenous opioid peptide which originates by proteolytic cleavage from POMC. Sevaral studies have shown that β-E producing neurons are located in the mediobasal hypothalamus (arcuate and retrochiasmatic nuclei and, possibly, in the brain stem (Liotta and Krieger 1983). For several reasons there emerged very early the assumption that β-E might be prominently involved in the pathophysiology of psychiatric disorders. Reports showing increased levels of the peptide in the serum and the CSF of schizophrenics led to the "β-E excess hypothesis" of the disease, while other work supports quite the opposite, the "β-E deficiency hypothesis". Contradictory results were also found with regard to depression. And finally, investigations revealing a deviant metabolism of β-E in psychiatric patients have founded the hypothesis that not β-E *per se* but its proteolytic fragments γ-endorphin and (probably) α-endorphin are of crucial importance (for overview see Bernstein et al. 2002b). Estimating the number of β-E immunoreactive arcuate nucleus neurons in schizophrenics, depressives and matched controls we found significantly fewer neurons in neuropsychiatric patients than in control cases. Moreover, when counting the number of endorphinoceptive (i.e. receiving a β-E innervation) PVN neurons (figure 5 A-H) we determined an apparent reduction of such cells in schizophrenia and depression (figure 6). Knowing that β-E is an inhibitory neuromodulator on PVN neurons (Kiss et al. 1984), reduced β-E innervation on PVN neurons might also contribute to the disinhibiton of the neurons in disease, which finally leads to overproduction of CRH, vasopressin and oxytocin.

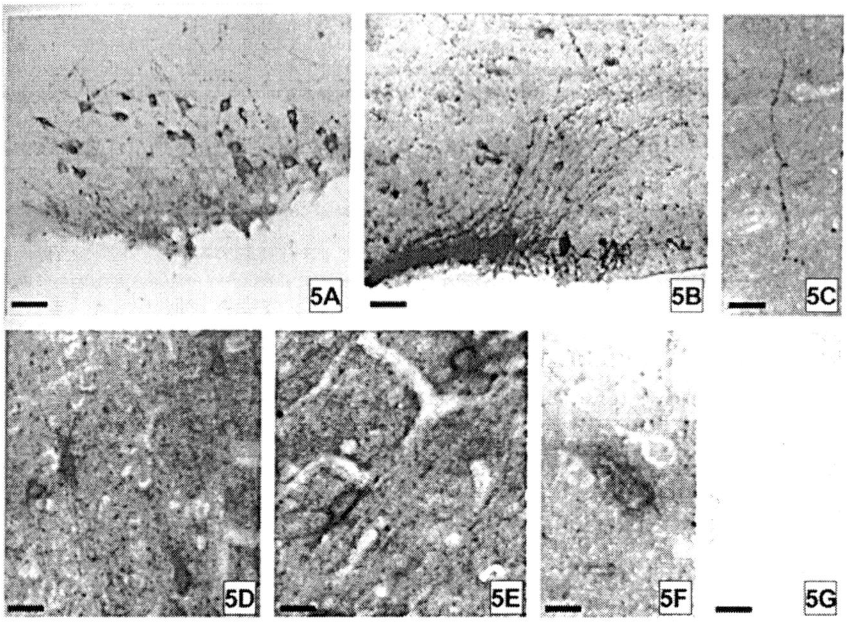

Figure 5. A. Beta-endorphin immunoreactive neurons in the human arcuate nucleus (control case). B. Beta-endorphin containing fibers entering the hypothalamus (control case), C. A single immunoreactive axon near the third ventricle (depressive patient), D. Beta-endorphinoceptive (receiving) neurons in the preoptic area (control case), E. A group of supraoptic neurons surrounded by β-E containing terminals (control case). F. A single PVN neuron receiving β-E input (control case). G. Immunohistochemical control reaction (preabsorption of the antiserum with pure peptide). Reproduced with from Cell Mol. Biol. 2002, with permission.

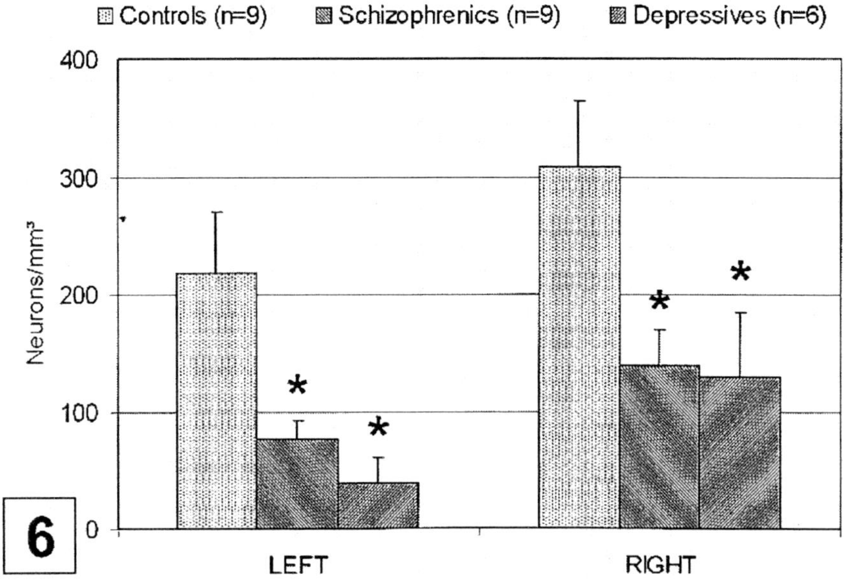

Figure 6. Quantitative estimation of the number of β-E receiving PVN neurons in controls, schizophrenics and depressives. Reproduced from Cell Mol. Bol., with permission.

7. GABA – An Underestimated Player in the Game?

In the introductory remarks we have shortly mentioned that biogenic amines are prominently involved in the regulation of peptide production in, and release from, hypothalamic neurons. Amongst them is GABA. When immunohistochemically analyzing the regional distribution and cellular localization of GABA synthetizing enzyme, glutamic acid decarboxlyase (GAD), one can easily see that this enzyme is abundantly present in nerve endings targeting human hypothalamic neurons. This holds especially true for the PVN (figure 7). Hence it is imaginable that GABA/GABA receptors might prominently be involved in the regulation of PVN neurons, and that the GABAergic system contributes to the disinhition of the neurons in depression and schizophrenia. Indeed, some recent data are in favor of this conjecture (summarized in Giordano et al. 2006). We are currently about to estimate the density of GAD-immunopositive synaptic contacts at PVN neurons of schizophrenics, depressives and controls. Our preliminary data show that at least in major depression there is a strong reduction in GAD terminal density in comparison to control cases (in preparation). Thus, GABA is obviously another hypothalamic transmitter system capable of influencing PVN neurons in neuropsychiatric disorders. And again, if this is true, a disinhibtion would be the result.

Figure 7. Double immunolabeling for nNOS (golden-brown) and glutamic acid decarboxylase (purple) of PVN neurons (case of a depressive patient).

8. Conclusions

We have tried to demonstrate that besides glucocorticoids other factors might also be involved in the overactivation of the HPA axis in depression (and most probably, in schizophrenia). Amongst them is NO. The reduced expression of the NO-generating enzyme NOS in several hypothalamic nuclei seems to be an important local factor to contribute to the disinhibiton of PVN neurons. Since NOS inhibitors have antidepressant effects in animals ( ) and long-term treatment of patients with antidepressant medication increases the number of

NOS neurons in the SCN (although by far not reaching the level of controls, Bernstein et al. 2005) we are convinced that the observed alterations in the "nitrergic" system are of clinical relevance. Apart from NO (and in cross-talk with it) we have identified β-E and GABA as new players in the disinbition game.

9. Acknowledgements

The authors wish to thank Drs. G. Keilhoff, A. Heinemann, A. Stanarius and H. Bielau for fruitful cooperation. This research was supported by BMBF of Germany (NBL2 and 3) and the Stanley consortium.

10. References

Abbott, L. C., Nahm, S. S. (2004) Neuronal nitric oxide synthase expression in cerebellar mutant mice. *Cerebellum.* 3, 141-151.

Akbarian, S., Kim, J. J. , Potkin, S. G., Hetrick, W. P., Bunney, W .E .Jr, Jones, E.G. (1996) Maldistribution of interstitial neurons in prefrontal white matter of the brains of schizophrenic patients. *Arch. Gen. Psychiatry.* 53, 425-436.

Akyol, O., Zoroglu, S. S., Armutcu. F., Sahin, S., Gurel, A. (2004) Nitric oxide as a physiopathological factor in neuropsychiatric disorders. *In vivo.* 18, 377-390.

Bernstein. H.-G., Stanarius, A., Baumann, B., Henning, H., Krell, D., Danos, P., Falkai, P., Bogerts, B. (1998a) Nitric oxide synthase-containing neurons in the human hypothalamus: reduced number of immunoreactive cells in the paraventricular nucleus of depressed patients and schizophrenics. *Neuroscience.* 83, 867-875.

Bernstein, H.-G., Keilhoff, G., Seidel, B., Stanarius, A., Huang, P. L., Fishman, M. C., Reiser, M., Bogerts, B., Wolf, G. (1998b) Expression of hypothalamic peptides in mice lacking neuronal nitric oxide synthase: reduced beta-END immunoreactivity in the arcuate nucleus. *Neuroendocrinology.* 68, 403-411.

Bernstein, H.-G., Jirikowski, G. F., Heinemann, A., Baumann, B., Hornstein, C., Danos, P., Diekmann, S., Sauer, H., Keilhoff, G., Bogerts, B. (2000) Low and infrequent expression of nitric oxide synthase/ NADPH-diaphorase in neurons of the human supraoptic nucleus: a histochemical study. *J. Chem. Neuroanat.* 20,177-183.

Bernstein, H.-G., Heinemann, A., Krell, D., Mawrin, C., Bielau, H., Danos, P., Diekmann, S., Keilhoff, G., Bogerts, B., Baumann, B. (2000a) Further immunohistochemical evidence for impaired NO signaling in the hypothalamus of depressed patients. *Ann. N. Y. Acad. Sci.* 973, 91-93.

Bernstein, H.-G., Krell, D., Emrich, H. M., Baumann, B., Dano, P., Diekmann, S., Bogerts, B. (2000b) Fewer beta-endorphin expressing arcuate nucleus neurons and reduced beta-endorphinergic innervation of paraventricular neurons in schizophrenics and patients with depression. *Cell Mol. Biol.* 48, OL 259-265.

Bernstein, H.-G., Heinemann, A., Bogerts, B. (2004) Electro-convulsive therapy, nitric oxide and HPA-axis: a closer look at the human hypothalamus. *Med. Hypotheses.* 62, 158-159.

Bernstein, H.-G., Bogerts, B., Keilhoff, G. (2005a) The many faces of nitric oxide in schizophrenia. A review. *Schizophr. Res.* 78, 69-86.

Bernstein, H.-G., Heinemann, A., Krell, D., Dobrowolny, H., Bielau, H., Keilhoff, G., Bogerts, B. (2005b) Hypothalamic nitric oxide synthase in affective disorder: focus on the suprachiasmatic nucleus. *Cell Mol. Biol.* 51, 279-284.

Boehning, D., Snyder, S. H. (2003) Novel neural modulators. *Annu. Rev. Neurosci.* 26, 105-131.

Bogdan, C. (2001) Nitric oxide and the regulation of gene expression. *Trends Cell Sci.* 11, 66-75.

Bredt, D. S., Snyder, H. (1990) Isolation of nitric oxide synthetase, a calmodulin-requiring enzyme. *Proc. Natl. Acad. Sci. USA.* 87 (1990) 682-685.

Bredt, D. S., Snyder, S. H. (1992) Nitric oxide, a novel neuronal messenger. *Neuron.* 8, 3-11.

Bremner, J. D., Vythilingam, M., Vermetten, E., Anderson, G., Newcomer, J. W., Charney, D. S. (2004) *Effects of glucocorticoids on declarative memory function in major depression.* 55, 811-815.

Buttenschon, H. N., Mors, O., Ewald, H., McQuillin, A., Kalsi, G., Lawrence, J., Gurling, H., Kruse, T. A. (2004) No association between a neuronal nitric oxide (NOS1) gene polymorphism on chromosome 12q24 and bipolar disorder. *Am. J. Med. Genet.* B 124, 73-75.

Elibol, B., Söylemezoglu, F., Ünal, I., Fujii, M., Hirt, L., Huang, P. L., Moskwit, M. A., Dalkara, T. (2001) Nitric oxide is involved in ischemia-induced apoptosis in brain: a study in neuronal NOS deficient null mice. *Neuroscience.* 105, 79-86.

Gold, P. W., Drevets, W. C., Charney, D. S., Drevets, W. C.(2002) New insights into the role of cortisol and the glucocorticoid receptor in severe depression. *Biol. Psychiatry.* 52, 381-385.

Hara, T., Waeber, P. L., Huan, P. L, Fujii, M., Fishman, M. C., Moskowitz, M. A. (1996) Brain distribution of nitric oxide synthase in neuronal and endothelial nitric oxide synthase mutant mice using [^{3}H]*L*-N^G-nitro-arginine autoradiography. *Neuroscience.* 75, 881-890.

Brzustowicz, M., Siomne, J., Mohseni, P., Hayter, J. E., Hodgkinsonl, K. A., Chow, E. W., Bassett, A. S. (2000) Linkage disequilibrium mapping of schizophrenia susceptibility to the CAPON region of chrosomome 1q22. *Am. J. Hum. Genet.* 74, 057-1063.

Das, I., Khan, N., Puri, S., Hirsch, S. R. (1996) Elevated endogenous nitric oxide synthase inhibitor in schizophrenic plasma may reflect abnormalities in brain nitric oxide production. *Neurosci. Lett.* 215, 209-211.

Giordano, R., Pellegrino, M., Picu, A., Monelli, L., Balbo, M., Berardelli, R., Lanfranco, F., Ghigo, E, Arvat, E. (2006) Neuroregulation of the hypothalamus-pituitary-adrenal (HPA) axis in humans: effects of GABA,-, mineralcorticoid-, and GH-secretagogues-receptor modulation. *ScientificWorld Journal.* 17, 1-11.

Hobbs, A. J, Ignarro, L. J.(1996) Nitric oxide-cyclic GMP signal transduction system. *Methods Enzymol.* 269, 134-148.

Holsboer, F. (2003) The corticosteroid hypothesis of depression. *Neuropsychopharmacology.* 23, 477-501.

Keilhoff, G., Seidel, B., Reiser, M., Stanarius, A., Huang, P. L., Bogerts, B., Wolf, G. (2001a) Lack of neuronal NOS has consequences for the expression for the expression of POMC and POMC-derived peptides in the mouse pituitary. *Acta Histochem.* 103, 397-412.

Keilhoff, G., Wolf, G., Bernstein, H.-G. (2001b) Altered laminar distribution of hippocampal zinc in mutant mice lacking neuronal nitric oxide synthase. A histochemical study. *Neurosci. Lett.* 305, 173-176.

Kirchner, L., Weizdörfer, R., Hoeger, H., Uri, A., Schmidt, P., Engelmann, M. Villar, S. R., Fountoulakis, M., Lubec, G., Lubec, C. (2004) Impaired cognitive performances in neuronal nitrc oxide synthase knockout mice is associated with hippocampal protein derangements. *Nitric Oxide.* 11, 316-330.

Kiss, J. Z., Cassell, M. D., Palkovits, M. (1984) Analysis of the ACTH/beta-End/alpha-MSH-immunoreactive afferent input in the hypothalamic nucleus of the rat. *Brain Res.* 324, 91-99.

Kiss, J. P., Vizi, E. S. (2001) Nitric oxide: A novel link between synaptic and nonsynaptic transmission. *Trends Neurosci.* 24, 211-215.

Klamer, D., Engel, J. A., Svensson, L. (2005) Effects of phencyclidine on acoustic startle and prpulse inhibition in neuronal nitric oxide synthase deficient mice. *Eur. Neuropsychopharmacol.* 15, 587-590.

Liou, Y.-J., Tsai, S.-J., Hon, C. .J., Liao, D.-L. (2003) Association analysis for the CA repeat polymorphism of the neuronal nitric oxide (NOS1) gene and schizophrenia. *Schizophr. Res.* 65, 57-59.

Liotta, A. S., Krieger, D. T. Pro-opiomelanocortin-related and other pituitary peptides in the central nervous system. In: *Brain Peptides.* (eds. Krieger D T, Brownstein M L, Martin J B). Wiley Interscience, New York 1983, pp. 613-659.

Lopez, J. F., Chalmers, D. ., Little, K. Y., Watson, J. (1998) Regulation of serotonin 1A, glucocorticoid, and mineralcorticoid receptor in rat and human hippocampus: implications for the neurobiology of depression. *Biol. Psychiatry.* 43, 547-573.

Nomura, M., Tsutsu, M., Shimo, H., Fujimoto, ., Ue, Y., Morishita, ., Yanagihara, N., Matsumoto, T. (2005) Effects of nitric oxide synthase isoforms deletion on oxytocin and vasopressin messenger RNA in mouse hypothalamus. *NeuroReport.* 15 , 413-417.

Norman, T. R. (2006) Prospects for the treatment of depression. *Aust. N. Z. J. Psychiatry.* 40, 379-380.

Orlando,. F., Langnaese, K., Landgraf, R., Spi, M.., Wolf, G., Engelmann, M. (2006) Neural nitric oxide gene inactivation affects the release profile of oxytocin into the blood in response to forced swimming. Nitric Oxide in press.

Purba, J. S., Hoogendijk, W. J. G., Hofmann, M. A., Swaab, D. F. (1996) Increased number of vasopressin-and oxytocin-expressing neurons in the paraventricular nucleus of the hypothalkamus in depression. *Arch. Gen. Psychiatry.* 53, 137-143.

Raadsheer, F. C., Van Heerikhuize, J. J., Lucassen, P. J,. Hoogendijk, W. J. G., Tilders, . J., Swaab, D. (1995) Increased corticotropin-releasing hormone (CRH) mRNA in paraventricular nuclei of depressed patients. *Am. J. Psychiatry.* 152, 1372-1376.

Raison, C. L., Miller, M. D. (2003) When not enough is too much : The role of insufficient glucocorticoid signaling in the pathophysiology of stress-related disorders. *Am. J. Psychiatry.* 160, 1554-1565.

Reif, A., Herterich, S., Strobel, A., Ehlis, A. C., Saur, D., Jacob, C. P., Wienker, T., Topner, T., Fritzen, S., Walter, U., Schmitt, A., Fallgatter, A. J., Lesch K. P. (2006) A neuronal nitric oxide (NOS-1) haplotype associated with schizophrenia modifies prefrontal function. *Mol. Psychiatry.* 11, 286-300.

Rosen, Y., Reznik, I., Sluvis, A., Kaplan, D., Mester, R. (2003) The significance of the nitric oxide in electro-convulsive therapy: a proposed neurophysiologcal mechanism. *Med. Hypotheses.* 60, 424-429.

Ryan, M.C., Sharifi, N., Condren, R., Thakore, J. H. (2004) Evidence of basal pituitary overactivity in first-episode, drug-naïve patients with schizophrenia. *Psychoneuroendocrinology.* 29, 4065-4070.

Scott, L. V., Dinan, T. G. (2002) Vasopressin as a target for antidepressant development: an assessment of the available evidence. *J. Affect. Disord.* 72, 113-124.

Selley, M. L. (2004) Increased (E)-4-hydroxy-nonenal and asymmetrical dimethylarginine concentrations and decreased nitric oxide concentrations in the plasma of patients with major depression. *J. Affect. Disord.* 80, 249-254.

Shinkai, T., Ohmori, O., Hori, H., Nakamura, J. (2002) Allelic association of the neuronal nitric oxide synthase (NOS1) gene with schizophrenia. *Mol. Psychiatry.* 7, 560-563.

Smart, J. L., Tolle, V., Otero-Corchon, V., Low, M. J. (2006) Central dysregulation of the hypothalamic-pituitray-adrenal axis in neuron-specific proopiomelanocortin-deficient mice. Endocrinology in press

Snyder S H, Bredt D S. Nitric oxide as a neuronal messenger. Trends. 12 (1991) 125-128.

Swaab, D. F., Bao, A. M., Lucassen, P. J. (2005) The stress system in the human brain in depression and neurodegeneration. *Ageing Res. Rev.* 4, 141-194.

Swanson, L., Simmons, D. (1989) Differential steroid hormone and neural influences on peptide mRNA levels in CRH cells aof the paraventricular nucleus : A hybrdization histochemical study. *J. Comp. Neurol.* 285, 413-435.

Volke, V., Wegener, G., Bourin, M., Vasar, E. (2003) Antidepressant-and anxiolytic-like effects of selective neuronal NOS inhibitor 1-(2-trifluoromethylphenyl)-imidazole in mice. *Behav. Brain Res.* 140, 141-147.

Walker, E. J., Difori, D. (1997) Schizophrenia: a neural diathesis-stress model. *Psychol. Rev.* 104, 667-685.

Wang, Y.C., Liou, Y. J., Liao, D. .L., Bai, Y. M., Li, C. C., Yu, S. C., Chen, J. Y. (2004) Association analysis of a neural nitric oxide synthase gene polymorphism and antipsychotics-induced tardive dyskincsia in Chinese schizophrenic patients. *J. Neural Transm.* 111, 623-629.

Yu, Y. W., Chen, T. J., Wang, Y. C., Liou, Y. L., Ho, C. J., Tsai, S. J. (2003) Association analysis for neuronal nitric oxide synthase gene polymorphism with major depression and fluorexitine response. *Neuropsychobiology.* 47, 137-140.

Zhou, J. N., Riemersma, R. F., Unmehopa, U. A., Hoogendijk, W. J., van Heerikhuize, J. J., Hofman, M. A., Swaab, D. F. (2001) Alterations in arginine vasopressin in the suprachiasmatic nucleus in depression. *Arch. Gen. Psychiatry.* 58, 655-662.

In: Neuropeptide Research Trends
Editor: Bernice A. Levine, pp. 229-241
ISBN: 978-1-60021-640-4

Chapter IX

Neurophysins in Mood Disorders

G. Scantamburlo[1], M. Ansseau[1] and J.J. Legros[2]
[1] Department of Psychiatry, [2] Department of Endocrinology; Psychoneuroendocrinology Unit; University of Liège, CHU of Liège, B35, B-4000 Liège, Belgium

Abstract

Neurophysins are polypeptides with molecular weights of about 10,000 and are synthesized within hypothalamic nuclei (mainly magnocellular supraoptic and paraventricular). Different forms of pituitary and circulating neurophysins have been described, but it is generally accepted in human subjects that two types of immonureactive neurophysins exist: one associated with AVP and one with OT synthesis. In plasma and cerebrospinal fluid (CSF), neurophysins are more stable than their related biological active nonapeptides. The neurophysins assay may therefore be considered a better measure of long-term AVP and OT secretion activity than isolated AVP or OT determinations. The radioimmunoassay for neurophysins has been proposed and used as an index of neuropituitary function in different physiological and pathological conditions. OT and AVP are two chemically very similar neurohypophyseal neuropeptides which could be involved in mood disorders. A few studies have been published about neurophysins in psychiatric pathology. In bipolar depressed patients, CSF AVP- and OT-neurophysins levels seems to be higher than in unipolar depressed patients or schizophrenic patients. Moreover, manic patients exhibit significantly higher plasma AVP-neurophysins values than depressed and schizophrenic patients. Following a switch from a depressive to a manic state, an increase in plasma AVP-neurophysins concentration has been observed. Depressed patients strive for a reduced AVP secretion. AVP-neurophysins have been also involved in the pathophysiology of the cortisol postdexamethasone nonsuppression and in suicidal behavior. The purpose of this paper is to propose an overview of neurophysins in mood disorders.

Introduction

The neurophysins are polypeptides which are synthesized along with vasopressin (AVP) and oxytocin (OT) in the suprachiasmatic, paraventricular and supraoptic nucleii (Acher, 1980). Neurophysins form part of a precursor of neurohypophyseal peptides. AVP and OT peptides were chemically identified in 1953 by Du Vigneaud, work for which he received the Nobel Prize in Chemistry in 1955. In 1955, Acher and his associates showed that the "mother-molecule" was not a pure substance but a complex comprising oxytocin, vasopressin and a non-active protein for which they have suggested the name neurophysin (review in Acher, 1990). The name for the neurophysin associated to the synthesis of AVP in human being changes from one author to another: hNpI, hNpII, AVP-Neur, MSEL-Neur, and those of an extremely similar composition, associated to OT: hNpII, hNpI, OT-Neur, VLDL-Neur (Legros and Geenen, 1996). Their main function is to transport the neurohyphysial hormones from their sites of synthesis, in the hypothalamus, to the neurohypophysis, the median eminence and the extrahypothalamic structures such as amygdala, septum, fornix, brain stem and spinal cord (Crine et al, 1982; Sofroniew, 1980). Neurophysins are also important for the package of the hormone in secretory granules and for correct trafficking of the prohormone throught the secretory pathway (Itoi et al, 2004). The wide distribution of vasopressinergic and oxytoninergic fibers in the brain and their connection with the limbic system strongly suggest important neuromodulatory and behavioral effects in humans. AVP- or OT-neurophysins assay may be an appropriate tool for investigating chronic vasopressinergic or oxytoninergic function in affective disorders.

Two Neurohypophysial Hormone Precursor

The nonapeptides vasopressin (AVP) and oxytocin (OT) and their associated proteins, the neurophysins, derive from a common precursor polypeptide. The vasopressin and oxytocin prohormones are highly conserved and also very similar in primary structure. The vasopressin gene encodes three polypeptides besides the signal peptide: vasopressin, neurophysin-vasopressin and the carboxy-terminal glycopeptide (Iwasaki et al, 2000). The main difference is that the oxytocin prohormone lacks the glycopeptide domain (de Bree, 2000). Since the two prohormones are virtually identical, with the exception of the glycocpeptide domain, they are often treated as equal in litterature. The vasopressin precursor protein, called preprovasopressin, is initially converted to provasopressin by removal the peptide signal. Provasopressin is packaged into secretory granules, and finally processed into three polypeptides by endopeptidases. It is assumed that neurophysin and glycopeptide as well as vasopressin are coreleased at the neurohypophysis into the peripheral circulation (Iwasaki et al, 2000).

The human AVP and OT genes are linked on the short arm of chromosome 20 (20p13) (2kb), separated by less than 12 kb of intervening DNA sequence. It has been assumed that the vasopressin and oxytocin genes have evolved by duplication of a common ancestor gene (Rinaman et al, 2000) roughly 450 million years ago. OT is structurally similar to AVP, presumably reflecting their common evolutionary derivation from the amphibian pituitary hormone, vasotocin. The two genes are transcribed from two opposite ends of the DNA. The structure of oxytocin is very similar to that of vasopressin also a nonapeptide with a sulfur bridge whose sequence differs from oxytocin by 2 amino acids.

The synthesis of the precursor occurs on ribosomes in the perikarya of neuron, then the neuropeptides and neurophysins are incorporated in neurosecretory granules in the Golgi complex and transported along the axon into the neural lobe where the content of the granules is released by exocytosis (Gonzalez et al, 1993) . The release of the nonapeptides and their neurophysins by exocytosis is a Ca^{2+} entry-dependent proces (McKelvy et al, 1980).

Hormone-Neurophysin Association

Hormone-neurophysin association itself lead to the purification of both vasopressin and oxytocin in the 1950s (Acher et al, 1958). When they were first discovered, neurophysins were thought to be carriers of active peptide, using low-affinity, noncovalent binding to prevent the passive distribution of AVP and OT in spaces outside the cells. They were then recognized as forming integral parts of the prepressophysin and preoxyphysin precursor (Legros and Geenen, 1996). A remarkable feature of the prohormone products is the tight association of hormone to neurophysin (de Bree et al, 2003). This association has been suggested to play a role during hormone biosynthesis. de Bree et al (2003) have demonstrated that the vasopressin domain is crucial for correct trafficking of the prohormone through the secretory pathway, and suggest that vasopressin-neurophysin association provides correct prohormone folding in the endoplasmic reticulum. Furthermore, it has been proposed that the correct disulphide linkage pattern in neurophysins relies entirely on the interaction between hormone domain and neurophysin domain (Huang et al, 1992).

Neurophysins as a Index of Neuropituitary Secretion

Different forms of pituitary and circulating neurophysins have been described, but it is generally accepted in human subjects that two types of immonureactive neurophysins exist: one associated with vasopressin (AVP) and one with oxytocin (OXT) synthesis (Legros and Louis, 1974). Neuropeptides and neurophysins are released simultaneously by exocytosis of neurosecretory granules in the peripheral circulation and in the cerebrospinal fluid (CSF): modifications of neurophysins concentrations therefor reflect the fluctuations of AVP or OT release in different physiological and pathological conditions. In plasma and cerebrospinal fluid, neurophysins are more stable than their related biological active nonapeptides (Legros 1978). For example, a dissociation exists between the CSF circadian profile of OT, which shows a daily fluctuation pattern, and OT-Neurophysin, wich reach a steady-state level and remains stable throughout the day (Amico et al, 1983). Investigating endocrine correlates of the hemodynamic changes induced by carbon dioxide pneumoperitoneum, Joris et al (1998) have shown that five minutes after the beginning of peritoneal inssuflation, it rapidly resulted in a marked release of AVP and AVP-neurophysin. Plasma concentrations of AVP then decreased, whereas plasma concentrations of neurophysin plateaued. The longer half-life of neurophysin explains this plateau. The neurophysins assay may therefore be considered a better measure of long-term AVP and OT secretion activity than isolated AVP or OT determinations (Legros et al, 2000). The radioimmunoassay (RIA) for neurophysins has been proposed and used as an index of neuropituitary function in different physiological and pathological conditions (Legros et al., 1975). Their assays have been performed to investigate the activity of the hypothalamoneurohyposial axis (Fawcett et al, 1968; Legros, 1975). For this reason, AVP- or OT-neurophysins assay may be an appropriate tool for investigating chronic vasopressinergic or oxytoninergic function in affective disorders.

Neurophysins Assays

Blood samples are taken by venipuncture in fasting, recumbent individuals at ± 8 a.m. Since the original neurophysins assay was no more available, we developed a personal neurophysisn assay. In summary, neurophysins are purified according to previously described method (Legros et al, 1970) followed by a high-performance liquid chromatography step. The labeling of neurophysins are realised as described previously for bovine and human neurophysins (Legros et al, 1969), the specific activity being 225 ± 25 µCi/µg. Anti-neurophysins serum is used at the initial dilution of 1/10000. Separation of free from bound hormones was achieved by double-antibody techniques. At this dilution, ± 30% of the tracer (± 9000 cpm) is bound, the limit of detection being 0,02 ng/tube, and the sensitivity (50 % inhibition binding) being 0,2 ng/tube. The volume of known or unknown samples is usually 100µl. Cross-reaction between OT-neurophysin and AVP-neurophysin is detectable at the 0,01 % level; no cross-reactivity is detected using synthetic nonapeptide (OT or AVP) or purified extracted human antehypophyseal hormone. Intraassay variability is 6,5 % and interassay variability is 10,7%.

Neurohypophyseal Hormones in Mood Disorders

A number of studies have reported abnormalities of neurohypophyseal secretions in major depressive disorder. Several lines of evidences suggest a role of AVP and OT in mood disorders. They act as neuromodulators of the stress response. AVP is known as an ACTH stimulating factor synergistic to CRF while OT could act as an antagonist of AVP on ACTH secretion. The ACTH-releasing properties occur via the V1b (V3) receptor subtype, expressed in the majority of pituitary corticotropes, in multiple brain regions and a number of peripheral tissues (Claes, 2004) . Human studies have shown that plasma concentrations of AVP and/or OT increase or decrease depending on different conditions of stress (Edelson and Roberston, 1986; Sanders et al, 1990). In 1982(a), Legros et al have suggested that neuropituitary could be involved in the ACTH regulation either to as modulatory to the CRF release through positive vasopressinergic and negative ocytoninergic influence or as a site of synthesis. Moreover, central AVP functions in animal are consistent with the assumption that impaired AVP transmission in the central nervous system could produce at least some symptoms of the depressive state, such as cognitive impairment, or disorganisation of circadian patterns. There are speculation that some of the symptoms commonly reported in depression (social withdrawal, reduced appetite, cognitive impairment) may reflect central OT function (Bell et al, 2006). OT has also revealed antidepressive (Arletti et al., 1987) and a anti-stress properties (Peterson et al., 1996; Uvnäs-Moberg and al., 1994, 1997). Hypothalamo-pituitary-adrenal axis overactivity is frequently observed in major depression CRH and AVP are the major secretagogues of the HPA/stress system (Scott and Dinan, 2002). Genes, childhood adversity and life events in adulthood contribute to the development of mood disorders, possibly via effects on HPA axis function (Watson et al, 2004).

AVP, OT and Neurophysins in Major Depression

There are relatively few data on plasma AVP levels in depression. Previous reports suggested that they are unchanged (Raskin et al., 1979; Gjerris et al., 1985; Sörensen et al., 1985; Ritchie et al., 1991). In contrast, van Londen et al. (1997) showed elevated plasma vasopressin concentrations in patients with depression when compared with controls. Concerning CSF concentrations of AVP, depressed patients showed significantly lower values than controls in two studies (Gold et al., 1984; Gjerris et al., 1985) but no difference in one study performed in depressive subjects (Heuser et al., 1998). In post-mortem studies, the numbers of AVP-neurons in the paraventricular nucleus levels of the hypothalamus has been reported to be increased (Purba et al., 1996). With regards to OT plasma concentrations, van Londen et al. (1997) found no significant difference between depressed patients and controls. Similar results were obtained regarding CSF OT (Demitrack and Gold 1988). In post-mortem studies, the numbers of AVP- and OT-expressing neurons in the paraventricular nucleus levels of the hypothalamus has been reported to be increased (Purba et al., 1996). Zhou et al (2001) have found an increase in the number of AVP-immunoreactive (AVP-IR) neurons in the suprachiasmatic nucleus in depression, together with a decrease in AVP-mRNA.

In 1984, Linkowski et al. demonstrated a reduction in the CSF concentration of AVP-neurophysins in a group of eight patients with unipolar depression compared with a group of 12 patients without psychiatric disorders. Basal levels of AVP-neurophysins were reported to be lower in major depressive patients than in normal controls (Laruelle et al., 1990; Legros et al., 1993; Scantamburlo et al., 2005). Few studies have investigated the influence of acute apomorphine or clonidine injection on AVP- and OT-neurophysin release in humans. Apormophine, a dopamine agonist, stimulates the release of various anterior pituitary hormones. Dopamine also has been reported to influence the secretion of posterior pituitary hormones. However stimulatory as well as inhibitory actions have been reported. With regards to noradrenergic receptors, clonidine (a α-2-adrenoceptor) has been widely used in psychiatric patients. Noradrenaline is known to inhibit AVP release from the hypothalamo-neurohypophyseal system in culture and to attenuate acetylcholine-stimulated AVP release in a concentration dependent manner (Armstrong et al, 1982). Dopamine and noradrenergic dysfunctions have been widely suggested in major depression, and neurophysins response to apomorphine and clonidine challenge could represent an indirect way to test this hypothesis. We previously reported a stimulatory influence of acute apomorphine injection on AVP-neurophysin release in one manic patient (Legros and Ansseau, 1986). Nussey et al (1988) also described a stimulatory action of apomorphine on AVP and OT in human. In schizophrenic patients, the basal level AVP-neurophysin was found to be reduce to those in the controls. In contrast, the basal level of OT-neurophysin was increased (Legros et al, 1992). The response to apomorphine was blunted, with no significant release of AVP- and OT-Neurophysin. Recently, we did not find any influence of acute apomorphine or clonidine injections on AVP- and OT-neurophysin (Scantamburlo et al, 2005).

HPA dysfunction is one of the best-documented neurophysiological changes seen in major depression (Claes, 2004). Hypercortisolemia observed in depressed patients could be due to an hyperproduction of corticotropin-releasing hormone at the hypothalamic level. Several animal and human studies suggested a possible role of AVP in ACTH release (Scott and Dinan, 1998). A relationship between HPA axis overactivity and AVP has been reported in depression (Legros et al, 1982; Gispen-de Wied et al., 1992; Dinan and Scott, 2005). In, on average, 30-50 % of depressed patients the HPA axis is not suppressed in the DST, with the highstes rates of nonsuppression in patients with melancholic, endogenous, familial or psychotic subcategory of depression (Nelson and Davis, 1997). Legros et al (1982b) have suggested that, beside ACTH, posterior pituitary function must be investigated to explain the origin of the weak suppressibility of plasma cortisol to DST in endogenously patients. Preliminary results have shown a correlation between neurophysins and cortisol in the non suppressor group both before and after DST. In contrast, Laruelle et al (1990) found that AVP-neurophysins values were not affected by DST administration and no difference in AVP-neurophysins levels between escapers and nonescapers. Some data suggest that increased AVP is related to a specific subcategory of depression in which the HAP axis is not suppresed during the DST, rather than to depression in wich response are increased in Dex-CRH test (de Winter et al, 2003). Indeed, plasma AVP levels are elevated in patients with major depression and in patients with the melancolic subcategory compared with healthy controls (van Londen et al, 1997). Moreover, plasma AVP and cortisol levels are positively correlated in depressed patients (Inder et al, 1997). They also showed a significant correlation between plasma AVP and plasma ACTH and cortisol. Recently, Watson et al (2006) found that post-DST AVP levels were significantly elevated in patients with bipolar disorder and chronic depression compared with controls. The incresead AVP plasma concentration may have been present before administration of dexamethasone, may reflect an increased release of AVP from parvocellular neurones of the paraventricular system or from the magnocellular system (Aguilera and Rabadan-Dielh, 2000; Scott and Dinan, 2002). Dinan and Scott (2005) suggest an important role for AVP in the pathophysiology of major depression. Extrapolating from animal models of chronic stress, the constant forward drive of the HPA axis may depend on AVP activity unrestrained by elevated ambient cortisol levels.

Increased plasma AVP concentrations have been reported in depressed suicide attempters (Inder et al, 1997). Brunner et al (2002) investigated CSF and plasma AVP in nineteen depressed patients. They found no difference between depressed suicide attempters, depressed nonsuicidal patients and neurological control subjects. Recently, Merali et al (2006) have shown several neuropeptide changes in relation to suicide. Levels of AVP-IR were elevated at the paraventricular nucleus, locus coeruleus and dorsolateral prefrontal cortex, and reduced at the dorsovagal complex. Depressed patients with a history of suicide attempt exhibited higher plasma AVP concentrations than depressed controls without history of suicidal behavior.

Neurophysins in Bipolar Disorder

Neuroendocrine studies in bipolar disorder have found abnormal HPA axis function during affective episodes (Rybakowski and Twardowska, 1999; Watson et al, 2004). Recently, Watson et al (2006) found post-dexamethasone AVP levels significantly elevated in patients with bipolar disorder compared with matched healthy controls. The difference was also present in remitted bipolars. They also found that post-dexamethasone AVP remain abnormal to a similar degree in remission (Watson et al, 2004). This supports the hypothesis of an allostatic shift in the function of the HPA axis in bipolar disorder.

Linkowski et al. (1984) demonstrated a reduction in the CSF concentration of AVP-neurophysins in a group of eight patients with unipolar depression compared with a group of 12 patients without psychiatric disorders. We also previously described an increase of CSF neurophysins in bipolar as compared to unipolar depressed patients (Legros et al, 1983b).

We published the first report af an increased in the basal blood level of a vasopressin-related peptide in a drug-free manic patient. This difference from previous studies might be due to the fact that plasma neurophysins concentrations better reflect chronic vasopressinergic function than isolated AVP assay itself (Legros and Ansseau, 1986). The increase of vasopressinergic activity in the manic compared to the depressed phase is fully consistent with studies that show either an increased plasma AVP release following hypertonic saline infusion in manic compared to depressed patients (Gold et al, 1980) or increased CSF AVP release in manics compared to controls or depressed patients (Gold et al, 1983).

The pathophysiological meaning of this increased vasopressinergic activity is still uncertain; it may be either a non specific consequence of the gross behavioural disturbances in mania or a part of the specific central mechanisms leading to mania. It is of interest to note that exogenous AVP given as a treatment for memory disturbances has been reported to sometimes induce manic symptoms. As AVP has been shown to potentiate catecholaminergic neurotransmisssion, and as catecholamine overactivity could be implicated in the pathogenesis of mania (Post, 1980), an increased vasopressin modulatory action might be responsible for an increased catecholamine activity. Lithium, which could act as an antagonist to AVP at the peripheral receptor level, may increased the sensitivity of the plasma AVP response to osmotic stimuli (Gold et al, 1983) and could share its antimanic properties through a similar action at the central AVP receptor system. Moreover, Legros and Ansseau showed that manic patients exhibit significantly higher plasma AVP-neurophysins values than depressed and schizophrenic patients (Legros et Ansseau, 1989).

Conclusion

Activity of the vasopressinergic or oxytoninergic systems is measured by assessing the neurophysin blood level which is considered as a better method to investigate neuropituitary function. Indeed, neurophysins are coreleased equimolecularly with the active nonapeptide and their stability in the blood is better than the one of AVP (Legros and Ansseau, 1986).

An important limitation of studies in litterature is that AVP in plasma reflects neuropituitary AVP release and not AVP activity in the magnocellular portion of the paraventricular nucleus. AVP that acts at the anterior lobe of the pituitary is co-released with CRH from these specific parvocellular neurons. Moreover, the use of a plasma measure of OT rather than one from the CSF although other studies have suggested that the production of central OT may be paralleled by release into the systemic circulation (van Londen et al, 1997). It may be however that measures of plasma OT are too blunt a tool to detect changes in discrete clusters of cells in the CNS or variations in OT receptors genes that may underlie changes to social affiliative behaviour (Bel et al, 2006; Insel, 1997).

AVP seems to play an important role in the HPA dysregulation in depressive illness. Animal studies of chronic psychological stress indicate a shift from CRF to AVP controlled pituitary-adrenal activation. Given the fact that two V1b receptor antagonists are currently in clinical development for the treatment of depression, studies wich increase the understanding of the role of AVP in the pathophysiology of depression are required. Recently, van West et al (2006) have suggested a protective effect against developing major depression of a single-nucleotide polymorphism haplotype in the AVP V1b receptor gene.

Moreover, the neuropsychopathological, neuroendocrinological, and neuropharmacological data are consistent with the hypothesis that an increase in vasopressin synthesis and/or release can occur in manic patients and could initiate, or participate in, the genesis of the symptoms. HPA axis dysfunction may be involved in the disease process underlying bipolar disorder. Obviously, the opportunistic observation of this patient must be confirmed in larger studies of bipolar disorder subjects. The hypothetic peripheral metabolic consequences of such AVP modifications (dehydration, water intoxication) must be also studied carefully controlled psychoendocrine paradigms.

In summary, he radioimmunoassay for neurophysins has been proposed and used as an index of neuropituitary function in different physiological and pathological conditions. Their assays have been performed to investigate the activity of the hypothalamo-neurohyposial-axis. AVP- or OT-neurophysin assay may be an appropriate tool for investigating chronic vasopressinergic or oxytoninergic function in affective disorders.

References

Acher, R; Light A; Du Vigneaud V. Purification of oxytocin and vasopressin by way of a protein complex. *J. Biol. Chem,* 1958, 233, 116-120.

Acher, R. Molecular evolution of biologically active polypeptides. Proceedings of the Royal Society of London, 1980, 210, 21-43.

Aguilera, G; Rabadan-Dielh, C. Vasopressinergic regulation of the hypothalamic-pituitary-adrenal axis : implications for stress adaptation. *Regulatory Peptides,* 2000, 96, 23-9.

Amico JA; Tenicela R; Johnston J; Robinson A. A time-dependant peak of oxytocin exists in cerebrospinal fluid but not in plasma of humans. *J. Clin. Endocrinol. Metab,* 1983, 57, 947-951.

Arletti R; Bertolini A. Oxytocin acts as an antidepressant in two animal models of depression. *Life Sci,* 1987, 41, 1725-1730.

Armstrong WE; Sladek CD; Sladek JR. Characterization of noradrenergic control of vasopressin release by the organ cultured hyppothalamoneurohypophyseal system. *Endocrinology,* 1982, 111, 273-279.

Bell C; Nicholson H; Mulder R; Luty S; Joyce P. Plasma oxytocin levels in depression and their correlation with the temperament dimension of reward dependence. *J. Psychopharmacol,* 2006, 5, 656-60.

Brunner J; Keck M; Landgraf R; Uhr M; Namendorf C; Bronish T. Vasopressin in CSF and plasma in depressed suicide attempters: preliminary results. *European Neuropsychopharmacology,* 2002, 12, 489-494.

Claes, S. CRH, stress, and major depression: a psychobiological interplay. *Vitamines and Hormones,* 2004, 69, 117-150.

Crine, AF; Louis F; Legros JJ. Peripheral neurophysins change under mild stress conditions. *Biological Psychology,* 1982, 14, 311-315.

de Bree, FM. Trafficking of the vasopressin and oxytocin prohormone trough the regulated secretory pathway, *Journal of Neuroendocrinology,* 2000, 12, 589-594.

de Bree, FM; van der Kleij, AM; Nijnhuis, M; Zalm, R; Murphy, D; Burbach P. The hormone domain of the vasopressin prohormne is required for the correct prohormone trafficking through the secretory pathway. *Journal of Neuroendocrinology,* 2003, 15, 1156-1163.

de Winter, RF; van Hemert, AM; DeRijk, RH; Zwinderman, KH; Frankhuijzen-Sierevogel, AC; Wiegant, VM; Goekoop, JG. Anxious-retarded depression: relation with plasma vasopressin and cortisol. *Neuropsychopharmacology,* 2003, 28(1),140-147.

Edelson, JT; Roberston, GL. The effects of the cold pressor test on vasopressin secretion in man. *Psychoneuroendocrinology,* 1986, 3, 307-316.

Fawcett, CP; Powell, AE; Sachs, H. Biosynthesis and release of neurophysin. *Endocrinology,* 1968, 83, 1299-1310.

Geenen, V; Langer, G; Koining, G; Schonbeck, G; Ansseau, M; von Frenckell, R; Smitz, S; Legros JJ. Release of human neurophysin I during insulin-induced hypoglycemia in depressed patients is abolished after recovery with clomipramine treatment. *Psychoneuroendocrinology,* 1985, 10, 61-69.

Gispen-de Wied, CC; Westenberg, HG; Koppeschaar, HP; Thijssen, JH; van Ree, JM. Stimulation of the pituitary-adrenal axis with low dose Arg8-vasopressin in depressed patients and healthy subjects. *Eur. Neuropsychopharmacology.* 1992, 2(4), 411-419.

Gjerris, A; Hammer, M; Vendsborg, P; Christensen, NJ; Rafaelsen, OJ. Cerebrospinal fluid vasopressin changes in depression. *Br. J. Psychiatry.* 1985, 147696-701.

Gold, PW; Goodwin, FK; Ballenger, JC; Weintgarner, H; Robertson, GL; Post, RM. Central vasopressin function in affective illness. In de Wied D., van Keep, P. A. (eds), Hormones and the brain. Lancaster, England: MTP Press, 241-251.

Gold, PW; Robertson, GL; Post, RM; Kaye, W; Ballenger, J; Rubinow, D; Goodwin, FK. The effect of lithium on the osmoregulation of arginine vasopressin secretion. *J. Clin. Endocrinol. Metab.* 1983, 56, 295-299.

Gold, PW; Chrousos, G; Kellner, C; Post, R; Roy, A; Augerinos, P; Schulte, H; Oldfield, E; Loriaux, DL. Psychiatric implications of basic and clinical studies with corticotropin-releasing factor. *Am. J. Psychiatry.* 1984, 141(5), 619-627.

Gonzalez, CB; Caorsi, CE; Figueroa, CD. Structure of neurosecretory granules and the chemistry of exocytosis. *Ann. N Y Acad. Sci,* 1993, 689, 59-73.

Heuser, I; Bissette, G; Dettling, M; Schweiger, U; Gotthardt, U; Schmider, J; Lammers, CH; Nemeroff, CB; Holsboer, F. Cerebrospinal fluid concentrations of corticotropin-releasing hormone, vasopressin, and somatostatin in depressed patients and healthy controls: response to amitriptyline treatment. *Depress Anxiety,* 1998, 8(2), 71-79.

Huang, HB; Breslow, E. Identification of the unstable neurophysin disulfide and localization to the hormone-binding site. Relationship to folding-unfolding pathways. *J. Biol. Chem,* 1992, 267, 6750-6756.

Inder, WJ; Donald, RA; Prickett, TCR; Frampton, CM; Sullivan, PF; Mulder, RT; Joyce, PR. Arginine vasopressin is asociated with hypercortisolemia and suicide attempts in depression. *Biol. Psychiatry.* 1997, 42, 744-747.

Insel TR, A neurobiological basis of social attachment. *Am. J. Psychiatry,* 154, 726-735.

Itoi, K; Jiang, YQ; Iwasaki, Y; Watson J. Regulatory mechanisms of corticotropin-releasing hormone and vasopressin gene expression in the hypothalamus. *Journal of Neuroendocrinology.* 2004, 16, 348-355.

Joris, JL; Chiche, JD; Canivet, JL; Jacquet, NJ; Legros, JJ; Lamy ML. Hemodynamic changes induced by laparoscopy and their endocrine correlates: effects of clonidine. *J. Am. Coll. Cardiol.* 1998, 32(5), 1389-1396.

Laruelle, M; Seghers, A; Goffinet, S; Bouchez, S; Legros, JJ. Plasmatic vasopressin neurophysin in depression : basic levels and relations with HPA axis. *Biol. Psychiatry.* 1990, 27, 1249-1263.

Legros, JJ; Louis, F. Identification of a vasopressin-neurophysin and of an oxytocin-neurophysin in man. *Neuroendocrinology,* 1973-1974, 13, 371-375.

Legros JJ.The radioimmunoassay of human neurophysin: contribution to the understanding of the physiopathology of neurohypophyseal function. *Ann. N. Y. Acad. Sci,* 1975, 248, 281-303.

Legros JJ. The radioimmunoassay of neurophysins as a tool for the posterior pituitary investigation. *Biochem. Pharmacol.* 27, 1978, 1009-1018.

Legros, JJ; Louis, F; Chiodera P. Concomitant increase of neurophysin I and ACTH during insulin tolerance test in one out five normal volunteers; influence of psychologicla stress ? *Neuroendocrinol. Lett*, 1982a, 6, 367-376.

Legros, JJ; Charles, G; Meunier, JC; Smitz, S; Mendlewicz J. Neurophysins and cortisol levels during dexamethasone suppression test in depression. *Neuroendocrinol. Lett*, 1982b, 4295-298.

Legros, JJ; Ansseau, M; Doumont, A; Crine, AF; Geenen, V; Smitz, S; Charles, G; Demey-Ponsart, E; Sulon, J; Mens, WB; Van Wimmersma-Greidanus, TJB. Relationship between vasopressin, neurophhysin and ACTH, cortisol plasma levels in non-suppressor patients during dexamethasone suppression test. *Neuroendocrinol. Lett*, 1983a, 5, 297-302.

Legros, JJ; Geenen, V; Linkowski, P; Mendelwicz, J. Increased neurophysin I and II cerebrospinal fluid concentration from bipolar versus unipolar depressed patients. *Neuroendocrinol. Lett*, 1983b, 5, 201-205.

Legros, JJ and Ansseau, M. Vasopressin-neurophysin and bipolar depression : a case report. *Biol. Psychiatry*, 1986, 21, 1212-1216.

Legros, JJ and Ansseau, M. Increased basal plasma vasopressin-neurophysin in mania. *Horm. Res*, 1989, 31,55-58.

Legros, JJ and Ansseau, M. Neurophysins as markers in neuropsychiatry. In : Ansseau M, von Frenckell R and Franck G, eds. Elsevier Science Publishers, 1991.

Legros, JJ and Ansseau, M. Neurohypophyseal peptides and psychopathology. Progress in *Brain Research*, 1992, 93, 455-461.

Legros, JJ; Gazzotti, C; Carvelli, T; Franchimont, P; Timsit-Berthier, M; von Frenckell, R; Ansseau M. Apomorphine stimulation of vasopressin- and oxytocin-neurphysins. Evidence for increased oxytoninergic and decreased vasopressinergic functions in schizophrenics. *Pshychoneuroendocrinology*, 1992, 6, 611-617.

Legros, JJ; Ansseau, M; Timsit-Berthier, M. Neurohypophyseal peptides and psychiatric diseases. *Regul. Pept*, 1993, 45, 133-138.

Legros, JJ; Geenen, V. Neurophysins in central diabetes insipidus. *Horm. Res*, 1996, 45, 182-186.

Legros, JJ; Boudarene, M; Timsit-Berthier, M; Heindricks, M; Crasson, M; Petraglia, C. AVP-neurophysic but not AVP nor CRF is implicated in cortisol response to psychological stress in anxious healthy human. *Neuropsychopharmacology*, 2000, 13, 550.

Linkowski, P; Geenen, V; Kerkofs, M; Mendlewicz, J; Legros, JJ. Cerebrospinal fluid neurophysins in affective illness and schizophrenia. *Eur. Arch. Psychiat. Neurol. Sci*, 1984, 234, 162-165.

Merali, Z; Kent, P; Du, L; Hrdina, P; Palkovits, M; Faludi, G; Poulter, M; Bédard; Anisman, T. Corticotropin-Releasing Hormone, Arginine Vasopressin, Gastrin-Releasing Peptide, and Neuromedin B Alterations in Stress-Relevant Brain Regions of Suicides and Control Subjects. *Biol. Psychiatry*, 2006, 594-602.

Nelson, JC; Davis, JM. DST studies in psychotic depression: a meta-analysis. *Am. J. Psychiatry*. 1997, 154(11), 1497-1503.

Petersson, M; Alster, P; Lundeberg, T; Uvnas-Moberg, K. Oxytocin increases nociceptive thresholds in a long-term perspective in female and male rats. *Neurosci. Lett.* 1996, 212(2), 87-90.

Post, RM. Biochemical theories of mania. In Belmaker R.H., Van Praag N.M. (eds). Mania: an evolving concept. New York: spectrum, 217-265.

Purba, JS; Hoogendijk, WJ; Hofman, MA; Swaab, DF. Increased number of vasopressin- and oxytocin-expressing neurons in the paraventricular nucleus of the hypothalamus in depression. *Arch. Gen. Psychiatry,* 1996, 53(2), 137-143.

Raadsheer, FC; Hoogendijk, WJ; Stam, FC; Tilders, FJ; Swaab, DF. Increased numbers of corticotropin-releasing hormone expressing neurons in the hypothalamic paraventricular nucleus of depressed patients. *Neuroendocrinology,* 1994, 60(4), 436-444.

Raskin, A and Sathananthan, G. Depression in the elderly [proceedings]. *Psychopharmacol. Bull.* 1979, 15(2), 14-16.

Ritchie, JC; Bragg, DC; White, N; Krishnan, KR; Carroll, BJ. Plasma vasopressin and ACTH baseline levels in depression. *Neuroendocrinol. Lett,* 1991, 208.

Rybabowski, JK; Twardowska K. The dexamethasone/corticotropin-releasing hormone test in depression in bipolar and unipolar affective illness. *Journal of Psychiatric Research.* 1999, 33, 363-70.

Sanders, G; Freilicher, J; Lightman SL. Psychological stress of exposure to uncontrollable noise increases plasma oxytocin in high emotionnality women. *Psychoneuroendocrinology,* 1990, 15,47-58.

Scantamburlo, G; Hansenne, M; Fuchs, S; Pitchot, W; Pinto, E; Reggers, J; Ansseau, M; Legros, JJ. AVP- and OT-neurophysins response to apomorphine and clonidine in major depression. *Psychoneuroendocrinology,* 2005, 30(9), 839-845.

Scott, LV and Dinan, TG. Vasopressin and the regulation of hypothalamic-pituitary-adrenal axis function : implications for the pathophysiology of depression. *Life Sciences,* 1998, 62, 1985-1998.

Scott, LV and Dinan, TG. Vasopressin as a target for antidepressant development: an assessment of the available evidence. *J. Affect. Disord,* 2002, 72(2),113-124.

Scott, LV and Dinan, TG. Anatomy of melancolia: focus on hypothalamic-pituitary-adrenal axis overactivity and the role of vasopressin. *J. Anat,* 2005, 207, 259-264.

Sofroniew M. Projections from vasopressin, oxytocin, and neurophysin neurons to neural targets in the rat and human. *Journal of Histochemistry and cytochemistry,* 1980, 28, 475-478.

Sorensen, PS; Gjerris, F; Hammer, M. Cerebrospinal fluid and plasma vasopressin during short-time induced intracranial hypertension. *Acta Neurochir.* (Wien) 1985, 77(1-2), 46-51.

Uvnas-Moberg, K; Ahlenius, S; Hillegaart, V; Alster P. High doses of oxytocin cause sedation and low doses cause an anxiolytic-like effect in male rats. *Pharmacol. Biochem. Behav,* 1994, 49(1),101-106.

Uvnas-Moberg, K. Oxytocin linked antistress effects-the relaxation and growth response. *Acta Physiol. Scand. Suppl,* 199, 64038-42.

van Londen, L; Goekoop, JG; van Kempen, GM; Frankhuijzen-Sierevogel, AC; Wiegant, VM. Plasma levels of arginine vasopressin elevated in patients with major depression. *Neuropsychopharmacology,* 1997, 17(4), 284-292.

Watson, S; Gallagher, P; Ritchie, JC et al . Hypothalamic-pituitary-adrenal axis function in patients with chronic depression. *British Journal of Psychiatry,* 2004, 184, 496-502.

Watson, S; Gallagher, P; Nocol Ferrier, I; Young H . Post-dexamethasone arginine vasopressin levels in patiens with severe mood disorders. *Journal of Psychiatric research,* 2006, 40, 353-359.

Zhou, JN; Riemersma, RF; Unmehopa, UA; Hoogendijk, WJ; van Heerikhuize, JJ; Hofman, MA; Swaab DF. Alterations in arginine vasopressin neurons in the suprachiasmatic nucleus in depression. *Arch Gen Psychiatry,* 2001, 58(7), 655-662.

In: Neuropeptide Research Trends
Editor: Bernice A. Levine, pp. 243-260
ISBN: 978-1-60021-640-4

Chapter X

Pairs of Neurotransmitters Act Synergistically to Regulate Neuropeptide Release from the Hypothalamo-Neurohypophyseal System

Zhilin Song
Department of Physiology and Biophysics, University of Colorado at Denver and Health Sciences Center, Aurora, CO

Abstract

The hypothalamo-neurohypophyseal system (HNS) secrets two physiologically important neuropeptides: vasopressin (VP) and oxytocin (OT), which regulate water secretion in kidney/blood pressure and milk ejection/uterus contraction, respectively. Both of these neuropeptides are synthesized by magnocellular neurons (MCNs) in the supraoptic nucleus (SON) and paraventricular nucleus of the hypothalamus and released from the posterior pituitary into the systemic circulation. Their synthesis and release are regulated by numerous synaptic input signals with the primary regulation of VP secretion coming from osmotic and cardiovascular signals. Cardiovascular signals are relayed through the A1 pathway to SON, where several neurotransmitters are released. Among them, a pair of co-released neurotransmitters, ATP and norepinephrine (NE), exerts unique regulation of VP and OT release. In vitro studies with ATP and phenylephrine (PE), an agonist to α-1 adrenergic receptor (α-1R), demonstrated that ATP and PE acted synergistically to stimulate VP and OT release from perifused HNS explants. Either of these agents alone only induced a transient and small VP or OT release. Combined application of these two agents transformed the response to a much larger and sustained one. This synergistic action on neuropeptide release also exists between another pair of neurotransmitters, angiotensin II and glutamate, which mediate osmotic signals to SON. Depletion of intracellular calcium stores by thapsigargin disrupted the synergistic response to ATP and PE, indicating involvement of calcium as a mediator. Since activation of P2X purinergic receptors (P2XR, ATP is an agonist) induces influx of calcium whereas activation of α-1Rs releases calcium from intracellular stores, it was anticipated that co-application of this pair of neurotransmitters might generate a different pattern of intracellular calcium, which served as the driving force for their synergistic action on neuropeptide release. Further studies employing a calcium imaging technique with calcium sensitive fluorescence dye, Fura-2 AM, revealed that either ATP or PE alone induced a large, rapid peak increase in intracellular calcium, but ATP and PE together induced a much larger sustained calcium increase than either agent alone. This larger sustained calcium response to combined ATP and PE treatment is not dependent on either pre-synaptic inputs, or continuous exposure to the agents. These results demonstrate that the pair of neurotransmitters acts post-synaptically to generate a different pattern of intracellular calcium, which in turn drives a synergistic neuropeptide release from posterior pituitary. The synergistic action on neuropeptide release may be physiologically important under circumstances such as severe blood loss induced hypovolemia when large and sustained release of VP is required to maintain blood pressure until the volume loss is corrected.

Introduction

The A1 pathway plays a critical role in relaying hemodynamic regulatory signals from the periphery to the magnocellular vasopressin (VP) neurons of the hypothalamo-neurohypophyseal system (HNS). Information about decreases in blood volume and pressure associated with moderate hemorrhage is monitored by volume and baroreceptors in the atria and carotid sinus and transmitted to the A1 catecholamine neurons in the caudal ventrolateral medulla by way of the nucleus tractris solitarius (Raby and Renaud, 1989; Smith et al., 1995). The A1 neurons synapse directly on and send excitatory signals to stimulate VP synthesizing neurons of the supraoptic nuclei (SON) and paraventricular nuclei (PVN) in HNS (Alonso and Assenmacher, 1984; Day et al., 1984; Day and Renaud, 1984; Day et al., 1992; Shioda et al., 1992; Ginsberg et al., 1994; Smith et al., 1995). Norepinephrine (NE) was surmised to be the primary neurotransmitter mediating this response but this mechanism was not supported by experimental data since adrenoceptor antagonists did not block A1 activation of VP cells (Day et al., 1990). Furthermore, glutamate cannot account for VP cellular responses to A1 input since injections of the broad-spectrum excitatory amino acid receptor antagonist kynurenic acid were ineffective in blocking excitation induced by stimulation of the A1 region either (Day et al., 1990). It is most likely that other excitatory neurotransmitters co-localized and/or co-released with NE play a joint role in the A1 regulation of VP release.

Several neurotransmitters are co-localized and/or co-released with NE from the A1 neurons, including neuropeptide Y (NPY), substance P (SP), and ATP (Everitt et al., 1984; Sawchenko et al., 1985; Blessing et al., 1986; Beroukas et al., 1989; Lundberg et al., 1989; Bittencourt et al., 1991; Sperlagh et al., 1998). Among them, the nucleotide ATP, which is commonly co-localized in catecholamine vesicles (Fried, 1980; Whittaker, 1982), may be the co-neurotransmitter with NE to activate VP neurons in response to A1 input. In the periphery, ATP is co-localized with NE in synaptic vesicles in the sympathetic nervous system (Geffen and Livett, 1971; Whittaker, 1982). It is released after physiological nerve activity (Sperlagh and Vizi, 1996) and functions as a co-transmitter with NE in a number of tissues innervated by the sympathetic system (Burnstock, 1990; Von Kugelgen and Starke, 1991). Central noradrenergic neurons might use similar co-transmission. P2 purinoceptors (P2R) are expressed abundantly in the CNS, and a high density of receptors exists in the hypothalamus (Balcar et al., 1995). A role for ATP in mediating responses to activation of the A1 pathway is supported by the finding that application of the P2 receptor blocker suramin (10 mM) in the SON reversibly blocked excitation of VP cells by A1 stimulation without preventing the excitatory effect of locally applied NE (Day et al., 1992). Histological and electrophysiological evidence demonstrating purinergic receptors on SON neurons further supports a role for purinergic transmission (Hiruma and Bourque, 1995; Shibuya et al., 1999).

As described in the following sections, studies using an in vitro hypothalamo-neurohypophyseal system model characterized the functional consequence of this adrenergic and purinergic co-transmission on VP and OT hormone release, explored the possible sites where these two neurotransmitters may interact, and revealed the underlying intracellular mechanisms that is driving this action. We also discuss similar interactions between other pairs of neurotransmitters in the HNS.

1. ATP and NE Act Synergistically to Stimulate VP and OT Release from Posterior Pituitary of the HNS

The effect of ATP and phenylephrine (PE, an agonist to α-1 adrenergic receptor) on VP and OT hormone release was studied extensively in an in vitro HNS explant perifusion system (Kapoor and Sladek, 2000). This explant includes the SON, with some preserved synaptic afferents to SON and an intact axonal pathway to the posterior pituitary in which the VP and OT hormones are transported through the median eminence for storage and from which they are released when the HNS is stimulated. ATP induced a rapid, but un-sustained, increase in VP or OT release. Since in other experiments, explants retained elevated VP/OT release throughout 6 hr of perifusion with 25 mM KCl (Ludwig et al., 1997), termination of the response is unlikely to reflect depletion of releasable VP or OT stores in posterior pituitary. Termination of the response is not due to the inhibitory effect of adenosine, the ATP metabolite either, since blockade of ATP breakdown to adenosine by a ecto-nucleotidase inhibitor did not prolong the response to ATP (Song and Sladek, 2005). Shortening of the response to ATP is rather most likely due to desensitization of the ligand-gated P2X ion channels. In electrophysiological studies, some SON neurons showed little-to-moderate desensitization, whereas others showed strong desensitization (Shibuya et al., 1999). The restoration of the ATP response after a washout period further supports the receptor desensitization mechanism. PE also induced a fast, un-sustained VP release. Interestingly, exposure of the explants to combined ATP and PE to mimic the in vivo situation where ATP and NE are presumed to be co-released from the A1 neurons innervating SON neurons dramatically changed the nature of VP and OT response (figure 1). Not only was the response several fold larger in amplitude compared with that resulting from exposure to the individual drugs, the response was also converted from a transient to a sustained one which lasted for hours. Thus, the response to combined exposure of ATP and PE apparently is not additive, but synergistic in nature. The synergistic response was blocked by 10 μM pyridoxal-phosphate-6-azophenyl-2', 4'-disulphonic acid (PPADS), indicating a role for P2XRs. It was also diminished by a protein kinase C (PKC) inhibitor and by actinomycin, which inhibits gene transcription thus blocking new protein synthesis.

Several intracellular mechanisms were postulated to underlie the synergistic increase in VP release. The first is recruitment of P2YRs by α-adrenergic receptors. This was considered feasible, because these receptors couple to the same class of G-proteins as the α-adrenergic receptors [$G_{q/11}$ (Zimmermann, 1994)] and it seems possible that co-activation of these receptors might amplify α-adrenergic-mediated activation of the phospholipase C (PLC)-PKC signal cascade. Functional P2Y2R has been demonstrated on SON neurons electrophysiologically (Hiruma and Bourque, 1995) and P2Y1R, P2Y2/4R, and P2Y6R have been shown to increase intracellular calcium in SON neurons (Song et al., 2007). However, recruitment of P2YRs alone could not account for the synergism because PPADS, which does not block P2YRs at 10 μM, prevented the ATP and PE-mediated synergism (Kapoor and Sladek, 2000). Thus a second hypothesis is that activation of α-adrenergic receptors activates PLC which in turn activates PKC by the production of diacylglycerol (DAG). Activation of PKC may increase conductance of the ATP-gated P2XRs or decrease their desensitization by inducing direct phosphorylation of these channels or indirect phosphorylation of some accessory proteins (Boue-Grabot et al., 2000; Paukert et al., 2001; Ennion and Evans, 2002; Brown and Yule, 2007). Although this is an intriguing possibility since a PKC inhibitor suppressed the synergistic hormone release, alone it is not an adequate explanation for the synergism, because phosphorylation events occur within seconds to minutes, while the ATP and PE synergism was delayed by about 40 min. A third possibility is that simultaneous opening of the ATP-gated P2X ion channel and activation of the PLC cascade by PE generates a different intracellular calcium pattern. An augmented calcium signal could in turn activate PKC as well as other Ca^{2+}-dependent kinases. Ca^{2+}-dependent phosphorylation of transcription factors might lead to synthesis of new receptors and thus result in a synergistic response. Thus, intracellular Ca^{2+} may be central to the augmentation phenomenon. Blockade of the synergism by the PKC inhibitor supports a role for Ca^{2+}-dependent phosphorylation in the synergism. Blockade of the synergism by actinomycin demonstrates its dependence on new protein synthesis and the delay in the occurrence of synergism is also in concert with this mechanism. We will discuss in more detail the importance of intracellular calcium in synergistic action in section 3.

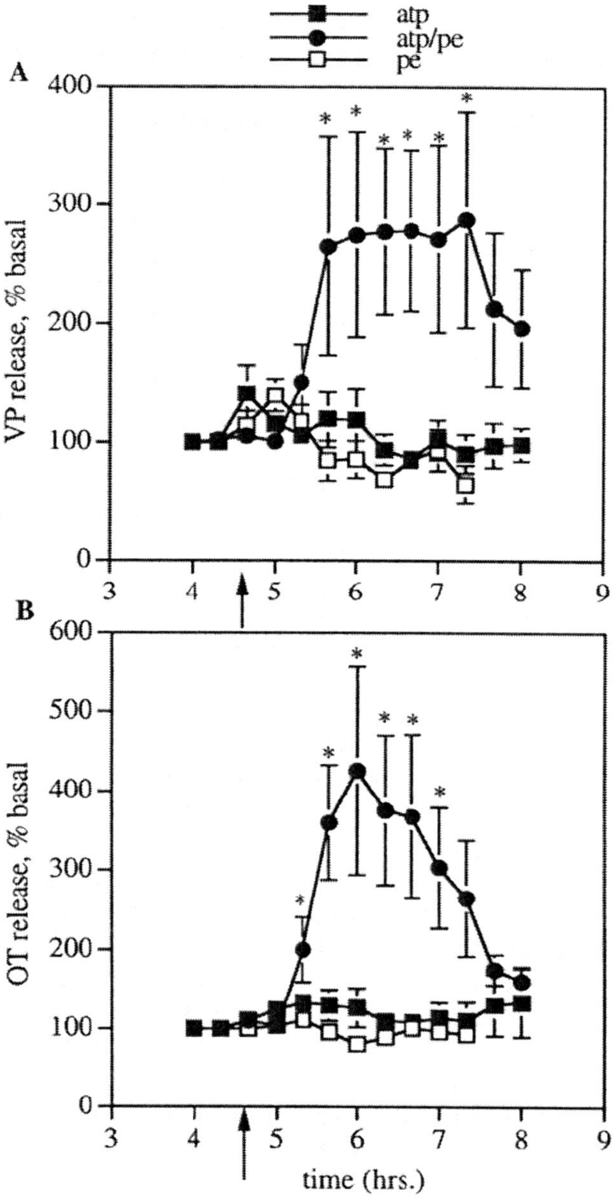

Figure 1. *A*, VP release in response to simultaneous exposure to ATP (100 μM) and PE (100 μM). The fast and un-sustained effects of ATP or PE on VP release are dwarfed by the augmented and sustained stimulation of VP release obtained by simultaneous exposure to ATP and PE (*$p \leq 0.002$). The synergistic response was delayed by ~40 min compared with peak responses observed with ATP (100 μM) or PE (100 μM) individually. Basal release for ATP plus PE (n=12), ATP (n=14), and PE (n=11) groups was 31.5±4.9, 42.6±8.5, and 35.1±14.1 pg/ml, respectively. *B*, OT release in response to simultaneous exposure to ATP (100 μM) and PE (100 μM). ATP and PE result in a synergistic stimulation of OT release when compared with the fast and un-sustained effects of ATP or the lack of an effect of PE on OT release (*$p \leq 0.002$). Basal release for ATP plus PE (n=8), ATP (n=11), and PE (n=12) groups was 67.8±19.8, 87.8±24.4, and 87±25.6 pg/ml, respectively. Explants were exposed to ATP and/or PE beginning at the time indicated by the *arrow*. (Reproduced with permission from: Kapoor and Sladek, J. Neuroscience, 2000, 20(23)).

2. The Synergistic Action of ATP and NE Is Mediated by a Postsynaptic Mechanism

A direct action of ATP on SON neurons is likely because these neurons express P2XR mRNA (Shibuya et al., 1999) and application of P2XR agonists elicits TTX-insensitive depolarization (Hiruma and Bourque, 1995) and increase in intracellular Ca^{2+} in dissociated SON neurons (Shibuya et al., 1999) or SON neurons in HNS explants (Song et al., 2006). Additionally, ATP may activate other neurons afferent to the SON neurons, and it may also act on SON neuronal terminals in the neural lobe. ATP is released from the neural lobe in a stimulus-dependent manner (Sperlagh et al., 1999), is co-stored with neuropeptides in the secretory granules of neurohypophyseal nerve endings in milli-molar concentrations (Gratzl et al., 1980), and increases VP, but not OT, release from the isolated neurohypophyseal terminals (Troadec et al., 1998). This effect of ATP in neural lobe may have contributed to the synergistic VP responses observed. However, neural lobe could not be the only site of ATP action since OT release was also stimulated.

Similar to ATP, NE could also be acting either post-synaptically on the SON neurons, in the neural lobe, or on afferent neurons adjacent to the SON. First, SON neurons express α1-adrenergic receptors. Several groups have demonstrated electrical activation of SON neurons following application of PE to HNS explants or hypothalamic slices (Armstrong et al., 1986; Randle et al., 1986; Yamashita et al., 1987) and intense immuno-reactivity for the α1b-adrenergic receptor subtype is present in the SON (Acosta-Martinez et al., 1999). The SON is richly innervated by noradrenergic fibers arising from the A1 and A2 catecholamine neurons in the medulla (Mcneill and Sladek, 1980; Sawchenko and Swanson, 1981). Direct synaptic innervation of SON neurons has also been demonstrated at the ultra-structural level (Shioda and Nakai, 1992). Furthermore, noradrenaline has been shown to increase intracellular Ca^{2+} in dissociated SON neurons via activation of α1 receptors (Shioda et al., 1997) and PE, the exclusive α1 receptor agonist, increases intracellular Ca^{2+} in SON neurons of HNS explants (Song et al., 2006). Thus, direct activation of postsynaptic α1-adrenergic receptors is clearly one possibility. Second, pre-synaptic effects of norepinephrine have also been reported. Norepinephrine induced increases in excitatory postsynaptic potentials (e.p.s.p.) in ~42% of magnocellular neurons in the PVN (Daftary et al., 1998) via an α1-mediated effect that was tetrodotoxin (TTX) sensitive. Since local glutamate-mediated excitatory inputs have been demonstrated in both the SON and PVN (Boudaba et al., 1997; Boudaba et al., 2003), it suggests that α1 receptors may be located on local glutamatergic neurons and induce glutamate release upon activation by NE. Finally, the neural lobe receives noradrenergic innervations and α1-adrenergic receptors are also present in the neural lobe (Desouza and Kuyatt, 1987). The noradrenergic innervation of the neural lobe is derived from the A2 catecholamine neurons in the brainstem (Garten et al., 1989) as well as sympathetic innervation from the superior cervical ganglion (Alper et al., 1980; Saavedra, 1985). The α1Rs are uniformly distributed in the neural lobe, are up-regulated by sympathetic ganglionectomy and are not decreased by pituitary stalk section suggesting that they are expressed on both glial and vascular elements (Desouza and Kuyatt, 1987).

Possible sites of interaction for ATP and NE in the HNS explants are summarized in figure 2. These include: (i) neurons afferent to the SON neurons that are included in the explant and may express these receptors on either their perikarya or nerve terminals, specifically, local glutamatergic neurons innervating SON; (ii) the VP/OT neurons in SON via postsynaptic receptors on the perikarya and/or dendrites; (iii) the neural lobe.

Several approaches were used to identify the sites of synergistic action of ATP and NE (PE) (Song and Sladek, 2006). Antagonists of glutamatergic receptors were used to address the possibility that the synergistic response requires activation of local glutamatergic afferents. TTX and isolated neural lobe preparations were used to address the possibility that the synergism derives from interactions of ATP and PE in the posterior pituitary gland.

Experiments to address these options derived the following results. First, the VP response to ATP+PE was not altered in the presence of either the CNQX/AP5 cocktail that blocked AMPA/NMDA ionotropic glutamate receptors or (*RS*)-α-methyl-4-carboxyphenylglycine (MCPG) that blocked the metabotropic glutamate receptors, thus eliminating the involvement of glutamatergic transmission. Second, synergistic stimulation of VP release by ATP+PE was not observed in the experiment with isolated neural lobe. Thus, the synergistic response observed in the whole HNS explant requires activation of P2Rs and adrenergic receptors other than those located in the neural lobe. Third, TTX completely eliminated the VP response to ATP+PE. This observation, taken together with the absence of a synergistic response in the isolated neural lobe experiments, suggests that action potentials propagated from the soma of VP neurons to the VP nerve terminals in the neural lobe are required for the synergistic VP release observed in response to ATP+PE stimulation.

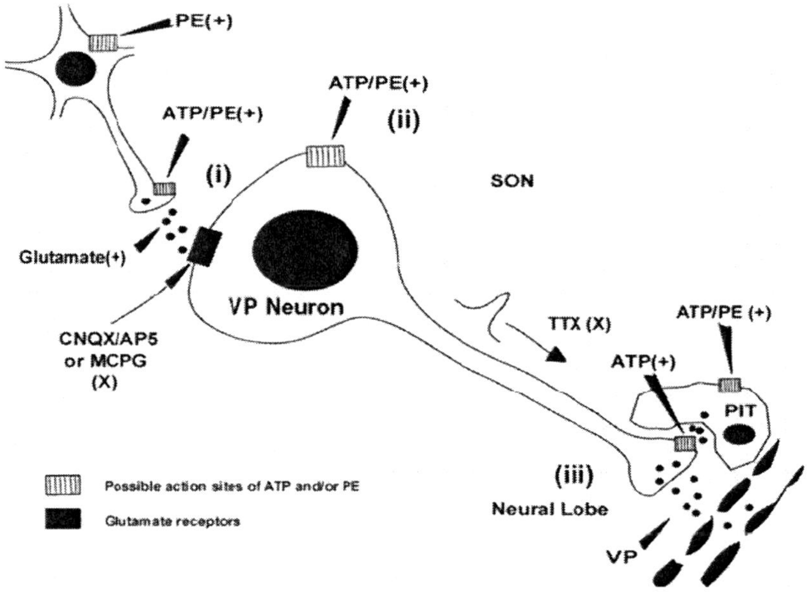

Figure 2. Diagram of potential sites for synergistic interactions of ATP and PE in regulation of VP release. These sites include: (i) activation of local afferent glutamatergic neurons, via actions of ATP and/or PE at the pre-synaptic terminals and/or on the perikarya of the afferent neurons; (ii) activation of receptors on the perikarya or dendrites of the SON neurons; and (iii) direct or indirect activation of receptors in the posterior pituitary gland. PIT, pituicyte. (Adapted with permission from: Song and Sladek, Journal of Neuroendocrinology, 2006, 18 (4)).

The combined results of the TTX, ionotropic and metabotropic glutamatergic receptor antagonists, and isolated neural lobe experiments strongly support a post-synaptic mechanism for the synergistic action of ATP+NE (PE): simultaneous activation of P2Rs and α1-adrenergic receptors other than those in the neural lobe or on the afferent glutamatergic neurons is required to obtain the potentiated and sustained stimulation of VP release elicited by simultaneous exposure of HNS explants to ATP and NE (PE). Although ATP and/or PE might induce release of some other excitatory neurotransmitters than glutamate, glutamate is the primary candidate for an excitatory afferent that might account for the synergistic response to ATP+PE since CNQX virtually eliminated e.p.s.p. in magnocellular neurons recorded in hypothalamic slices (van den Pol et al., 1990), making it the predominant local excitatory input to the SON. In addition, the soma of other excitatory afferents (e.g. cholinergic and histaminergic) are excluded from the HNS explants used in these experiments. Local GABAergic neurons have been shown to influence basal AVP release from these explants also (Sladek and Armstrong, 1987), but they are not responsible for the synergistic stimulation induced by ATP+PE since ATP has been shown to increase inhibitory postsynaptic potentials in SON neurons (Ponzio and Hatton, 2004) and there is no evidence that α1R agonists reduce GABA release either. Activation of ATP release by PE from hypothalamic astrocytes has also been demonstrated to have enduring effects on miniature excitatory postsynaptic currents in magnocellular neurons in the PVN by inducing insertion of new AMPA receptors into postsynaptic membrane (Gordon et al., 2005). However, this mechanism cannot account for the extended elevation of VP release in response to ATP+PE for two reasons. First, PE alone does not induce the sustained elevation in VP release indicating that PE-induced release of glial ATP is inadequate to produce the synergistic effect. Second, the synergistic effect is not blocked by CNQX and therefore does not reflect increased AMPA receptor activation. Thus, the reported findings have eliminated several possible mechanisms, leaving the postsynaptic mechanism the most feasible one for the synergistic effect on VP release by combined exposure to ATP+NE (PE).

3. Intracellular Calcium Is the Driving Force for Synergistically Stimulated VP and OT Release by ATP and NE

As discussed in section 1, a central role for intracellular calcium in ATP+PE stimulated synergistic release of VP and OT is supported by a substantial amount of evidence. But how could this happen? The possible intracellular signal cascades leading to the synergism are depicted in figure 3. Briefly, activation of the ligand-gated non-selective cation channel P2XR by ATP may allow Ca^{2+} as well as Na^{+} to enter the cell (Chen et al., 1995). Influx of Na^{+} depolarizes membrane, thus activating voltage-gated calcium channels and this in turn induces more calcium influx. On the other hand, the α1-adrenergic receptor is a G protein-coupled receptor that activates PLC inducing the formation of inositol triphosphate (IP3) which in turn stimulates Ca^{2+} release from intracellular stores (Graham et al., 1996). Activation of PLC also activates PKC in a calcium dependent manner and PKC could in turn modulate P2XRs. Thus, the two signal cascades interact and generate a different intracellular calcium pattern than either signal cascade alone. This unique pattern of intracellular calcium rise might in turn result in activation of Ca^{2+} sensitive kinases and subsequent phosphorylation of other proteins such as transcription factors. This could lead to a change in gene expression and new receptor synthesis, thus giving rise to the extended elevation in hormone release seen with co-exposure to ATP+PE. The absence of synergistic VP release by ATP+PE following thapsigargin (TG) depletion of intracellular calcium stores supports this hypothesis that convergence of intracellular calcium signals is required for synergistic stimulation of VP release by combined ATP and PE (Song et al., 2006).

A series of experiments were conducted to characterize the ATP+PE induced intracellular calcium change by directly measuring the intracellular calcium ($[Ca^{2+}]_i$) using the calcium sensitive dye, fura-2 AM (Song et al., 2006). The effect of ATP+PE on $[Ca^{2+}]_i$ was examined under four distinct conditions: Extended (4 min) exposure to ATP+PE in low extra-cellular calcium ($[Ca^{2+}]_o$; 0.3 mM), normal $[Ca^{2+}]_o$ (physiological condition, 2mM), and in the presence of TTX, and after short (1 min) exposure to ATP+PE. Low $[Ca^{2+}]_o$ was used to replicate the conditions of the hormone release studies. However, because low $[Ca^{2+}]_o$ increases the excitability of SON neurons by increasing the amplitude of the persistent Na^+ current (I_{NaP}) (Li and Hatton, 1996), it was important to determine whether the effects of ATP+PE were dependent on $[Ca^{2+}]_o$. TTX was used to determine whether the effects of ATP+PE were dependent on I_{NaP} or on transient Na^+ current (e.g., action potential generation). The latter could reflect activation of local neuronal circuitry by ATP and/or PE.

These results are summarized in figure 4. Both ATP and PE induced rapid, but transient, elevations in $[Ca^{2+}]_i$. In 0.3 mM Ca^{2+}, the peak response to ATP was greater than to PE but did not differ from the peak response to ATP+PE. A sustained elevation in $[Ca^{2+}]_i$ was induced by ATP+PE, that was greater than ATP or PE alone. In 2 mM Ca^{2+}, the peak response to ATP+PE was significantly greater than to either ATP or PE alone, and the sustained response to ATP+PE was greater than to either agent alone. Responses were comparable in the presence of TTX. The sustained elevation in $[Ca^{2+}]_i$ was also observed when ATP+PE was removed after 1 min, but it was eliminated either by TG pre-treatment to deplete the intracellular calcium store or by removing external calcium (Song et al., 2006), indicating that both calcium influx and calcium release from internal stores are required.

These observations support the hypothesis that the intracellular signaling cascades activated by ATP and PE converge to yield a greater sustained elevation in $[Ca^{2+}]_i$ than is achieved with either agent independently. The sustained increase in $[Ca^{2+}]_i$ induced by exposure to ATP+PE, although significantly greater than that observed with either agent alone, was not greater than the sum of the elevation observed to ATP or PE alone. Thus it probably reflects the additive effects of ATP+PE on $[Ca^{2+}]_i$. This and the observation that continued exposure to ATP+PE was not necessary to induce a sustained elevation in $[Ca^{2+}]_i$ indicate that an altered $[Ca^{2+}]_i$ signal persists that could initiate the events that lead to new gene transcription and synergistic stimulation of VP release.

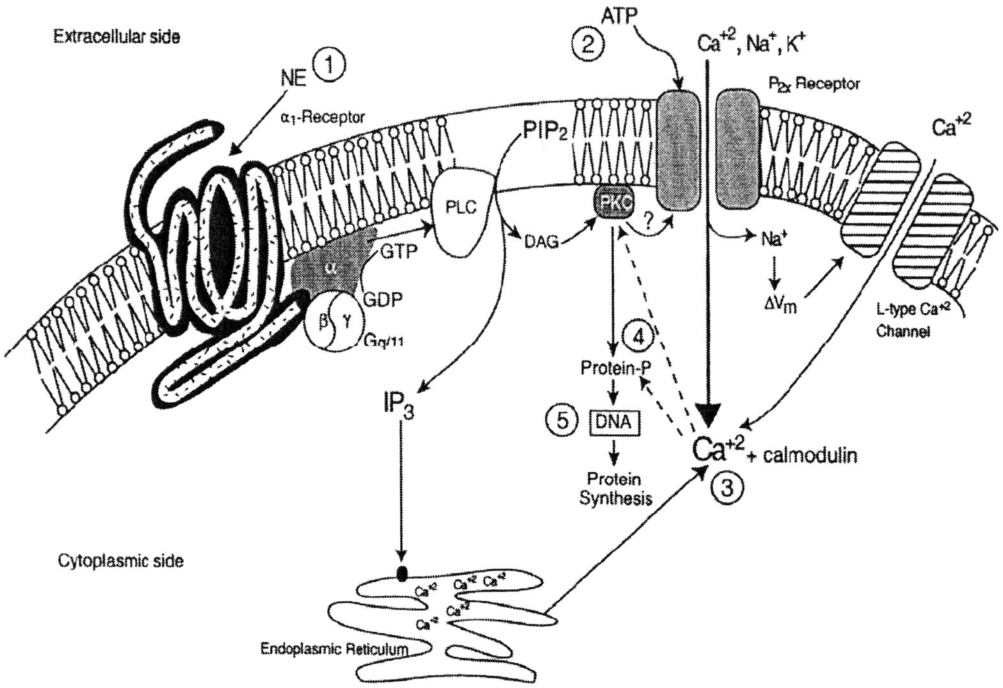

Figure 3. Schematic diagram showing possible points of convergence of intracellular signaling cascades that might lead to NE plus ATP-mediated synergism. *1*, NE activates the $\alpha 1$ receptor, which is coupled to $G_{q/11}$-protein. Activation of the receptor triggers an exchange of GDP for GTP. The trimeric G-protein then dissociates into the α and β/γ dimer. The GTP-bound form of the α subunit then activates the PLC/IP_3/PKC pathway. *2*, ATP binds to the P_{2X} receptor resulting in the opening of a nonspecific ion channel. *3*, both of these events increase intracellular Ca^{2+}. Activation of the IP_3 branch of the PLC pathway releases intracellular Ca^{2+} from the endoplasmic reticulum. Activation of P_{2X} receptors allows Ca^{2+} entry through the P_{2X} channel. In addition, Na^{+} influx through the channel depolarizes the membrane, probably opening voltage-sensitive Ca^{2+} channels and further increasing Ca^{2+} influx. Thus, increased intracellular Ca^{2+} represents one major point of convergence. *4*, the increase in intracellular Ca^{2+} increases the activity of kinases. Specifically, because PKC is Ca^{2+} dependent, higher internal Ca^{2+} levels increase the activity of PKC. Increases in PKC activity might lead to phosphorylation of the ATP-dependent P_{2X} ion channel, leading to a higher open probability for the channel. *5*, PKC and/or Ca^{2+}/calmodulin may phosphorylate transcription factors, thereby inducing transcription of other transcription factors (such as immediate-early genes) or perhaps additional receptors. The observed delay of ~40 min before the onset of the synergistic response as well as the inhibition of the synergism by actinomycin indicates that this final step is a requirement for synergism. *PIP_2*, Phosphatidylinositol 4, 5-bisphosphate. (Reproduced with permission from: Kapoor and Sladek, J. Neuroscience, 2000, 20(23))

Elevations in $[Ca^{2+}]_i$ alter a wide array of cellular functions that might initiate events leading to the extended elevation in VP release observed in response to exposure to ATP+PE. Specifically, in VP neurons, elevated $[Ca^{2+}]_i$ inactivates a Ca^{2+}-sensitive K^{+}-current ($I_{K,leak}$) causing membrane depolarization and excitation (Li and Hatton, 1996). $[Ca^{2+}]_i$ also regulates the activity of Ca^{2+}-sensitive kinases such as PKC and Ca^{2+}-calmodulin kinase. This could lead to alterations in K^{+}-channel activation (Li and Hatton, 1996), receptor trafficking, or gene expression by phosphorylation of transcription factors. The latter possibility is supported by the finding that actinomycin treatment blocked the extended elevation in VP release, which indicates that gene transcription is required for the response (Kapoor and

Sladek, 2000). Thus it is possible that the larger extended elevation in $[Ca^{2+}]_i$ although not in itself synergistic, initiates cellular processes, including new protein synthesis that convert a transient response into a response that, by virtue of its extended nature, is clearly synergistic(Kapoor and Sladek, 2000).

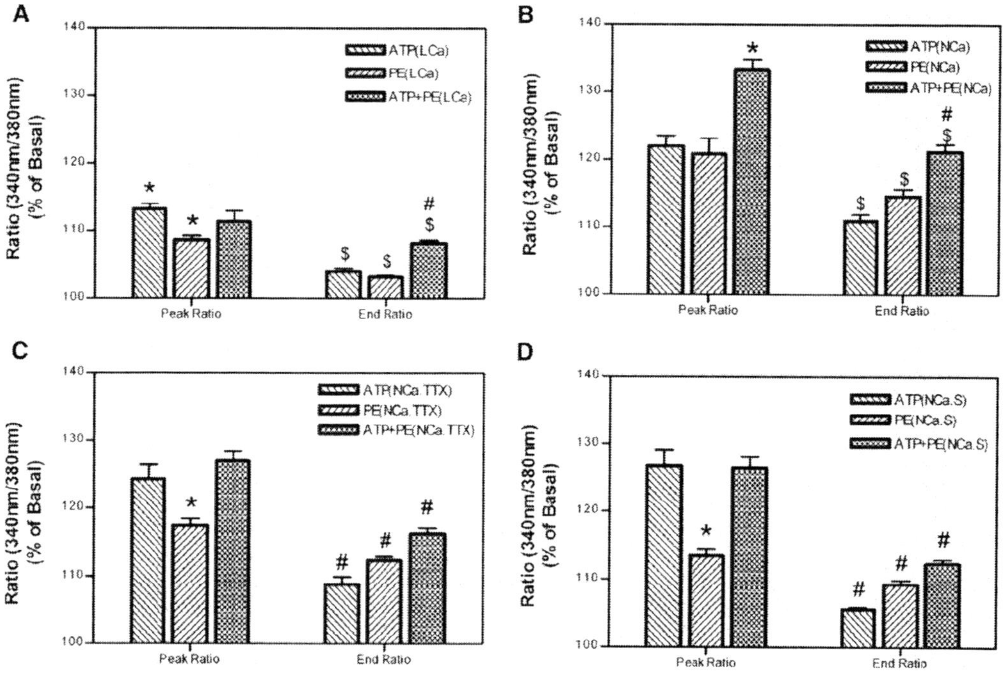

Figure 4. Comparisons of peak and end 340/380 ratios induced by ATP, PE, or ATP+PE in 0.3 mM Ca^{2+} (*A*), 2 mM Ca^{2+} (*B*), 2 mM Ca^{2+} with TTX (*C*), and after a short drug exposure (*D*). *A*: in low Ca^{2+} medium (0.3 mM, LCa), the ATP-induced peak ratio was greater than that induced by PE (*$P < 0.05$ ATP vs. PE) but not different from that induced by ATP+PE; the end ratio during continued exposure to ATP+PE was greater than that during exposure to either ATP or PE alone (#$P < 0.05$ ATP+PE vs. ATP or PE), although the end ratio remained significantly above basal for all three groups ($$P < 0.001$). *B*: in normal Ca^{2+} medium (2 mM, NCa), the peak response to ATP+PE was greater than that to either ATP or PE alone (*$P < 0.05$ ATP+PE vs. ATP or PE); the end ratio during continuous exposure to ATP+PE was greater than during exposure to either PE or ATP (#$P < 0.05$ ATP+PE vs. PE or ATP), although the end ratio remained significantly above basal for all three groups ($$P < 0.001$). Data were obtained from two explants per treatment. The total numbers of neurons analyzed were ATP, n=30; PE, n=42; ATP+PE, n=30. *C*: in 2 mM Ca^{2+} medium with TTX treatment (NCa.TTX), the peak responses to ATP and ATP+PE were greater than that to PE (*$P < 0.05$), but there was no difference between ATP and ATP+PE; the end ratio following ATP+PE was greater than PE or ATP; PE was also greater than ATP (#$P < 0.05$ vs. all other groups). Data were obtained from two explants per treatment. The total numbers of neurons analyzed were ATP, n=42; PE, n=41; ATP+PE, n=32. *D*: drug washout was initiated 1 min after the peak response (NCa.S). The peak response to ATP+PE was greater than that of PE but not different from that of ATP (*$P < 0.05$, PE vs. ATP or ATP+PE). The end ratio followed the same order as in *C*: ATP+PE > PE > ATP (#$P < 0.05$ vs. all other groups). Data were obtained from two explants per treatment. The total numbers of neurons analyzed were ATP, n=36; PE, n=35; ATP+PE, n=35. (Reproduced with permission from: Song and Sladek, Am J Physiol Regul Integr Comp Physiol 2006, 291)

A role for $[Ca^{2+}]_i$ in driving the synergistic response is further supported by experiments that have eliminated several other possible mechanisms. Alterations in ATP metabolism to adenosine is not likely to participate, because inhibition of ATP conversion to adenosine does not convert the transient response to ATP to a sustained response (Song and Sladek, 2005). Modification of stimulus secretion coupling at the nerve terminal was eliminated in experiments using isolated neural lobes, and involvement of glutamatergic afferents was eliminated in experiments demonstrating that neither ionotropic nor metabotropic excitatory amino acid receptor antagonists block the synergistic stimulation of VP release by ATP+PE (Song and Sladek, 2006).

4. Synergism between other Pairs of Neurotransmitters in HNS

In addition to ATP and NE, other neuropeptides are also co-expressed and therefore may be released from the A1 neurons. These include substance P (SP) and neuropeptide Y (NPY). Like PE, SP activates $G_{q/11}$ coupled receptors and thus producing IP3 via activation of PLCβ1; IP3 then releases calcium from intracellular calcium stores. Experiments have shown that SP synergizes with ATP, but not PE, to stimulate VP release from perifused HNS explants (Kapoor and Sladek, 2001). In contrast, NPY induces IP3-independent release of calcium stores via $G_{i/o}$ activation of PLCβ2. It synergizes with PE, but not ATP, to increase VP release. The synergism between ATP and SP might be due to convergence of intracellular calcium signals, similar to that of ATP and PE. On the other hand, the synergism between NPY and PE might be due to activation of multiple signals for the release of intracellular Ca^{++} stores, e.g., IP3 dependent and independent mechanisms.

Osmotic stimulation of VP secretion is transmitted to the VP neurons by glutamatergic synaptic inputs from the rostral hypothalamus. Angiotensin II (AII) is known to facilitate the osmotic response. Glutamate activation of N-methyl-D-aspartate (NMDA) receptors stimulates VP release. Like the P2X receptors, the NMDA receptors are Na^{+} and Ca^{++} permeable ion channels. Although the exact nature of the AII receptors in SON remains unclear, it is likely that AII acts on a G-protein related complex that could have similar signal transduction cascades to the α1-adrenergic receptors. Therefore, it is reasonable to hypothesize that AII and glutamate could have synergistic actions on the regulation of VP hormone release. Experiments performed to assess the effect of combined exposure to AII and NMDA (50 μM) on VP release from explants of the HNS demonstrated that small and transient VP response to AII or NMDA alone were converted to a significantly larger and sustained response when both agents were administered together (Song and Sladek, 2003). The interaction between AII and NMDA is also blocked by TG pretreatment, indicating dependence on release of intracellular calcium stores. This clear evidence for a synergistic interaction between AII and glutamate (NMDA), the pair of neurotransmitters that mediate osmotic signals to HNS suggests that induction of mechanisms leading to sustained response to neurotransmitters is a common theme in the regulation of VP secretion.

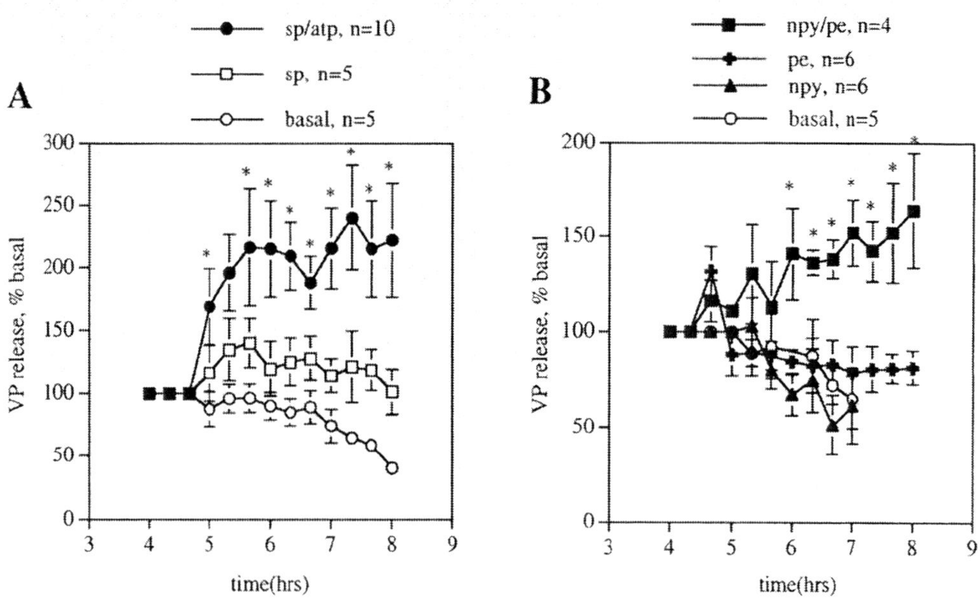

Figure 5. A: VP release in response to simultaneous exposure to ATP (100 μM) and SP (1 μM). Explants were exposed to agents beginning at about 4.7th hour. The combination elicits a larger and sustained response relative to the responses observed with ATP (100 μM) or SP (1 μM) individually. Basal release for SP-ATP, SP, and time control groups was 31.2±4.0, 28.2±18, and 27±2.9 pg/ml, respectively (*P<0.05, SP vs. SP-ATP and time control). B: VP release in response to the combined exposure to neuropeptide Y (NPY; 10 μM) and PE (100 μM). Explants were exposed to agents beginning at about 4.7th hour. VP release was significantly increased in the NPY-PE-treated explants compared with explants exposed to PE alone (*P<0.009). Basal release for NPY-PE, PE, NPY, and basal groups was 122.19±71, 79.06±39.77, 64.13±48.03, and 70.29±50.8 pg/ml, respectively. (Adapted with permission from: Kapoor and Sladek, Am J Physiol Regul Integr Comp Physiol, 2001, 280).

Conclusion

A synergistic action on VP and OT hormone release, which is driven by the convergence of signal cascades that are activated by ATP on purinergic receptors and by NE on adrenergic receptors, is clearly established. The synergistic effect of co-released ATP and NE is likely to have physiological significance for the hemodynamic regulation of VP release under the situations when maximal and sustained increases in plasma VP are either required to achieve homeostasis or may contribute to pathology. Such stimuli include severe hemorrhage, chronic hypotension, or congestive heart failure. Demonstration and characterization of the synergistic effect of purinergic-adrenergic co-transmission on VP release may also provide insight into the role of purinergic co-transmission at other synapses in the hypothalamus and throughout the CNS. The A1 cell group contributes up to 60% of the total noradrenergic input to the whole hypothalamus(Palkovits, 1981; Pacak et al., 1995), and the synergistic effect of ATP and NE on OT release demonstrates that the phenomenon is not limited to VP neurons. ATP also may be co-released with NE from locus coeruleus neurons (Shen and North, 1993) which are associated with receiving and distributing information from and to sites associated with a number of motor, sensory, and behavioral functions. Besides NE, SP and NPY that have been discussed in this chapter, ATP is also stored and co-released with other neurotransmitters, including acetylcholine, nitric oxide, and vasoactive intestinal peptide (Sneddon et al., 1999). We also show that a similar synergistic action between AII and glutamate neurotransmission exists in the osmotic pathway in hypothalamus. Probably, this kind of co-transmission in control of peptide release exists in a broad range in the central nervous system and is necessary for accomplishing important sustained physiological functions associated with maintenance of homeostasis or arousal.

Acknowledgement

Special thanks to Dr. Celia D Sladek, Department of Physiology, UCDHSC, for her editorial suggestions.

References

Acosta-Martinez M, Fiber JM, Brown RD, Etgen AM (1999) Localization of alpha(1B)-adrenergic receptor in female rat brain regions involved in stress and neuroendocrine function. *Neurochemistry International.* 35:383-391.

Alonso G, Assenmacher I (1984) Ultrastructural analysis of the noradrenergic innervation of rat supraoptic nucleus. *Neurosci. Lett.* 49:45-50.

Alper RH, Demarest KT, Moore KE (1980) Effects of Surgical Sympathectomy on Catecholamine Concentrations in the Posterior Pituitary of the Rat. *Experientia.* 36:134-135.

Armstrong WE, Gallagher MJ, Sladek CD (1986) Noradrenergic Stimulation of Supraoptic Neuronal-Activity and Vasopressin Release Invitro - Mediation by an Alpha-1-Receptor. *Brain Research.* 365:192-197.

Balcar VJ, Li Y, Killinger S, Bennett MR (1995) Autoradiography of P-2x Atp Receptors in the Rat-Brain. *British Journal of Pharmacology.* 115:302-306.

Beroukas D, Willoughby JO, Blessing WW (1989) Neuropeptide Y-like immunoreactivity is present in boutons synapsing on vasopressin-containing neurons in rabbit supraoptic nucleus. *Neuroendocrinology.* 50:222-228.

Bittencourt JC, Benoit R, Sawchenko PE (1991) Distribution and origins of substance P-immunoreactive projections to the paraventricular and supraoptic nuclei: partial overlap with ascending catecholaminergic projections. *J. Chem. Neuroanat.* 4:63-78.

Blessing WW, Howe PR, Joh TH, Oliver JR, Willoughby JO (1986) Distribution of tyrosine hydroxylase and neuropeptide Y-like immunoreactive neurons in rabbit medulla oblongata, with attention to colocalization studies, presumptive adrenaline-synthesizing perikarya, and vagal preganglionic cells. *J. Comp. Neurol.* 248:285-300.

Boudaba C, Schrader LA, Tasker JG (1997) Physiological evidence for local excitatory synaptic circuits in the rat hypothalamus. *Journal of Neurophysiology.* 77:3396-3400.

Boudaba C, Di S, Tasker JG (2003) Presynaptic noradrenergic regulation of glutamate inputs to hypothalamic magnocellular neurones. *Journal of Neuroendocrinology.* 15:803-810.

Boue-Grabot E, Archambault V, Seguela P (2000) A Protein Kinase C Site Highly Conserved in P2X Subunits Controls the Desensitization Kinetics of P2X2 ATP-gated Channels. *J. Biol. Chem.* 275:10190-10195.

Brown DA, Yule DI (2007) Protein kinase C regulation of P2X3 receptors is unlikely to involvedirect receptor phosphorylation. Biochimica et Biophysica Acta (BBA) - *Molecular Cell Research.* 1773:166-175.

Burnstock G (1990) Noradrenaline and Atp as Cotransmitters in Sympathetic-Nerves. *Neurochemistry International.* 17:357-368.

Chen ZP, Levy A, Lightman SL (1995) Nucleotides as Extracellular Signaling Molecules. *Journal of Neuroendocrinology.* 7:83-96.

Daftary SS, Boudaba C, Szabo K, Tasker JG (1998) Noradrenergic excitation of magnocellular neurons in the rat hypothalamic paraventricular nucleus via intranuclear glutamatergic circuits. *Journal of Neuroscience.* 18:10619-10628.

Day TA, Renaud LP (1984) Electrophysiological evidence that noradrenergic afferents selectively facilitate the activity of supraoptic vasopressin neurons. *Brain Res.* 303:233-240.

Day TA, Ferguson AV, Renaud LP (1984) Facilitatory influence of noradrenergic afferents on the excitability of rat paraventricular nucleus neurosecretory cells. *J. Physiol.* 355:237-249.

Day TA, Renaud LP, Sibbald JR (1990) Excitation of supraoptic vasopressin cells by stimulation of the A1 noradrenaline cell group: failure to demonstrate role for established adrenergic or amino acid receptors. *Brain Research.* 516:91-98.

Day TA, Sibbald JR, Smith DW (1992) A1 neurons and excitatory amino acid receptors in rat caudal medulla mediate vagal excitation of supraoptic vasopressin cells. *Brain Res.* 594:244-252.

Desouza EB, Kuyatt BL (1987) Alpha-1-Adrenergic Receptors in the Neural Lobe of the Rat Pituitary - Autoradiographic Identification and Localization. *Endocrinology.* 120:2227-2233.

Ennion SJ, Evans RJ (2002) P2X1 Receptor Subunit Contribution to Gating Revealed by a Dominant Negative PKC Mutant. *Biochemical and Biophysical Research Communications.* 291:611-616.

Everitt BJ, Hokfelt T, Terenius L, Tatemoto K, Mutt V, Goldstein M (1984) Differential co-existence of neuropeptide Y (NPY)-like immunoreactivity with catecholamines in the central nervous system of the rat. *Neuroscience.* 11:443-462.

Fried G (1980) Small noradrenergic storage vesicles isolated from rat vas deferens--biochemical and morphological characterization. *Acta Physiol. Scand. Suppl.* 493:1-28.

Garten LL, Sofroniew MV, Dyball REJ (1989) A Direct Catecholaminergic Projection from the Brain-Stem to the Neurohypophysis of the Rat. *Neuroscience.* 33:149-155.

Geffen LB, Livett BG (1971) Synaptic vesicles in sympathetic neurons. *Physiol. Rev.* 51:98-157.

Ginsberg SD, Hof PR, Young WG, Morrison JH (1994) Noradrenergic innervation of vasopressin- and oxytocin-containing neurons in the hypothalamic paraventricular nucleus of the macaque monkey: quantitative analysis using double-label immunohistochemistry and confocal laser microscopy. *J. Comp. Neurol.* 341:476-491.

Gordon GRJ, Baimoukhametova DV, Hewitt SA, Rajapaksha WRAKJS, Fisher TE, Bains JS (2005) Norepinephrine triggers release of glial ATP to increase postsynaptic efficacy. *Nature Neuroscience*. 8:1078-1086.

Graham RM, Perez DM, Hwa J, Piascik MT (1996) alpha(1)-Adrenergic receptor subtypes - Molecular structure, function, and signaling. *Circulation Research*. 78:737-749.

Gratzl M, Torppedersen C, Dartt D, Treiman M, Thorn NA (1980) Isolation and Characterization of Secretory Vesicles from Bovine Neurohypophyses. *Hoppe-Seylers Zeitschrift Fur Physiologische Chemie*. 361:1615-1628.

Hiruma H, Bourque CW (1995) P2 purinoceptor-mediated depolarization of rat supraoptic neurosecretory cells in vitro. *Journal of Physiology*. 489:805-811.

Kapoor JR, Sladek CD (2000) Purinergic and adrenergic agonists synergize in stimulating vasopressin and oxytocin release. *Journal of Neuroscience*. 20:8868-8875.

Kapoor JR, Sladek CD (2001) Substance P and NPY differentially potentiate ATP and adrenergic stimulated vasopressin and oxytocin release. *Am. J. Physiol. Regul. Integr. Comp. Physiol.* 280:R69-78.

Li ZH, Hatton GI (1996) Oscillatory bursting of phasically firing rat supraoptic neurones in low-Ca2+ medium: Na+ influx, cytosolic Ca2+ and gap junctions. *Journal of Physiology*-London 496:379-394.

Ludwig M, Morris M, Sladek CD (1997) Effects of antisense oligodeoxynucleotides on peptide release from hypothalamoneurohypophysial explants. *American Journal of Physiology-Regulatory Integrative and Comparative Physiology*. 41:R1441-R1446.

Lundberg JM, Pernow J, Lacroix JS (1989) Neuropeptide Y: Sympathetic Cotransmitter and Modulator? *News Physiol. Sci.* 4:13-17.

Mcneill TH, Sladek JR (1980) Simultaneous Monoamine Histofluorescence and Neuropeptide Immunocytochemistry .2. Correlative Distribution of Catecholamine Varicosities and Magnocellular Neurosecretory Neurons in the Rat Supraoptic and Paraventricular Nuclei. *Journal of Comparative Neurology*. 193:1023-1033.

Pacak K, Palkovits M, Kopin IJ, Goldstein DS (1995) Stress-Induced Norepinephrine Release in the Hypothalamic Paraventricular Nucleus and Pituitary-Adrenocortical and Sympathoadrenal Activity - in-Vivo Microdialysis Studies. *Frontiers in Neuroendocrinology*. 16:89-150.

Palkovits M (1981) Catecholamines in the hypothalamus: an anatomical review. *Neuroendocrinology*. 33:123-128.

Paukert M, Osteroth R, Geisler H-S, Brandle U, Glowatzki E, Ruppersberg JP, Grunder S (2001) Inflammatory Mediators Potentiate ATP-gated Channels through the P2X3 Subunit. *J. Biol. Chem.* 276:21077-21082.

Ponzio TA, Hatton GI (2004) Activation of presynaptic P2X receptors causes glutamate and GABA release in the supraoptic nucleus. *Soc. Neurosci. Abstract Viewer*. 422:4.

Raby WN, Renaud LP (1989) Dorsomedial medulla stimulation activates rat supraoptic oxytocin and vasopressin neurones through different pathways. *J. Physiol.* 417:279-294.

Randle JC, Bourque CW, Renaud LP (1986) Alpha 1-adrenergic receptor activation depolarizes rat supraoptic neurosecretory neurons in vitro. *Am. J. Physiol.* 251:R569-574.

Saavedra JM (1985) Central and Peripheral Catecholamine Innervation of the Rat Intermediate and Posterior Pituitary Lobes. *Neuroendocrinology*. 40:281-284.

Sawchenko PE, Swanson LW (1981) Central Noradrenergic Pathways for the Integration of Hypothalamic Neuroendocrine and Autonomic Responses. *Science.* 214:685-687.

Sawchenko PE, Swanson LW, Grzanna R, Howe PR, Bloom SR, Polak JM (1985) Colocalization of neuropeptide Y immunoreactivity in brainstem catecholaminergic neurons that project to the paraventricular nucleus of the hypothalamus. *J. Comp. Neurol.* 241:138-153.

Shen KZ, North RA (1993) Excitation of rat locus coeruleus neurons by adenosine 5'-triphosphate: ionic mechanism and receptor characterization. *J. Neurosci.* 13:894-899.

Shibuya I, Tanaka K, Hattori Y, Uezono Y, Harayama N, Noguchi J, Ueta Y, Izumi F, Yamashita H (1999) Evidence that multiple P2X purinoceptors are functionally expressed in rat supraoptic neurones. *Journal of Physiology.* 514:351-367.

Shioda S, Nakai Y (1992) Noradrenergic Innervation of Vasopressin-Containing Neurons in the Rat Hypothalamic Supraoptic Nucleus. *Neuroscience Letters.* 140:215-218.

Shioda S, Shimoda Y, Nakai Y (1992) Ultrastructural studies of medullary synaptic inputs to vasopressin-immunoreactive neurons in the supraoptic nucleus of the rat hypothalamus. *Neurosci. Lett.* 148:155-158.

Shioda S, Yada T, Muroya S, Takigawa M, Nakai Y (1997) Noradrenaline activates vasopressin neurons via alpha(1)-receptor-mediated Ca2+ signaling pathway. *Neuroscience Letters.* 226:210-212.

Sladek CD, Armstrong WE (1987) Gamma-Aminobutyric-Acid Antagonists Stimulate Vasopressin Release from Organ-Cultured Hypothalamoneurohypophyseal Explants. *Endocrinology.* 120:1576-1580.

Smith DW, Sibbald JR, Khanna S, Day TA (1995) Rat vasopressin cell responses to simulated hemorrhage: stimulus-dependent role for A1 noradrenergic neurons. *Am. J. Physiol.* 268:R1336-1342.

Sneddon P, Westfall TD, Todorov LD, Mihaylova-Todorova S, Westfall DP, Kennedy C (1999) Modulation of purinergic neurotransmission. *Prog. Brain Res.* 120:11-20.

Song Z, Sladek CD (2006) Site of ATP and phenylephrine synergistic stimulation of vasopressin release from the hypothalamo-neurohypophyseal system. *Journal of Neuroendocrinology.* 18:266-272.

Song Z, Vijayaraghavan S, Sladek CD (2006) Simultaneous Exposure to ATP and Phenylephrine Induces a Sustained Elevation in the Intracellular Calcium Concentration in Supraoptic Neurons. *Am. J. Physiol. Regul. Integr. Comp. Physiol.* 291:R37-R45.

Song Z, Vijayaraghavan S, Sladek CD (2007) ATP increases intracellular calcium in supraoptic neurons by activation of both P2X and P2Y purinergic receptors. *Am. J. Physiol. Regul. Integr. Comp. Physiol.* 292:R423-431.

Song ZL, Sladek CD (2003) Synergistic stimulation of vasopressin (VP) release by angiotensin II and glutamate: involvement of NMDA receptors. *Faseb. Journal.* 17:A930-A931.

Song ZL, Sladek CD (2005) Does conversion of ATP to adenosine terminate ATP-stimulated vasopressin release from hypothalamo-neurohypophyseal explants? *Brain Research.* 1047:105-111.

Sperlagh B, Vizi ES (1996) Neuronal synthesis, storage and release of ATP. *Seminars in the Neurosciences.* 8:175-186.

Sperlagh B, Sershen H, Lajtha A, Vizi ES (1998) Co-release of endogenous ATP and [3H]noradrenaline from rat hypothalamic slices: origin and modulation by alpha2-adrenoceptors. *Neuroscience.* 82:511-520.

Sperlagh B, Mergl Z, Juranyi Z, Vizi ES, Makara GB (1999) Local regulation of vasopressin and oxytocin secretion by extracellular ATP in the isolated posterior lobe of the rat hypophysis. *Journal of Endocrinology.* 160:343-350.

Troadec JD, Thirion S, Nicaise G, Lemos JR, Dayanithi G (1998) ATP-evoked increases in [Ca2+]i and peptide release from rat isolated neurohypophysial terminals via a P2X2 purinoceptor. *Journal of Physiology.* 511:89-103.

van den Pol AN, Wuarin JP, Dudek FE (1990) Glutamate, the dominant excitatory transmitter in neuroendocrine regulation. *Science.* 250:1276-1278.

Von Kugelgen I, Starke K (1991) Release of Noradrenaline and Atp by Electrical-Stimulation and Nicotine in Guinea-Pig Vas-Deferens. *Naunyn-Schmiedebergs Archives of Pharmacology.* 344:419-429.

Whittaker VP (1982) The synaptic vesicle. In: *Handbook Neurochem.* 7, pp 41-69.

Yamashita H, Inenaga K, Kannan H (1987) Depolarizing Effect of Noradrenaline on Neurons of the Rat Supraoptic Nucleus Invitro. *Brain Research.* 405:348-352.

Zimmermann H (1994) Signalling via ATP in the nervous system. *Trends Neurosci.* 17:420-426.

In: Neuropeptide Research Trends
Editor: Bernice A. Levine, pp. 261-271
ISBN: 978-1-60021-640-4

Chapter XI

Role of the Angiotensin II-Type 1 Receptor System in Cardiometabolic Disorders

Sho-ichi Yamagishi [*a], ***Kazuo Nakamura*** [a], ***Takanori Matsui*** [a], ***and Hiroyoshi Inoue*** [b]

[a] Department of Internal Medicine, [b] Radioisotope Institute for Basic and Clinical Medicine, Kurume University School of Medicine, Kurume 830-0011, Japan

Abstract

There is widespread agreement that the renin-angiotensin system (RAS) plays a pivotal role in the pathogenesis of insulin resistance and cardiovascular disease (CVD) in diabetes. Indeed, large clinical trials have demonstrated substantial benefit of the blockade of this system for end-organ protection. Thus the blockade of the RAS may be a promising strategy for the treatment of the patients with the metabolic syndrome, which consists of a constellation of coronary risk factors such as hypertension, insulin resistance, obesity and hyperlipidemia. Although several types of angiotensin II type 1 (AT_1) receptor blockers (ARBs) are commercially available for the treatment of patients with hypertension, we have recently found that telmisartan (Micardis®) could have the strongest binding affinity to AT_1 receptor. Telmisartan will be a promising cardiometabolic sartan due to its unique peroxisome proliferator-activated receptor-γ (PPAR- γ)-inducing properties as well. In this paper, we reviewed the role of angiotensin II-type 1 receptor system in cardiometabolic disorders, especially focusing on telmisartan. We discussed its potential therapeutic implication in these devastating disorders.

Key Words: AGEs, angiotensin II, cardiovascular disease, insulin resistance, PPAR-γ.

---

[*] Reprints and correspondence to Sho-ichi Yamagishi MD, PhD, Department of Internal Medicine, Kurume University School of Medicine, Kurume 830-0011, Japan; Tel; +81-942-31-7580; Fax; +81-942-31-7707; E-mail; shoichi@med.kurume-u.ac.jp

Introduction

There is a growing body of evidence that blood pressure (BP) level is one of the major determinants of cardiovascular morbidity and mortality in individuals [1,2]. Further, a recent analysis by the Blood Pressure Lowering Treatment Trialists' Collaboration (BPLTTC) revealed that any commonly-used BP-lowering regimen reduced the risk of total major cardiovascular events, and larger reductions in BP produced larger reductions in the risk [3]. These observations suggest that most of the differences among treatment regimens in their effects on cardiovascular outcomes could be explained by the differences in achieved BP level. However, it may also be true that some treatment regimen is superior or inferior to others with regard to the risk reduction of cardiovascular events.

Recently, there is accumulating evidence to show that the renin-angiotensin system (RAS) plays an important role in the development and progression of cerebro-cardiovascular diseases in patients with hypertension [4]. Angiotensin II (AT II), a physiologically active major substance of the RAS, acts as a vasopressor by inducing vasoconstriction and elicits water and sodium absorption in the proximal renal tubule by stimulating secretion of aldosterone [5]. Thus, inhibition of pathophysiological effects of AT II is considered beneficial for the treatment of hypertension.

Effects of AT II are mediated by binding of AT II to the AT II receptors. Four subtypes of AT II receptors are known– AT_1, AT_2, AT_3 and AT_4–although details of AT_3 and AT_4 subtypes remain to be elucidated. The AT_1 receptor exists in the blood vessels, liver, kidneys, adrenal cortex, and heart, and cardiovascular effects of AT II are mainly mediated by this receptor [5, 6]. A physiological role of the AT_2 receptor is not well understood; however, it also exists in the blood vessels, kidneys, adrenal glands, heart, and brain, and it is generally thought that the AT_2 receptor may have physiologically opposing effects on AT_1 receptor-mediated actions [6, 7]. Thus, selective blockade of the AT_1 receptor by ARBs (AT II type 1 receptor blockers) may allow free AT II to stimulate unblocked AT_2 receptors, providing possible beneficial effects on cardiovascular systems [8].

ARBs suppress the effects of AT II generated by all pathways, including chymase, whose activity is not inhibited by angiotensin converting enzyme (ACE) inhibitors. Furthermore, ARBs have less adverse reactions; they are unlikely to cause dry cough and angioedema associated with ACE inhibitors [9]. Therefore, it is plausible that inhibition of the RAS by ARBs may become a promising strategy for the organ protection in patients with hypertension [10-12]. In this review, we focused on telmisartan (Micardis®), a selective AT_1 receptor blocker newly developed by Boehringer Ingelheim GmbH, Germany, and discussed its potential therapeutic implication in cardiometabolic disorders.

Binding Affinity of Telmisartan to Angiotensin II Type 1 Receptor

Although several types of ARBs are commercially available for the treatment of patients with hypertension, comparisons of the binding affinity to AT_1 receptor among them remain to be elucidated. Therefore, we first examined the dissociation rate of several ARBs from AT_1

receptor *in vitro*. Angiotensin II time-dependently dissociated telmisartan, olmesartan, candesartan, valsartan, losartan, and an active metabolite of losartan, EXP3174 from membrane components containing human AT_1 receptor; the dissociation rate constant of each ARB was 0.003248, 0.004171, 0.005203, 0.009946, 0.01027, and 0.008561 min^{-1}, with corresponding half-lives ($t_{1/2}$) of 213, 166, 133, 70, 67, and 81 min, respectively [13]. These results demonstrate that telmisartan could have the strongest binding affinity to AT_1 receptor among various ARBs examined here; the rank order of affinity is telmisartan > olmesartan > candesartan > EXP3174 $\geq$ valsartan $\geq$ losartan. Telmisartan may have superior cardioprotective properties in patients with hypertension due to its long-lasting angiotensin II type 1 receptor antagonistic ability. Indeed, we have very recently found that the cerebroprotective effects of telmisartan in angiotensin II-infused spontaneously hypertensive rats stroke-prone (SHR-SP) were stronger than those of losartan although there was no significant difference of BP levels between the two groups. Cumulative neurological deficit and death rates were significantly less in telmisartan-treated group than in non-treated SHR-SP. These findings suggest that endogenous RAS activation could be involved in neuronal cell damage in SHR-SP and blockade of the RAS by telmisartan may become a promising strategy for the treatment of high-risk stroke-prone hypertensive patients.

Beneficial Aspects of Telmisartan on Insulin Resistance

The metabolic syndrome is strongly associated with insulin resistance and consists of a constellation of factors such as hypertension and hyperlipidemia that raise the risk for cardiovascular diseases and diabetes mellitus [14]. Hypertension occurs approximately twice as frequently in patients with diabetes compared with in non-diabetic controls [15-18]. Conversely, recent data suggest that hypertensive patients are more likely to develop diabetes than normotensive persons [15-18]. The association of diabetes with hypertension increases its risk of cardiovascular morbidity and mortality. Indeed, up to 75% of cardiovascular disease (CVD) in diabetic patients can be attributed to hypertension [15-18]. Therefore, the primary goals of treating the metabolic syndrome are prevention of type 2 diabetes and cardiovascular events.

Recently, telmisartan was found to act as a partial agonist of peroxisome proliferator-activated receptor-γ (PPAR-γ) [19-21]. Molecular modeling studies suggest that telmisartan might influence PPAR-γ activity by interacting with regions of the ligand-binding domain that are not typically engaged by full agonists of the receptor. Furthermore, none of the commercially available ARBs were found not to activate PPAR-γ when tested at the concentrations of therapeutic ranges. Telmisartan also induced PPAR-γ activity in AT_1 receptor-deficient cell models, thus further supporting the concept that telmisartan could stimulate PPAR-γ activity independent of its AT_1 receptor blocking actions [21].

PPAR-γ activity influences the gene expression involved in carbohydrate and lipid metabolism, and pioglitazone and rosiglitazone, ligands for PPAR-γ, improve insulin resistance in diabetic patients [22]. Moreover, there is a growing body of evidence that activators of PPAR-γ exert anti-inflammatory, anti-oxidative and anti-proliferative effects on vascular wall cells, thus decreasing the risks for atherosclerosis [22,23]. These observations suggest that due to its unique PPAR-γ-modulating activity, telmisartan may become a promising 'cardiometabolic sartan', that targets both diabetes and CVD in hypertensive patients [24].

In vitro, telmisartan augmented glucose transporter isoform 4 expression and 2-deoxy glucose uptake both in basal and insulin-stimulated state of adipocytes [25]. In animal study, telmisartan administration caused a significant attenuation of weight gain and reduced glucose, insulin, and triglyceride levels in rats fed a high-fat, high-carbohydrate diet, compared with treatments of losartan, another type of ARB [19]. Further, recently, telmisartan, but not valsartan, promoted increases in caloric expenditure in this model [26]. Telmisartan reduced the accumulation of visceral fat and decreased adipocyte size to a much greater extent than valsartan and was also associated with a significant reduction in hepatic triglyceride levels. In addition, telmisartan treatment decreased the weight of visceral adipose tissue and triglyceride content in the liver and muscle, and increased serum adiponectin levels with a concomitant improvement of insulin resistance in diet-induced obese mice [27]. These observations suggest that due to its unique PPAR-γ-modulating activity, telmisartan may help protect against dietary-induced visceral obesity and insulin resistance.

Furthermore, recently, some clinical papers also reported the insulin-sensitizing effects of telmisartan in hypertensive patients [28, 29]. Replacement of valsartan and candesartan by telmisartan in hypertensive patients with type 2 diabetes resulted in a significant decrease in fasting insulin, triglyceride, and high sensitive-C-reactive protein levels. Serum adiponectin levels were also increased by the treatment with telmisartan. No adverse effects such as weight gain and edema were noted in the replacement by telmisartan. These clinical observations further support the concept that telmisartan could retain the metabolic efficacy of PPAR-γ activation with reduction in adverse effects in parallel angiotensin II type 1 receptor blockade. This may offer a new idea for the treatment of CVD and insulin resistance in patients with the metabolic syndrome by telmisartan. Ongoing clinical trial (ONTARGET) has been designed the efficacy of telmisartan with an ACE inhibitor, ramipril, alone or in combination [30]. This randomized, double-blind, multicenter international studies will provide further information whether telmisartan can improve insulin resistance and subsequently reduce the development of diabetes and CVD in high-risk hypertensive patients.

Beneficial Effects of Telmisartan on Diabetic Retinopathy

Diabetic retinopathy is one of the most important microvascular complications in diabetes and is a leading cause of acquired blindness among the people of occupational age [31]. The prevalence of diabetic retinopathy increases with duration of diabetes. After 30 years of diabetes, nearly all patients in the former group have some degree of retinopathy and the prevalence of proliferative retinopathy is about 60 %.

The lesions of diabetic retinopathy can be grouped into those associated with background, preproliferative and proliferative retinopathy. The earliest histopathological hallmark of diabetic retinopathy is loss of pericytes [32]. Normally, the ratio of endothelial cells (ECs) to pericytes in the retinal capillaries is 1:1, but its levels are reported to decrease to 1:4 after several years of diabetes and eventually to 1:10 with even longer diabetic exposure. In parallel with loss of pericytes, several characteristic changes including thickening of the basement membrane, hyperpermeability, and formation of microaneurysm are observed [31]. These structural and functional abnormalities are followed by microvascular occlusion in the retinas, which ultimately progresses to proliferative changes associated with neovascularization [31]. It has been postulated that many of these changes are the consequent of the loss of pericytes.

Pericytes are elongated cells of the mesodermal origin, wrapping around and along ECs of small vessels [33]. As pericytes contain contractile muscle filaments on their EC side, they have been regarded for a long time just as microvascular counterparts of smooth muscle cells, and implicated in the maintenance of capillary tone [34]. In 1983, D'Amore developed a procedure for isolating pericytes from small vessels, and enabled us to elucidate the functional roles and biological characteristics of pericytes [34]. By using pericyte-EC co-culture systems, we have found that pericytes not only regulate the growth, but also preserve the prostacyclin-producing ability and protect against lipid-peroxide-induced injury of ECs, thus playing an important role in the maintenance of microvascular homeostasis [34-36]. Therefore, the loss of pericytes could predispose the vessels to angiogenesis, thrombogenesis, and EC injury, leading to full-blown clinical expression of diabetic retinopathy. Further, recently, Hammes *et al.* showed that retinal capillary coverage with pericytes was crucial for the survival of ECs, particularly under stress conditions such as diabetes, and that pericyte deficiency leads to reduced inhibition of EC proliferation, thus promoting angiogenesis in the retinopathy of prematurity model [37]. Taken together, our *in vitro* and their recent *in vivo* observations provide a basis for understanding why diabetic retinopathy develops consequent to pericyte loss, the earliest histopathological hallmarks of diabetic retinopathy.

The local RAS is activated under diabetes [38]. We have recently found that AT II stimulates intracellular reactive oxygen species (ROS) generation in retinal pericytes through an interaction with type 1 receptor. Further, AT II decreased DNA synthesis and simultaneously up-regulated vascular endothelial growth factor (VEGF) mRNA levels in pericytes, both of which were blocked by treatment with telmisartan or an anti-oxidant, *N*-acetylcysteine [39, 40]. These results suggest that AT II-type 1 receptor interaction could induce pericyte loss through intracellular ROS generation, thus being involved in diabetic retinopathy, and that telmisartan could be a promising therapeutic strategy for the treatment of early diabetic retinopathy.

VEGF is a specific mitogen to ECs, also known as vascular permeability factor, and is thought a pivotal factor in the pathogenesis of proliferative diabetic retinopathy. Indeed, some clinical observations have demonstrated that VEGF level in ocular fluid is positively correlated with the activity of neovascularization in diabetic retinopathy [41, 42]. Recently, VEGF level was also found to be associated with the breakdown of the blood-retinal barrier, thus being involved in microvascular hyperpermeability in background retinopathy [43]. These observations suggest that telmisartan might play a protective role against the development and progression of diabetic retinopathy by blocking VEGF overexpression in pericytes as well.

Blockade by Telmisartan of the Advanced Glycation End Product-Their Receptor (AGE-RAGE) System

Reactive derivatives from non-enzymatic glucose-protein condensation reactions, as well as lipids and nucleic acids exposed to reducing sugars, form a heterogeneous group of irreversible adducts called "AGEs". AGEs were originally characterized by a yellow-brown fluorescent color and by an ability to form cross-links with and between amino groups [44], but the term is now used for a broad range of advanced products of the glycation process (also called the "Maillard reaction"), including N-carboxymethyllysine and pyrraline, which show neither color nor fluorescence and do not cross-link proteins [45-47].

AGEs are formed by the Maillard process, a non-enzymatic reaction between reducing sugars and the amino groups of proteins that contributes to the aging of proteins and to the pathological complications of diabetes [48-52]. The formation of AGEs *in vitro* and *in vivo* is dependent on the turnover rate of the chemically modified target, the time available, and the sugar concentration. In the hyperglycemia elicited by diabetes, this process begins with the conversion of reversible Schiff base adducts to more stable, covalently-bound Amadori rearrangement products. Over the course of days to weeks, these Amadori products undergo further rearrangement reactions to form the irreversibly-bound moieties known as AGEs.

A variety of molecular mechanisms underlying the actions of AGEs and their contribution to CVD have been proposed [53-57]. AGEs formed on the extracellular matrix results in decreased elasticity of vasculatures, and quench nitric oxide, which could mediate defective endothelium-dependent vasodilatation in diabetes [58]. AGE modification of low-density lipoprotein (LDL) exhibits impaired plasma clearance and contributes significantly to increased LDL *in vivo*, thus being involved in atherosclerosis [59]. Binding of AGEs to RAGE results in generation of intracellular ROS generation and subsequent activation of the redox-sensitive transcription factor NF-κB in vascular wall cells, which promotes the expression of a variety of atherosclerosis-related genes, including intercellular adhesion molecule-1, vascular cell adhesion molecule-1, monocyte chemoattractant protein-1, plasminogen activator inhibitor-1, tissue factor, VEGF, and RAGE [60-69]. AGEs have the ability to induce osteoblastic differentiation of microvascular pericytes, which would contribute to the development of vascular calcification in accelerated atherosclerosis in diabetes as well [70].

The interaction of the RAS and AGE-RAGE system has also been proposed. We have found that AT II potentiates the deleterious effects of AGEs on pericytes by inducing RAGE protein expression [71]. *In vivo*, AGE injection stimulated RAGE expression in the eye of spontaneously hypertensive rats, which was blocked by telmisartan. *In vitro*, AT II-type 1 receptor-mediated ROS generation elicited RAGE gene expression in retinal pericytes through NF-□B activation. Further, AT II augmented AGE-induced pericyte apoptosis, the earliest hallmark of diabetic retinopathy. Further, we have recently found that telmisartan blocks the AT II-induced RAGE expression in ECs as well [72]. Telmisartan could decrease endothelial RAGE levels in patients with essential hypertension. Taken together, these observations provide the functional interaction between the AGE-RAGE system and the RAS in the pathogenesis of diabetic vascular complications, thus suggesting a novel beneficial aspect of telmisartan on these devastating disorders.

Perspectives

We reviewed here the potential therapeutic implication of telmisartan in cardiometabolic disorders. The clinical relevance of this unique sartan should be further studied in patients with the metabolic syndrome. To develop novel therapeutic strategies that specifically target insulin resistance and CVD may be helpful for most patients with hypertension.

References

[1] Black HR. The burden of cardiovascular disease: following the link from hypertension to myocardial infarction and heart failure. *American Journal of Hypertension.* 2003; 9: 4S-6S.

[2] Staessen JA, Wang JG, Thijs L. Cardiovascular prevention and blood pressure reduction: a quantitative overview updated until 1 march 2003. *J. Hypertension.* 2003; 21: 1055-76.

[3] Blood Pressure Lowering Treatment Trialists' Collaboration. Effects of different blood-pressure-lowering regimens on major cardiovascular events: results of prospectively-designed overviews of randomized trials. *Lancet.* 2003; 362: 1527-35.

[4] Higaki J, Baba S, Katsuya T, et al. Deletion Allele of Angiotensin-Converting Enzyme Gene Increases Risk of Essential Hypertension in Japanese Men. The Suita Study. *Circulation.* 2000; 101: 2060-5.

[5] Burnier M, Brunner HR. Angiotensin II receptor antagonists. *Lancet.* 2000; 355: 637-45.

[6] de Gasparo M, Catt KJ, Inagami T, Wright JW, Unger T. International Union of Pharmacology, XXIII. The Angiotensin II Receptors. *Pharmacol. Rev.* 2000; 52: 415-72.

[7] Stoll M, Unger T. Angiotensin and its AT_2 receptor: new insights into an old system. *Regul. Pept.* 2001; 99: 175-82.

[8] Siragy HM. Angiotensin receptor blockers: How important is selectivity? *Am. J. Hypertens.* 2002; 15: 1006-14.

[9] Pylypchuk GB. ACE inhibitor- versus angiotensin II blocker-induced cough and angioedema. *Ann. Pharmacother.* 1998; 32: 1060-6.

[10] RENAAL Study Investigators: Effects of losartan on renal and cardiovascular outcomes in patients with type 2 diabetes and nephropathy. *N. Engl. J. Med.* 2001; 345: 861-9.

[11] Dahlöf B, Devereux RB, Kjeldsen SE, et al. Cardiovascular morbidity and mortality in the Losartan Intervention for Endpoint reduction in hypertension study (LIFE): a randomized trial against atenolol. *Lancet.* 2002; 359: 995-1003.

[12] CHARM Investigators and Committees: Effects of candesartan in patients with chronic heart failure and reduced left-ventricular systolic function intolerant to angiotensin-converting-enzyme inhibitors: the CHARM-Alternative trial. *Lancet.* 2003; 362: 772-6.

[13] Kakuta H, Sudoh K, Sasamata M, Yamagishi S. Telmisartan has the strongest binding affinity to angiotensin II type 1 receptor. Comparison with other angiotensin II type 1 receptor blockers. *Int. J. Clin. Pharm. Res.* 2005; 25: 41-6.

[14] Scheen AJ. Management of the metabolic syndrome. *Minerva Endocrinol.* 2004; 29: 31-45.

[15] Khamaisi M, Wexler ID, Skrha J, Strojek K, Raz I, Milicevic Z. Cardiovascular disease in type 2 diabetics: epidemiology, risk factors and therapeutic modalities. *Isr. Med. Assoc. J.* 2003; 5: 801-6.

[16] Sowers JR. Insulin resistance and hypertension. *Am. J. Physiol. Heart Circ. Physiol.* 2004; 286: H1597-602.

[17] Sowers JR, Frohlich ED. Insulin and insulin resistance: impact on blood pressure and cardiovascular disease. *Med. Clin. North A.* 2004; 88: 63-82.

[18] Watson KE, Peters Harmel AL, Matson G. Atherosclerosis in type 2 diabetes mellitus: the role of insulin resistance. *Cardiovasc. Pharmacol. Ther.* 2003; 8: 253-60.

[19] Benson SC, Pershadsingh HA, Ho CI, et al. Identification of telmisartan as a unique angiotensin II receptor antagonist with selective PPARgamma-modulating activity. *Hypertension.* 2004; 43: 993-1002.

[20] Yamagishi S, Nakamura K. Telmisartan, its potential therapeutic implications in cardiometabolic disorders. *Recent Patents on Cardiovascular Drug Discovery.* 2006; 1: 79-83.

[21] Schupp M, Janke J, Clasen R, Unger T, Kintscher U. Angiotensin type 1 receptor blockers induce peroxisome proliferator-activated receptor-gamma activity. *Circulation.* 2004; 109: 2054-57.

[22] Takano H, Hasegawa H, Zou Y, Komuro I. Pleiotropic actions of PPAR gamma activators thiazolidinediones in cardiovascular diseases. *Curr. Pharm. Des.* 2004; 10: 2779-86.

[23] Marx N, Duez H, Fruchart JC, Staels B. Peroxisome proliferator-activated receptors and atherogenesis. Regulators of gene expression in vascular cells. *Cir. Res.* 2004; 94: 1168-78.

[24] Yamagishi S, Takeuchi M. Telmisartan is a promising cardiometabolic sartan due to its unique PPAR-γ-inducing property. *Med. Hypothese.* 2005; 64: 476-8.

[25] Fujimoto M, Masuzaki H, Tanaka T et al. An angiotensin II AT_1 receptor antagonist, telmisartan augments glucose uptake and GLUT4 protein expression in 3T3L-1 adipocytes. *FEBS Lett.* 2004; 576: 492-7.

[26] Sugimoto K, Qi NR, Kazdova L, et al. Telmisartan but not valsartan increases caloric expenditure and protects against weight gain and hepatic steatosis. *Hypertension.* 2006; 48: 1003-9.

[27] Araki K, Masaki T, Katsuragi I, et al. Telmisartan prevents obesity and increases the expression of uncoupling protein 1 in diet-induced obese mice. *Hypertension.* 2006; 48: 51-57.

[28] Pershadsihgh HA, Kurtz TW. Insulin-sensitizing effects of telmisartan. *Diabetes Care.* 2004; 27, 1015.

[29] Miura Y, Yamamoto N, Tsunekawa S, Taguchi S, Eguchi Y, Ozaki N, Oiso Y. Replacement of valsartan and candesartan by telmisartan in hypertensive patients with diabetes. *Diabetes Care.* 2005; 28, 757-8.

[30] Zimmermann M, Unger T. Challenges in improving prognosis and therapy: the Ongoing Telmisartan Alone and in Combination with Ramipril Global End point Trial programme. *Expert Opin. Pharmacother.* 2004; 5: 1201-8.

[31] L'Esperance FA, James WA, Judson PH. In: Lifkin H, Porte D Ed, The eye and diabetes mellitus. Ellenberg and Rifkin's Diabetes Mellitus, Theory and Practice. New York, Elsevier, 1990; 661-83.

[32] Cogan DG, Toussaint D, Kuwabara T. Retinal vascular patterns. IV. Diabetic retinopathy. *Arch. Ophthalmo.* 1961; 66: 366-78.

[33] Sims DE. Recent advances in pericyte biology--implications for health and disease. *Can. J. Cardiol.* 1991; 7: 431-43.

[34] Gitlin JD, D'Amore PA. Culture of retinal capillary cells using selective growth media. *Microvasc. Res.* 1983; 26: 1455-62.

[35] Yamagishi S, Kobayashi K, Yamamoto H. Vascular pericytes not only regulate growth, but also preserve prostacyclin-producing ability and protect against lipid peroxide-induced injury of co-cultured endothelial cells. *Biochem. Biophys. Res. Commun.* 1993; 190: 418-25.

[36] Yamagishi S, Hsu CC, Kobayashi K, Yamamoto H. Endothelin 1 mediates endothelial cell-dependent proliferation of vascular pericytes. *Biochem. Biophys. Res. Commun.* 1993; 191: 840-6.

[37] Hammes HP, Lin J, Renner O, et al. Pericytes and the pathogenesis of diabetic retinopathy. *Diabetes.* 2002; 51: 3107-12.

[38] Anderson S. Role of local and systemic angiotensin in diabetic renal disease. Kidney *Int. Suppl.* 1997; 63: S107-10.

[39] Yamagishi S, Amano S, Inagaki Y, et al. Angiotensin II-type 1 receptor interaction upregulates vascular endothelial growth factor messenger RNA levels in retinal pericytes through intracellular reactive oxygen species generation. *Drugs Exp. Clin. Res.* 2003; 29: 75-80.

[40] Amano S, Yamagishi S, Inagaki Y, Okamoto T. Angiotensin II stimulates platelet-derived growth factor-B gene expression in cultured retinal pericytes through intracellular reactive oxygen species generation. *Int. J. Tissue React.* 2003; 25: 51-5.

[41] Adamis AP, Miller JW, Bernal MT, et al. Increased vascular endothelial growth factor levels in the vitreous of eyes with proliferative diabetic retinopathy. *Am. J. Ophthalmol.* 1994; 118: 445-50.

[42] Aiello LP, Avery RL, Arrigg PG, et al. Vascular endothelial growth factor in ocular fluid of patients with diabetic retinopathy and other retinal disorders. *N. Engl. J. Med.* 1994; 331: 1480-7.

[43] Murata T, Ishibashi T, Khalil A, et al. Vascular endothelial growth factor plays a role in hyperpermeability of diabetic retinal vessels. *Ophthalmic Res.* 1995; 27: 48-52.

[44] Brownlee M, Cerami A, Vlassara H. Advanced glycosylation end products in tissue and the biochemical basis of diabetic complications. *N. Engl. J. Med.* 1988; 318: 1315-21.

[45] Dyer DG, Blackledge JA, Thorpe SR, Baynes JW. Formation of pentosidine during nonenzymatic browning of proteins by glucose. Identification of glucose and other carbohydrates as possible precursors of pentosidine in vivo. *J. Biol. Chem.* 1991; 266: 11654-60.

[46] Grandhee SK, Monnier VM. Mechanism of formation of the Maillard protein cross-link pentosidine. Glucose, fructose, and ascorbate as pentosidine precursors. *J. Biol. Chem.* 1991; 266: 11649-53.

[47] Takeuchi M, Kikuchi S, Sasaki N, et al. Involvement of advanced glycation end-products (AGEs) in Alzheimer's disease. *Curr. Alzheimer Research.* 2004; 1:39-46.

[48] Bucala, Cerami A. Advanced glycosylation: chemistry, biology, and implications for diabetes and aging. *Adv. Pharmacol.* 1992; 23: 1-34.

[49] Vlassara H, Bucala R, Striker L. Pathogenic effects of advanced glycosylation: biochemical, biologic, and clinical implications for diabetes and aging. *Lab. Invest.* 1994; 70: 138-51.

[50] Brownlee M. Advanced protein glycosylation in diabetes and aging. *Ann. Rev. Med.* 1995; 46: 223-34.

[51] Yamagishi S, Takeuchi M, Inagaki Y, Nakamura K, Imaizumi T. Role of advanced glycation end products (AGEs) and their receptor (RAGE) in the pathogenesis of diabetic microangiopathy. *Int. J. Clin. Pharmcol. Res.* 2003; 23: 129-34.

[52] Matsumura T, Yamagishi S, Brownlee M. In: Leroith D, Taylor SI, Olefsky JM Ed, Advanced glycation end products and the pathogenesis of diabetic complications. Diabetes Mellitus: A Fundamental and Clinical Text. New York, Lippincott-Raven Publishers, 2000; 983-91.

[53] Vlassara H, Palace MR. Diabetes and advanced glycation endproducts. *J. Intern. Med.* 2002; 251: 87-101.

[54] Bierhaus A, Hofmann MA, Ziegler R, Nawroth PP. AGEs and their interaction with AGE-receptors in vascular disease and diabetes mellitus. I. The AGE concept. *Cardiovasc. Res.* 1998; 37: 586-600.

[55] Wendt T, Bucciarelli L, Qu W, et al. Receptor for advanced glycation endproducts (RAGE) and vascular inflammation: insights into the pathogenesis of macrovascular complications in diabetes. *Curr. Atheroscler. Rep.* 2002; 4: 228-37.

[56] Schmidt AM, Stern D. Atherosclerosis and diabetes: the RAGE connection. *Curr. Atheroscler. Rep.* 2000; 2: 430-6.

[57] Stitt AW, Bucala R, Vlassara H. Atherogenesis and advanced glycation: promotion, progression, and prevention. *Ann. N Y Acad. Sci.* 1997; 811: 115-27.

[58] Bucala R, Tracey KJ, Cerami A. Advanced glycosylation products quench nitric oxide and mediate defective endothelium-dependent vasodilatation in experimental diabetes. *J. Clin. Invest.* 1991; 87: 432-8.

[59] Bucala R, Mitchell R, Arnold K, et al. Identification of the major site of apolipoprotein B modification by advanced glycosylation end products blocking uptake by the low density lipoprotein receptor. *J. Biol. Chem.* 1995; 270: 10828-32.

[60] Inagaki Y, Yamagishi S, Okamoto T, Takeuchi M, Amano S. Pigment epithelium-derived factor prevents advanced glycation end products-induced monocyte chemoattractant protein-1 production in microvascular endothelial cells by suppressing intracellular reactive oxygen species generation. *Diabetologia.* 2003; 46: 284-7.

[61] Yamagishi S, Fujimori H, Yonekura H, Yamamoto Y, Yamamoto H. Advanced glycation endproducts inhibit prostacyclin production and induce plasminogen activator inhibitor-1 in human microvascular endothelial cells. *Diabetologia.* 1998; 41: 1435-41.

[62] Yamagishi S, Yonekura H, Yamamoto Y, et al. Advanced glycation end products-driven angiogenesis in vitro. Induction of the growth and tube formation of human microvascular endothelial cells through autocrine vascular endothelial growth factor. *J. Biol. Chem.* 1997; 272: 8723-30.

[63] Lander HM, Tauras JM, Ogiste JS, et al. Activation of the receptor for advanced glycation end products triggers a p21(ras)-dependent mitogen-activated protein kinase pathway regulated by oxidant stress. *J. Biol. Chem.* 1997; 272: 17810-4.

[64] Schmidt AM, Hori O, Chen JX, et al. Advanced glycation endproducts interacting with their endothelial receptor induce expression of vascular cell adhesion molecule-1 (VCAM-1) in cultured human endothelial cells and in mice. A potential mechanism for the accelerated vasculopathy of diabetes. *J. Clin. Invest.* 1995; 96: 1395-403.

[65] Vlassara H, Fuh H, Donnelly T, Cybulsky M. Advanced glycation endproducts promote adhesion molecule (VCAM-1, ICAM-1) expression and atheroma formation in normal rabbits. *Mol. Med.* 1995; 1: 447-56.

[66] Bierhaus A, Illmer T, Kasper M, et al. Advanced glycation end product (AGE)-mediated induction of tissue factor in cultured endothelial cells is dependent on RAGE. *Circulation.* 1997; 96: 2262-71.

[67] Tanaka N, Yonekura H, Yamagishi S, et al. The receptor for advanced glycation end products is induced by the glycation products themselves and tumor necrosis factor-alpha through nuclear factor-kappa B, and by 17beta-estradiol through Sp-1 in human vascular endothelial cells. *J. Biol. Chem.* 2000; 275: 25781-90.

[68] Yamagishi S, Nakamura K, Taakeuchi M, Imaizumi T. Molecular mechanism for accelerated atherosclerosis in diabetes and its potential therapeutic intervention. *Int. J. Clin. Pharm. Res.* 2004; 24: 129-134.

[69] Yamagishi S, Nakamura K, Imaizumi T. Advanced glycation end products (AGEs) and diabetic vascular complications. *Curr. Diabetes Rev.* 2005; 1: 93-106.

[70] Yamagishi S, Fujimori H, Yonekura H, Tanaka N, Yamamoto H. Advanced glycation endproducts accelerate calcification in microvascular pericytes. *Biochem. Biophys. Res. Commun.* 1999; 258: 353-7.

[71] Yamagishi S, Takeuchi M, Matsui T et al. Angiotensin II augments advanced glycation end product-induced pericyte apoptosis through RAGE overexpression. *FEBS Lett.* 2005: 579: 4265-70.

[72] Nakamura K, Yamagishi S, Nakamura Y, et al. Telmisartan inhibits expression of a receptor for advanced glycation end products (RAGE) in angiotensin-II-exposed endothelial cells and decreases serum levels of soluble RAGE in patients with essential hypertension. *Microvasc. Res.* 2005; 70: 137-41.

Index

A

B

C

D

E

F

G

H

I

J

K

L

M

N

O

P

Q

R

S

T

X

Y

Z